Canberra International Physics Summer Schools

The New Cosmology

Proceedings of the

16th International Physics Summer School, Canberra

Canberra International Physics Summer Schools

The New Cosmology

Proceedings of the

16th International Physics Summer School, Canberra

Canberra, Australia 3 – 14 February 2003

editor

Matthew Colless

Anglo-Australian Observatory, Australia

World Scientific

NEW JERSEY • LONDON • SINGAPORE • BEIJING • SHANGHAI • HONG KONG • TAIPEI • CHENNAI

Published by

World Scientific Publishing Co. Pte. Ltd.
5 Toh Tuck Link, Singapore 596224
USA office: 27 Warren Street, Suite 401-402, Hackensack, NJ 07601
UK office: 57 Shelton Street, Covent Garden, London WC2H 9HE

British Library Cataloguing-in-Publication Data
A catalogue record for this book is available from the British Library.

Cover image by the 2dF Galaxy Redshift Survey Team and Swinburne University Centre for Astrophysics and Supercomputing.

THE NEW COSMOLOGY
Proceedings of the 16th International Physics Summer School

ISBN 981-256-066-1

Printed by FuIsland Offset Printing (S) Pte Ltd, Singapore

PREFACE

Since 1988 the Canberra International Physics Summer Schools sponsored by the Australian National University have provided intensive courses in topical areas of physics not covered in most undergraduate programs. The 2003 Summer School brought together students from around Australia and beyond to hear lectures by leading international experts on the topic of *The New Cosmology*. The lectures encompassed a treatment of the classical elements of cosmology and an introduction to the new cosmology of inflation, the cosmic microwave background, the high-redshift universe, dark matter, dark energy and particle astrophysics. These lecture notes, which are aimed at senior undergraduates and beginning postgraduates, therefore provide a comprehensive overview of the broad sweep of modern cosmology and entry points for deeper study.

Matthew Colless

CONTENTS

THE EXPANDING AND ACCELERATING UNIVERSE

BRIAN P. SCHMIDT

Research School of Astronomy and Astrophysics, Mt. Stromlo Observatory, The Australian National University, via Cotter Rd, Weston Creek, ACT 2611, Australia
E-mail: brian@mso.anu.edu.au

Measuring distances to extragalactic objects has been a focal point for cosmology over the past 100 years, shaping (sometimes incorrectly) our view of the Universe. I discuss the history of measuring distances, briefly review several popular distance measuring techniques used over the past decade, and critique our current knowledge of the current rate of the expansion of the Universe, H_0, from these observations. Measuring distances back to a significant portion of the look back time probes the make-up of the Universe, through the effects of different types of matter on the cosmological geometry and expansion. Over the past five years two teams have used type Ia supernovae to trace the expansion of the Universe to a look back time more than 70% of the age of the Universe. These observations show an accelerating Universe which is best explained by a cosmological constant, or other form of dark energy with an equation of state near $w = p/\rho = -1$. There are many possible lurking systematic effects. However, while difficult to completely eliminate, none of these appears large enough to challenge current results. However, as future experiments attempt to better characterize the equation of state of the matter leading to the observed acceleration, these systematic effects will ultimately limit progress.

1. An Early History of Cosmology

Cosmology became a major focus of astronomy and physics early in the 20th century when technology and theory had developed sufficiently to start asking basic questions about the Universe as a whole. The state of play of cosmology in 1920 is well summarised by the "Great Debate", which took place between Heber Curtis and Harlow Shapley. This debate was hosted by the United States Academy of Science, and featured the topic, "Scale of the Universe?" — and, in addition to debating the size and extent of the Universe, it tried to address the question, "Is the Milky Way an island universe, or just one of many such galaxies". With the benefit of 80 years of progress, the arguments made in favour of the island universe by Shapley, and those made by Curtis in favour of other galaxies existing alongside the Milky Way, serve modern day cosmology as a lesson on how various pitfalls can lead to wrong conclusions (see Hoskin 1976 for a nice review of the debate[37]).

1.1. *Curtis Shapley Debate*

Harlow Shapley, the young director of Harvard College Observatory, believed the evidence favoured the island universe hypothesis, and argued that spiral nebulae were part of our own galaxy, the Milky Way. His own work, using the positions of globular clusters, indicated that the Milky Way was very large — extending out to 100,000 parsecs (316,000 light years). He made the measurements by observing variable stars (RR Lyrae) in these objects, and comparing their brightnesses to closer objects. These same observations also indicated that we were not located in the centre of the Milky Way, as the measurement showed we were clearly displaced from the centre of the distribution of globular clusters. Novae — the sudden explosions of certain stars — were oftentimes seen in the Milky Way, and Shapley argued further, that these same objects had been seen in spiral nebulae such as the Andromeda nebula, and had the same apparent brightness as those seen in the middle of the Milky Way. If, as Curtis was arguing, these spiral nebulae were distant copies of the Milky Way, the novae should appear much fainter. To Shapley this was proof that these nebulae were not distant, but rather part of our own Galaxy. Next, Shapley appealed to the measurement of the rotation of the spiral M101 by van Maanen [62] — one of the largest of the spiral nebulae. If this galaxy were as distant as required for it to be beyond the Milky Way, then it could not be physically rotating as fast as van Maanen's measurement indicated without exceeding the speed of light. Shapley then noted Slipher's measurements of the recession of the nebulae, and the fact that they avoided a plane through the centre of the Milky Way. He suggested that this observation showed association of the objects with the Milky Way because these objects were somehow repulsed away from the Milky Way by some as yet unknown physical mechanism. Finally, Shapley argued that his colour measurement of the spiral nebulae indicated they had colours bluer than any objects in the Milky Way, further arguing that these were objects unlike anything we were familiar with, and not copies of the Milky Way, which was essentially a conglomeration of stars.

Heber Curtis, the wizened Director of the Allegheny Observatory, argued that spiral nebulae were distant objects, and like our own Milky Way. Curtis appealed to measurements of stars and star counts in the different parts of the sky to argue the Milky Way is more like 10,000 parsec in diameter, with the sun near the centre, and therefore it is hard to see what is going on. Curtis, while unable to explain the few bright novae in the spiral nebulae, also noted that many novae in the Andromeda nebula were faint — about the right brightness to be the same novae seen in our own Galaxy at a much greater distance. He noted that despite the colour measurements of Shapley, the spectra of spiral nebulae looked like the integrated spectrum of many stars, arguing that these were not unknown physical entities. Furthermore, he pointed to observations of many spiral nebulae that showed they had a dark ring of occulting material which explained why galaxies avoided the central plane of the Milky Way — they were obscured — although Curtis didn't have an explanation for the galaxies' mass exodus away from our galaxy. Finally, Curtis pointed to evidence that the Milky Way had spiral structure just like the other spiral nebulae.

The debate was solved in October 1923 (although the world didn't find out about it until some time later) when Hubble, using the new Hooker 100 inch telescope, discovered some of Shapley's variable stars (this time Cepheid variable stars) in the Andromeda Galaxy (and two other galaxies), indicating that these galaxies were at a great distance — well beyond the Milky Way — and had an expanse similar to that of the Milky Way.

The take home message from this debate is that cosmology is full of red herrings, bad observations, and missing information. Shapley appealed to his wrong measurements of the colour of spiral galaxies, as well as van Maanen's flawed measurement of the rotation of the spirals. The expanse of the Milky Way was a red herring — Shapley was more or less correct, but it wasn't very important to the argument in the end (Shapley had intended for the huge distances required for Shapley's argument to simply not be plausible). And finally, both the dust we now know is scattered throughout the plane of spiral galaxies, and supernovae, the incredibly bright explosions of stars, were both missing information — although Curtis had realised this, it was hard for him to prove in 1920. Definitive observations, coupled with sound theory, still provide a way through the fog today as they did in the 1920s.

1.2. *The Emergence of Relativity and the Expanding Universe*

Einstein first published his final version of general relativity in 1916, and within the first year, de Sitter had already investigated the cosmological implications of this new theory. While relativity took the theoretical physics world by storm, especially after Eddington's eclipse expedition in 1919 confirmed the first independent predictions of the theory, not all of science was so keen. In 1920, when George Ellery Hale was attempting to set up the great debate, the home secretary of the National Academy of Sciences, Abbot, remarked, "As to relativity I must confess that I would rather have a subject in which there would be a half dozen members of the Academy competent enough to understand at least a few words of what the speakers were saying... I pray the progress of science will send relativity to some region of space beyond the 4th dimension, from whence it will never return to plague us"

Theoretical progress was swift in cosmology after Eddington's confirmation of general relativity. In 1917 Einstein published his cosmological constant model, where he attempted to balance gravity with a negative pressure inherent to space, to create a static model seemingly needed to explain the Universe around him. In 1920 de Sitter published the first models that predicted spectral redshift of objects in the Universe, dependent on distance, and in 1922, Friedmann published his family of models for an isotropic and homogenous Universe.

The contact between theory and observations at this time, appears to have been mysteriously poor. Hubble had started to count galaxies to see the effects of non-Euclidean geometry, possible with general relativity, but failed to find the effect as late as 1926 (in retrospect, he wasn't looking far enough afield). In 1927,

Lemaitre, a Belgian monk with a newly received PhD from MIT, independently derived Friedmann universes, predicted the Hubble Law, noted that the age of the Universe was approximately the inverse of the Hubble Constant, and suggested that Hubble's/Slipher's data supported this conclusion[60] — his work was not well known at the time. In 1928, Robertson, at CalTech (just down the road from Hubble), in a very theoretical paper predicted the Hubble law and claimed to see it (but not substantiated) if he compared Sliper's redshift versus Hubble's galaxy brightness measurements[91]. Finally, in 1929, Hubble presented data in support of an expanding universe, with a clear plot of galaxy distance versus redshift[36]. It is for this paper that Hubble is given credit for discovering the expanding universe. Within two years, Hubble and Humason had extended the Hubble law out to 20000 km/s using the brightest galaxies, and the field of measuring extragalactic distance, from a 21st century perspective, made little substantive progress for the next 30 and some might argue even 60 years.

2. The Cosmological Paradigm

Astronomers use a standard model for understanding the Universe and its evolution. The assumptions of this standard model, that general relativity is correct, and the Universe is isotropic and homogenous on large scales, are not proven beyond a reasonable doubt — but they are well tested, and they do form the basis of our current understanding of the Universe. If these pillars of our standard model are wrong, then any inferences using this model about the Universe around us may be severely flawed, or irrelevant.

The standard model for describing the global evolution of the Universe is based on two equations that make some simple, and hopefully valid, assumptions. If the universe is isotropic and homogenous on large scales, the Robertson-Walker metric,

$$ds^2 = dt^2 - a(t)\left[\frac{dr^2}{1 - kr^2} + r^2 d\theta^2\right]. \tag{1}$$

gives the line element distance (s) between two objects with coordinates r,θ and time separation, t. The Universe is assumed to have a simple topology such that if it has negative, zero, or positive curvature, k takes the value $-1, 0, 1$, respectively. These universes are said to be open, flat, or closed, respectively. The dynamic evolution of the Universe needs to be input into the Robertson-Walker metric by the specification of the scale factor $a(t)$, which gives the radius of curvature of the Universe over time — or more simply, provides the relative size of a piece of space at any time. This description of the dynamics of the Universe is derived from general relativity, and is known as the Friedman equation

$$H^2 \equiv (\dot{a}/a)^2 = \frac{8\pi G\rho}{3} - \frac{k}{a^2}. \tag{2}$$

The expansion rate of the universe (H), is called the Hubble parameter (or the Hubble constant H_0 at the present epoch) and depends on the content of the Universe. Here we assume the Universe is composed of a set of matter components, each having a fraction Ω_i of the critical density

$$\Omega_i = \frac{\rho_i}{\rho_{crit}} = \frac{\rho_i}{\frac{3H_0^2}{8\pi G}}, \tag{3}$$

with an equation of state which relates the density ρ_i and pressure p_i as $w_i = p_i/\rho_i$. For example w_i takes the value 0 for normal matter, +1/3 for photons, and -1 for the cosmological constant. The equation of state parameter does not need to remain fixed; if scalar fields are at play, the effective w will change over time. Most reasonable forms of matter or scalar fields have $w_i >= -1$, although nothing seems manifestly forbidden. Combining equations 1 – 3 yields solutions to the global evolution of the Universe[12].

In cosmology, there are many types of distance, the luminosity distance, D_L, and angular size distance, D_A, being the most useful to cosmologists. D_L, which is defined as the apparent brightness of an object as a function of its redshift z — the amount an object's light has been stretched by the expansion of the Universe — can be derived from equations 1 – 3 by solving for the surface area as a function of z, and taking into account the effects of energy diminution and time dilation as photons get stretched travelling through the expanding universe. The angular size distance, which is defined by the angular size of an object as a function of z, is closely related to D_L, and both are given by the numerically integrable equation,

$$D_L = D_A(1+z)^2 = \frac{c}{H_0}(1+z)\kappa_0^{-1/2} S\{\kappa_0^{1/2} \int_0^z dz' [\sum_i \Omega_i (1+z')^{3+3w_i} - \kappa_0 (1+z')^2]^{-1/2}\}. \tag{4}$$

We define $S(x) = \sin(x)$, x, or $\sinh(x)$ for closed, flat, and open models respectively, and the curvature parameter κ_0, is defined as $\kappa_0 = \sum_i \Omega_i - 1$.

Historically, equation 4 has not been easily integrated, and was expanded in a Taylor series to give

$$D_L = \frac{c}{H_0}\{z + z^2 \left(\frac{1-q_0}{2}\right) + \mathcal{O}(z^3)\}, \tag{5}$$

where the deceleration parameter, q_0 is given by

$$q_0 = \frac{1}{2}\sum_i \Omega_i (1 + 3w_i). \tag{6}$$

From equation 6, we can see that in the nearby universe, the luminosity distance scales linearly with redshift, with H_0 serving as the constant of proportionality. In the more distant Universe, D_L depends to first order on the rate of acceleration/deceleration (q_0), or equivalently, the amount and types of matter that it is made up of. For example, since normal gravitating matter has $w_M = 0$ and the cosmological constant has $w_\Lambda = -1$, a universe composed of only these two forms of matter/energy has $q_0 = \Omega_M/2 - \Omega_\Lambda$. In a universe composed of these two types of matter, if $\Omega_\Lambda < \Omega_M/2$, q_0 is positive, and the Universe is decelerating. These decelerating Universes have D_Ls that are smaller as a function of z (for low z) than their accelerating counterparts. If distance measurements are made at a low-z and a small range of redshift at higher redshift, there is a degeneracy between Ω_M and Ω_Λ; it is impossible to pin down the absolute amount of either species of matter (only their relative fraction which at $z = 0$ is given by equation 6). However, by observing objects over a range of high redshift (e.g. $0.3 > z > 1.0$), this degeneracy can be broken, providing a measurement of the absolute fractions of Ω_M and Ω_Λ[24].

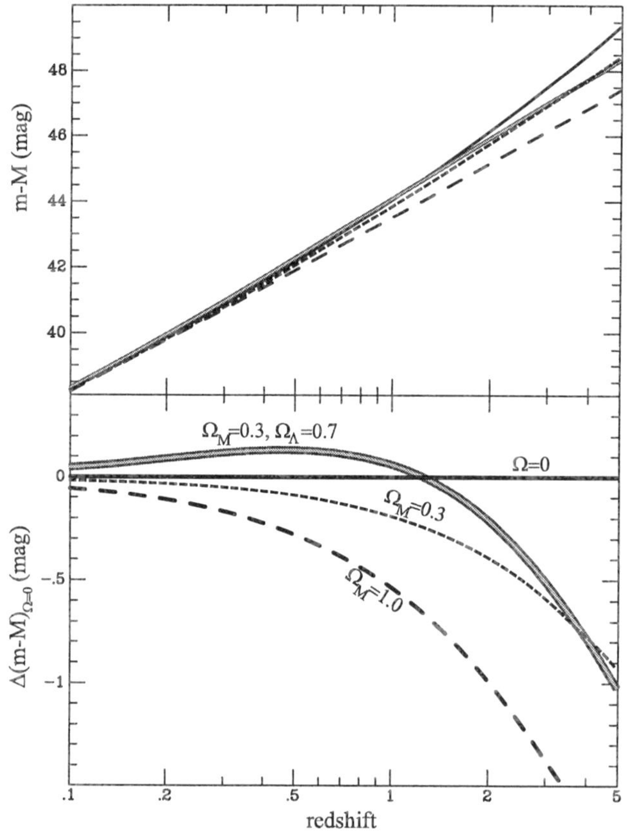

Figure 1. D_L expressed as distance modulus $(m - M)$ for four relevant cosmological models; $\Omega_M = 0$, $\Omega_\Lambda = 0$ (empty Universe); $\Omega_M = 0.3$, $\Omega_\Lambda = 0$; $\Omega_M = 0.3$, $\Omega_\Lambda = 0.7$; and $\Omega_M = 1.0$, $\Omega_\Lambda = 0$. In the bottom panel the empty universe has been subtracted from the other models to highlight the differences.

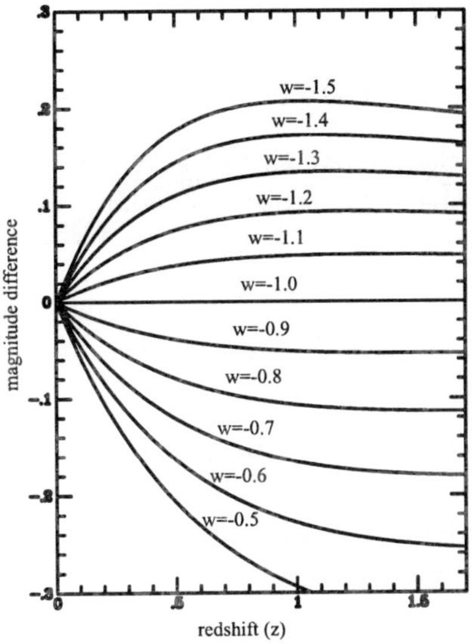

Figure 2. D_L for a variety of cosmological models containing $\Omega_M = 0.3$ and $\Omega_x = 0.7$ with equation of state w_x. The $w_x = -1$ model has been subtracted off to highlight the differences of the various models.

To illustrate the effect of cosmological parameters on the luminosity distance, in Figure 1 we plot a series of models for both Λ and non-Λ Universes. In the top panel, the various models show the same linear behaviour at $z < 0.1$ with models with the same H_0 indistinguishable to a few percent. By $z = 0.5$, the models with significant Λ are clearly separated, with distances that are significantly further than the zero-Λ universes. Unfortunately, two perfectly reasonable universes, given our knowledge of the local matter density of the Universe ($\Omega_M \sim 0.25$), one with a large cosmological constant, Ω_Λ=0.7, $\Omega_M = 0.3$, and one with no cosmological constant, $\Omega_M = 0.2$, show differences of less than 10%, even to redshifts of $z > 5$. Interestingly, the maximum difference between the two models is at $z \sim 0.8$, not at large z.

Figure 2 illustrates the effect of changing the equation of state of the non $w = 0$ matter component, assuming a flat universe $\Omega_{tot} = 1$. If we are to discriminate a dark energy component that is not a cosmological constant, measurements better than 5% are clearly required, especially since the differences in this diagram include the assumption of flatness, and also fix the value of Ω_M.

Other tests of cosmology are also possible within the standard model. These have been less widely used because of the difficulty in implementing them observationally. For example, if the absolute age difference of objects were known (for example, by radioactive dating of stars), then this could be compared to the modelled cosmological age

$$t_0 - t_1 = \frac{1}{H_0}\int_0^z dz' \left((1+z)\sqrt{(1+z)^2(1+\Omega_M z) - z(2+z)\Omega_\Lambda}\right)^{-1}. \quad (7)$$

Or, following Hubble, if the relative size of a volume of space were known as a function of z (*e.g.* via numbers of galaxies), then this provides another cosmological test

$$V(D_L) = \frac{1}{2H_0^3\kappa_0}\left[\frac{H_0 D_L}{(1+z)}\sqrt{1+\kappa_0[H_0^2 D_L^2(1+z)^2]} - \frac{1}{\sqrt{\kappa_0}}S^{-1}\left[H_0 D_L(1+z)\sqrt{\kappa_0}\right]\right] \quad (8)$$

where S and κ_0 have the same definitions as for equation (4).

Other ways to learn about our Universe include the density test — simply counting up how much mass there is in the Universe by its gravitational effect, and structure evolution tests, where the evolution of structure of the Universe is compared to a model. These two tests have become very powerful with the advent of large galaxy redshift surveys, and even larger cosmic simulations of the large scale structure growth in the Universe.

3. The Extragalactic Distance Toolbox

Since the 1930s Astronomers have developed a range of methods for measuring extragalactic distances. None is perfect, and none can be used in all situations, and this has made progress very slow in measuring distances. Here is a brief description of some of the most popular and influential distance methods over the past two decades in alphabetical order, excluding supernovae which I will give special attention to at the end of the section.

3.1. *Brightest Cluster Galaxies*

The Brightest Cluster Galaxy method has been popular since the 1950s because the objects used as standard candles — the brightest galaxy in a cluster of galaxies — are so bright[38]. The method has been most recently exploited by Lauer & Postman who found, by including a parameter related to the diffuseness of the galaxy, they could increase the precision of the method to roughly $\sigma \sim 0.25$mag[49]. Evolution of the galaxies[104] precludes using these as anything but local tracers, and the poor physical basis of the method plus some unexplained results (e.g. Lauer and Postman 1994)[50] has caused this method to fall out of favour with the general community.

3.2. *Cepheids*

The Period Luminosity (P-L) relationship of Cepheid variable stars has been exploited since it was first recognised by Leavitt through looking at stars in the LMC[51,52]. The method has a strong theoretical basis, and although theoretical calibrations of the P-L relationship exist, the empirical relationships derived from the Large Magellanic Cloud are still used to measure distances by the community. The Cepheids have gained special notoriety over the past decade because the Hubble Space Telescope is able to observe these objects in a large number of galaxies at distances beyond 20 Mpc. It is sometimes assumed that Cepheids are problem free, but they have many of the problems that other methods face. As massive stars, Cepheids are often highly extinguished (and this is difficult to remove with optical data alone). There is a poorly constrained relationship versus metallicity, and photometry of these faint objects on complex backgrounds is very difficult, even with the Hubble Space Telescope. Even so, Cepheids, with their good theoretical understanding, and distance uncertainties of roughly $\sigma \sim 0.1$ mag per galaxy, are a cornerstone of extragalactic distance indicators, and are used to calibrate most other methods.

3.3. *Fundamental Plane (aka $D_n - \sigma$)*

Elliptical galaxies exhibit a correlation between their surface brightness within a half-light radius and their velocity dispersion. This relationship, often called the $D_n - \sigma$ or Fundamental Plane, is observationally cheap, and has been used to discover the "Great Attractor"[61], as well as measure the Hubble constant. The method, while a favourite for building up large distance data sets in early type galaxies, has a poor physical basis, is imprecise ($\sigma \sim 0.4$ mag per galaxy) and there are some questions as to environmental effects leading to systematic errors in the distances derived.

3.4. *Lensing Delay*

It was suggested by Einstein that it was possible for a galaxy or star to act as a gravitational lens, bending light from a distant object over multiple paths, and magnifying the background object. Refsdal[85] realised, well before the discovery of the first lens, that the measurement of the time delay between light travelling on two or more of the different paths would enable the absolute distance to the lens to be measured. Many attempts were made at measuring the time delay for this first QSO lens 0957+561, with different groups getting different answers, depending on the analysis techniques. An unambiguous result was obtained by Kundic *et al.* in 1997 who observed the delay to be 417 ± 3 days[48]. At least 10 lenses with the necessary information to measure distances are currently available, and the results are summarised by Kochanek & Schechter[46]. The principal uncertainty in the method is knowing the mass distribution of the lensing galaxy, and this requires significant further work.

3.5. *Sunyaev-Zeldovich*

The Sunyaev-Zeldovich Effect (SZE) was first proposed in 1970 as a distance measuring technique[98],[99]. The SZE occurs when photons in the Cosmic Microwave Background undergo inverse Compton scattering off of hot electrons in the intracluster gas of galaxy clusters (seen as thermal emission in X-rays). By comparing measurements of the SZE effect with X-ray measurements of the cluster gas through a model. Since the X-ray emission is proportional to the electron density squared, and the SZE is linearly proportional to the electron density, it is possible to solve simultaneously for the electron density and distance, using a simple model (isothermal sphere) of the X-ray emitting gas. Complications arise because the few clusters examined in detail show deviations from the usual simple isothermal spheres assumed, through asphericity, and much worse, clumping. As the X-ray data improves, so will the modelling. We can expect, in the next decade, to have detailed distances to hundreds, and possibly, orders of magnitude more clusters.

3.6. *Surface Brightness Fluctuations*

Images of elliptical galaxies show brightness variations from pixel to pixel caused by the position fluctuations of the number of stars in each resolution element. These so called "Surface Brightness Fluctuations" (SBF) depend on the ratio of resolution and distance, because as more and more stars fall into a resolution element, the $\sqrt{N}$ fluctuations become a smaller and smaller fraction of the light within this area. Nearby galaxies appear highly mottled, whereas their more distant cousins appear as smoother objects under the same conditions. The method is explained in detail in Jacoby *et al.* [39], with the most comprehensive implementation of the method given by Tonry *et al.* [105]. This I-band implementation to several hundred objects shows the method provides distances with a precision of approximately 6–7%, making the method among the most precise available to astronomy. The method is limited on the ground to approximately $z < 0.015$, and using the Hubble Space Telescope to $z < 0.03$, although it appears possible to extend the range of the method observing in the near-IR[40] , possibly to cosmological distances using diffraction limited 30m telescopes.

3.7. *Tully-Fisher*

The empirical relationship between the luminosity of a spiral galaxy and its rotational velocity dates back to Opik[67], but gained acceptance as a useful method of measuring distances after the work of Tully and Fisher[107], and the method is usually referred to now as the Tully-Fisher method. The method is explained in detail within the review of Jacoby *et al.* [39], and the method has been applied to thousands of galaxies, using rotational velocities measured either from radio HI 21cm emission, or optical Hα emission. The method is relatively imprecise (20% uncertainty), but this is made up for by the relative ease of measuring distances. Measurements of

10 objects can beat down the uncertainty to a level as good as any indicator. The method has been used to a redshift of $z \sim 0.1$, and with current instrumentation, it should be possible to extend the method to objects at higher redshift. Unfortunately, because this is an empirical relationship being applied to a class of objects that show evolution even at $z < 0.5$, it is unlikely that the Tully-Fisher relationship can be used to probe cosmological parameters other than H_0.

3.8. *Type II Supernovae*

Massive stars come in a wide variety of shapes and sizes, and would seemingly not be useful objects for making distance measurements under the standard candle assumption — however, from a radiative transfer standpoint, these objects are relatively simple, and can be modelled with sufficient accuracy to measure distances to approximately 10%. The expanding photosphere method (EPM), was developed by Kirshner and Kwan in 1974[45], and implemented on a large number of objects by Schmidt *et al.* in 1994[92] after considerable improvement in the theoretical understanding of type II supernovae (SN II) atmospheres[113,14,15].

EPM assumes that SN II radiate as dilute blackbodies

$$\theta_{ph} = \frac{R_{ph}}{D} = \sqrt{\frac{f_\lambda}{\zeta^2 \pi B_\lambda(T)}}, \tag{9}$$

where θ_{ph} is the angular size of the photosphere of the SN, R_{ph} is the radius of the photosphere, D is the distance to the SN, f_λ is the observed flux density of the SN, and $B_\lambda(T)$ is the Planck function at a temperature T. Since SN II are not perfect blackbodies, we include a correction factor, ζ, which is calculated from radiative transfer models of SN II. Supernovae freely expand, and

$$R_{ph} = v_{ph}(t - t_0) + R_0, \tag{10}$$

where v_{ph} is the observed velocity of material at the position of the photosphere, t is the time elapsed since the time of explosion, t_0. For most stars, the stellar radius at the time of explosion, R_0, is negligible, and equations 9 and 10 can be combined to yield

$$t = D\left(\frac{\theta_{ph}}{v_{ph}}\right) + t_0 \tag{11}$$

By observing a SN II at several epochs, measuring the flux density and temperature of the SN (via broad band photometry) and v_{ph} from the minima of the weakest lines in the SN spectrum, we can solve simultaneously for the time of explosion and distance to the SN II. The key to successfully measuring distances via EPM is an accurate calculation of $\zeta(T)$. Requisite calculations were performed by Eastman, Schmidt and Kirshner[15], but, unfortunately, no other calculations of $\zeta(T)$ have yet been published for typical SN II-P progenitors.

Hamuy *et al.* [25] and Leonard *et al.* [57] have both measured the distances to SN 1999em, and have investigated other aspects of the implementation of EPM. Hamuy *et al.* [25] challenged the prescription of measuring velocities from the minima of weak lines, and developed a framework of cross-correlating spectra with synthesised spectra to estimate the velocity of material at the photosphere. This different prescription does lead to small systematic differences in estimated velocity using weak lines, but provided the modelled spectra are good representations of real objects, this method should be more correct. As yet, a revision of the EPM distance scale using this method of estimating v_{ph} has not been made.

Leonard *et al.* [56] have obtained spectropolarimetry of SN 1999em at many epochs, and see polarization intrinsic to the SN which is consistent with the SN having asymmetries of 10 to 20 percent. Asymmetries at this level are found in most SN II[115], and may ultimately limit the accuracy EPM can achieve on a single object (10% RMS) — however, the mean of all SN II distances should remain unbiased.

Type II supernovae have played an important role in measuring the Hubble constant independently of the rest of the extragalactic distance scale. In the next decade, it is quite likely that surveys will begin to turn up significant numbers of these objects at $z \sim 0.5$, and therefore the possibility exists that these objects will be able to make a contribution to the measurement of cosmological parameters beyond the Hubble Constant. Since SN II do not have the precision of the SN Ia (next section), and are significantly harder to obtain relevant data from, they will not replace the SN Ia, but they are an independent class of object which have the potential to confirm the interesting results that have emerged from the SN Ia objects.

3.9. *Type Ia Supernovae*

SN Ia have been used as extragalactic distance indicators since Kowal first published his Hubble diagram ($\sigma = 0.6$ mag) for SNe I in 1968[43]. We now recognize that the old SNe I spectroscopic class is comprised of two distinct physical entities: SN Ib/c which are massive stars that undergo core collapse (or in some rare cases might undergo a thermonuclear detonation in their cores) after losing their hydrogen atmospheres, and the SN Ia which are most likely thermonuclear explosions of white dwarfs. In the mid-1980s, it was recognized that studies of the Type I supernova sample had been confused by these similar-appearing supernovae, which were henceforth classified as Type Ib[116,108,68] and Type Ic[33]. By the late 1980s/early 1990s, a strong case was being made that the vast majority of the true Type Ia supernovae had strikingly similar lightcurve shapes[10,53,54,55], spectral time series[7,71,31,17], and absolute magnitudes[63,55]. There were a small minority of clearly peculiar Type Ia supernovae, e.g. SN 1986G[72], SN 1991bg[18,59], and SN 1991T[18,83], but these could be identified and "weeded out" by unusual spectral features. A 1992 review by Branch & Tammann[8] of a variety of studies in the literature concluded that the intrinsic dispersion in B and V maximum for Type Ia supernovae must be less than 0.25 mag, making them "the best standard candles known so far."

In fact, the Branch & Tammann review indicated that the magnitude dispersion was probably even smaller, but the measurement uncertainties in the available datasets were too large to tell. Realising the subject was generating a large amount of rhetoric despite not having a sizeable well-observed data set, a group of astronomers based in Chile started the Calan/Tololo Supernova Search in 1990[28]. This work took the field a dramatic step forward by obtaining a crucial set of high-quality supernova lightcurves and spectra. By targeting a magnitude range that would discover Type Ia supernovae in the redshift range between 0.01 and 0.1, the Calan/Tololo search was able to compare the peak magnitudes of supernovae whose relative distance could be deduced from their Hubble velocities.

The Calan/Tololo Supernova Search observed some 25 fields (out of a total sample of 45 fields) twice a month for over $3\frac{1}{2}$ years with photographic plates or film at the CTIO Curtis Schmidt telescope, and then organized extensive follow-up photometry campaigns primarily on the CTIO 0.9m telescope, and spectroscopic observation on either the CTIO 4m or 1.5m. The search was a major success; with the cooperation of many visiting CTIO astronomers and CTIO staff, it created a sample of 30 new Type Ia supernova lightcurves, most out in the Hubble flow, with an almost unprecedented (and unsuperseded) control of measurement uncertainties[27].

In 1993 Phillips, in anticipation of the results he could see coming in as part of the Calan/Tololo search (he was a member of this team), looked for a relationship between the rate at which the Type Ia supernova's luminosity declines and its absolute magnitude. He found a tight correlation between these parameters using a sample of nearby objects, where he plotted the absolute magnitude of the existing set of nearby SN Ia which had dense photoelectric or CCD coverage, versus the parameter $\Delta m_{15}(B)$, the amount the SN decreased in brightness in the B band over the 15 days following maximum light[73]. For this work, Phillips used a heterogenous mixture of other distance indicators to provide relative distances, and while the general results were accepted by most, scepticism about the scatter and shape of the correlation remained. The Calan/Tololo search presented their first results in 1995 when Hamuy *et al.* showed a Hubble diagram of 13 objects at $cz > 5000$ km/s that displayed the generic features of the Phillips (1993) relationship[27]. It also demonstrated that the intrinsic dispersion of SN Ia using the $\Delta m_{15}(B)$ method was better than 0.15 mag.

As the Calan/Tololo data began to become available to the broader community, several methods were presented that could select for the "most standard" subset of the Type Ia standard candles, a subset which remained the dominant majority of the ever-growing sample[6]. For example, Vaughan *et al.* presented a cut on the $B - V$ colour at maximum that would select what were later called the "Branch normal" SN Ia, with an observed dispersion of less than 0.25 mag[111].

The community more or less settled on the notion that including the effect of lightcurve shape was important for measuring distances with SN Ia when in 1996 Hamuy *et al.* showed the scatter in the Hubble diagram dropped from $\sigma \sim 0.38$ mag in B to $\sigma \sim 0.17$ mag for their sample of nearly 30 SN Ia at $cz > 3000$ km/s using the $\Delta m_{15}(B)$ correlation[30] .

Impressed by the success of the $\Delta m_{15}(B)$ parameter, Riess, Press and Kirshner developed the multi-colour lightcurve shape method (MLCS), which parameterizes the shape of SN lightcurves as a function of their absolute magnitude at maximum[90]. This method also included a sophisticated error model, and fitted observations in all colours simultaneously, allowing a colour excess to be included. This colour excess, which we attribute to intervening dust, enables the extinction to be measured.

Another method that has been used widely in cosmological measurements with SN Ia is the "stretch" method, described by Perlmutter *et al.* [82,80]. This method is based on the observation that the entire range of SN Ia lightcurves, at least in the B and V bands, can be represented with a simple time-stretching (or shrinking) of a canonical lightcurve. The coupled stretched B and V lightcurves serve as a parameterized set of lightcurve shapes, providing many of the benefits of the MLCS method, but as a much simpler (and constrained) set. This method, as well as recent implementations of $\Delta m_{15}(B)$[74,22], and template fitting[106] also allows extinction to be directly incorporated into the SN Ia distance measurement. Other methods that correct for intrinsic luminosity differences or limit the input sample by various criteria have also been proposed to increase the precision of SNe Ia as distance indicators[102,16,4,109]. While these latter techniques are not as developed as the $\Delta m_{15}(B)$, MLCS, and stretch methods, they all provide distances that are comparable in precision, roughly $\sigma = 0.18$ mag about the inverse square law, equating to a fundamental precision of SN Ia distance being 6% (0.12 mag), once photometric uncertainties and peculiar velocities are removed.

4. Measuring the Hubble Constant

To measure H_0, most methods must still be externally calibrated with Cepheids, and this calibration is the major limitation to measuring H_0. The Key Project has used Hubble Space Telescope observations of Cepheid variable stars in many galaxies to calibrate several of the distance methods described above. From their analysis, using SN Ia, Tully-Fisher, Fundamental Plane, and Surface Brightness Fluctuations, the Key Project concludes that $H_0 = 72 \pm +3 \pm 7$ where the first error bar is statistical, and the second, systematic (Figure 3).

The current nearby SN Ia sample[27,87,22,41] contains >100 objects (Figure 4), and accurately defines the slope in the Hubble diagram from $0 < z < 0.1$ to 1%. A team competing with the Key Project has also used the Hubble Space Telescope to independently calibrate several SN Ia supernovae. Two separate teams' analysis of the Cepheids and SN Ia have yielded surprisingly divergent values for the Hubble constant: Saha *et al.* find $H_0 = 59 \pm 6$[94] while Freedman *et al.* find $H_0 = 71 \pm 2 \pm 6$[19].

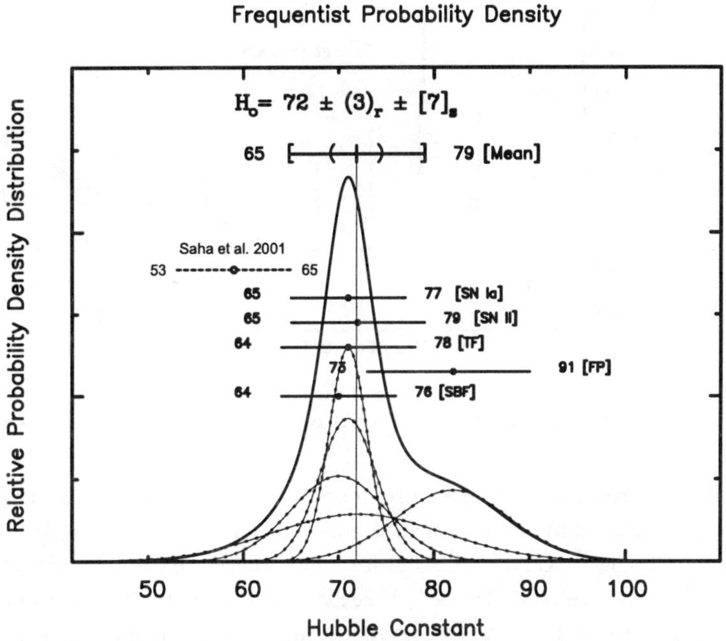

Figure 3. The derived values and uncertainties of the Key Project's Cepheid calibration (Freedman *et al.* 2001) of a variety of distance indicators. Overlaid is the Saha *et al.* 2001 SN Ia calibration. Figure adapted from Freedman 2001

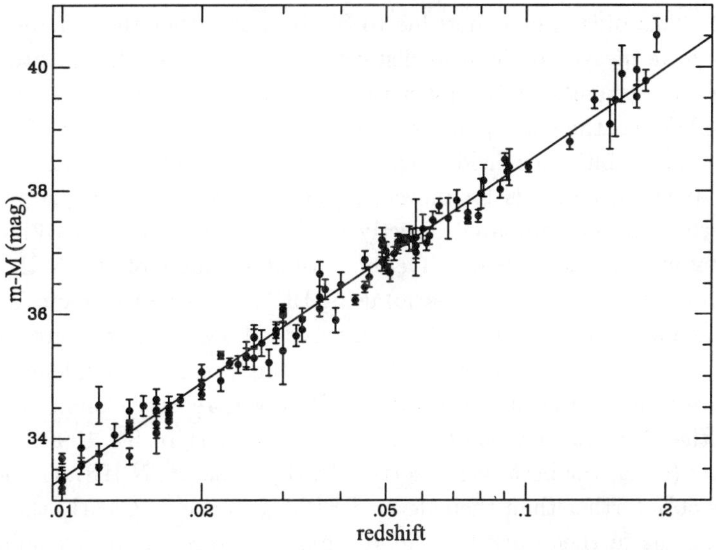

Figure 4. The Hubble diagram for High-Z SN Ia from $0.01 > z > 0.2$. The 102 objects in this redshift range have a residual about the inverse square line of approximately 10%.

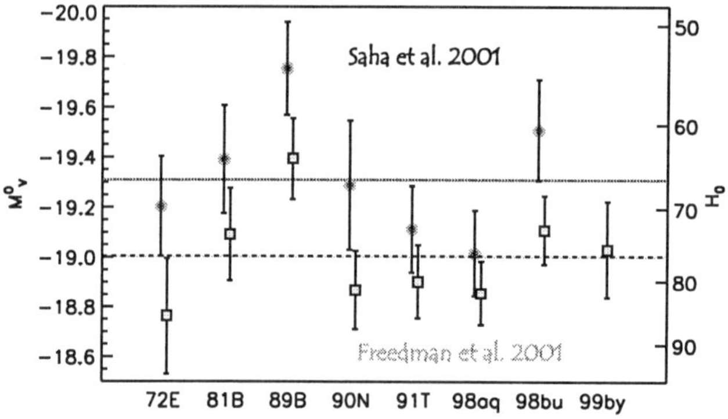

Figure 5. The derived values and uncertainties of each SN Ia's absolute magnitude using the Key Project's Cepheid calibration and the SN Ia Project calibration. Figure adapted from Jha 2002.

Jha has compared the SN Ia measurements using an updated version of MLCS to the distances measured by the two *HST* teams that have obtained Cepheid distances to SN Ia host galaxies[41]. Of the 12 SN Ia for which there are Cepheid distances to the host galaxy: 1895B*, 1937C*, 1960F*, 1972E, 1974G*,1981B, 1989B, 1990N, 1991T, 1998eq, 1998bu, 199by, four were observed by non-digital means marked by *, and are best excluded from analysis on the basis that non-digital photometry routinely has systematic errors far greater than 0.1 mag. Using the digitally observed SN Ia only, he finds, using distances from the SN Ia project[94], $H_0 = 66 \pm 3 \pm 7$ km/s/Mpc. The same analysis to the Key Project distances[19] give $H_0 = 76 \pm 3 \pm 8$ km/s/Mpc (Figure 5). This difference is not due to SN Ia, but rather the different ways the two teams have measured Cepheid distances with *HST*. While the two values do overlap in the extremes of the estimates of systematic error, it is none-the-less uncomfortable that the discrepancies are as large as this, when most of the claimed systematic uncertainties are held in common between the two teams.

Of the physical methods for measuring H_0, the SN II are arguably the most useful, as they can be compared directly to the Cepheids, and provide their own Hubble flow measurement. Schmidt *et al.* , using a sample of 16 SN II, estimated $H_0 = 73 \pm 6$(statistical)± 7 (systematic) using EPM[92]. Using this paper's distances, the Cepheid and EPM distance scales, compared galaxy to galaxy, agree within 5%, and are consistent within the errors[15,57], and this provides confidence that both methods are providing accurate distances. However, recently Leonard *et al.* have measured the Cepheid distance to NGC 1637, the host of SN 1999em. For this single object (albeit the best ever observed SN II-P besides SN 1987A), the Cepheid distance is 50% further than their derived EPM distance[58]. Clearly this large discrepancy signals further work (and more objects) are required to confidently use EPM distances in this age of precision cosmology.

The S-Z effect and Lensing both provide distance measurement to objects in the Hubble flow, however, concerns are still large for systematic modelling uncertainties for both these methods. Kochanek & Schechter have used lensing to derive distance to 10 objects, and find a surprisingly low value of $H_0 = 48 \pm 3$ km/s/Mpc if they assume isothermal mass distributions for the lensing galaxies[46]. This current work needs to assume the form of the mass distributions of the lensing galaxies, but future work should place better constraints on these inputs. With this information, it should become more obvious if there is indeed a conflict between the value of H_0 measured via lensing at $z = 0.3$ and the more local measurements. In general, future work on measuring H_0 lies not with the secondary/tertiary distance indicators, but with the Cepheid calibrators, or using other primary distance indicators such as EPM, Sunyaev-Zeldovich effect, or Lensing.

5. The Measurement of Acceleration by SN Ia

The intrinsic brightness of SN Ia allow them to be discovered to $z > 1.5$. Figure 1 shows that the differences in luminosity distances due to different cosmological models at this redshift are roughly 0.2 mag. For SN Ia, with a dispersion 0.2 mag, 10 well observed objects should provide a 3σ separation between the various cosmological models. It should be noted that the uncertainty described above in measuring H_0 is not important in measuring other cosmological parameters, because it is only the relative brightness of objects near and far that is being exploited in equation 4 — the value of H_0 scales out.

The first distant SN search was started by a Danish team. With significant effort and large amounts of telescope time spread over more than two years, they discovered a single SN Ia in a $z = 0.3$ cluster of galaxies (and one SN II at $z = 0.2$)[65,32]. The SN Ia was discovered well after maximum light, and was only marginally useful for cosmology itself.

Just before this first discovery in 1988, a search for high-redshift Type Ia supernovae was begun at the Lawrence Berkeley National Laboratory (LBNL) and the Center for Particle Astrophysics, at Berkeley. This search, now known as the Supernova Cosmological Project (SCP), targeted SN at $z > 0.3$. In 1994, the SCP brought on the high-Z SN Ia era, developing the techniques which enabled them to discover 7 SN at $z > 0.3$ in just a few months.

The High-Z SN Search (HZSNS) was conceived at the end of 1994, when this group of astronomers became convinced that it was both possible to discover SN Ia in large numbers at $z > 0.3$ by the efforts of Perlmutter[78], and also use them as precision distance indicators as demonstrated by the Calan/Tololo group[27]. Since 1995, the SCP and HZSNS have both been working feverishly to obtain a significant set of high-redshift SN Ia.

5.1. *Discovering SN Ia*

The two high-redshift teams both used this pre-scheduled discovery-and-follow-up batch strategy pioneered by Perlmutter's group in 1994. They each aimed to use the observing resources they had available to best scientific advantage, choosing, for example, somewhat different exposure times or filters.

Quantitatively, type Ia supernovae are rare events on an astronomer's time scale — they occur in a galaxy like the Milky Way a few times per millennium[11,69,70,106]. With modern instruments on 4 meter-class telescopes, which scan 1/3 of a square degree to $R = 24$ magnitude in less than 10 minutes, it is possible to search a million galaxies to $z < 0.5$ for SN Ia in a single night.

Since SN Ia take approximately 20 days to rise from nothingness to maximum light[89], the three-week separation between "before and after" observations (which equates to 14 restframe days at $z = 0.5$) is a good filter to catch the supernovae on the rise. The supernovae are not always easily identified as new stars on galaxies — most of the time they are buried in their hosts, and we must use a relatively sophisticated process to identify them. In this process, the imaging data that we take in a night is aligned with the previous epoch, with the image star profiles matched (through convolution) and scaled between the two epochs to make the two images as identical as possible. The difference between these two images is then searched for new objects which stand out against the static sources that have been largely removed in the differencing process[82,93]. The dramatic increase in computing power in the 1980s was thus an important element in the development of this search technique, as was the construction of wide-field cameras with ever-larger CCD detectors or mosaics of such detectors.

This technique is very efficient at producing large numbers of objects that are, on average, at near maximum light, and does not require obscene amounts of telescope time. It does, however, place the burden of work on follow-up observations, usually with different instruments on different telescopes. With the large number of objects able to be discovered (50 in two nights being typical), a new strategy is being adopted by both teams, as well as additional teams like the CFHT Legacy survey, where the same fields are repeatedly scanned several times per month in multiple colours, for several consecutive months. This type of observing program provides both discovery of objects and their follow up, integrated into one efficient program. It does require a large block of time on a single telescope — a requirement which was apparently not politically feasible in years past, but is now possible.

5.2. *Obstacles to Measuring Luminosity Distances at High-Z*

As shown above, the distances measured to SN Ia are well characterized at $z < 0.1$, but comparing these objects to their more distant counterparts requires great care. Selection effects can introduce systematic errors as a function of redshift, as can uncertain K-corrections, and an evolution of the SN Ia progenitor population as a function of look-back time. These effects, if they are large and not constrained or corrected with measurements, will limit our ability to accurately measure relative luminosity distances, and have the potential to undermine the potency of high-z SN Ia at measuring cosmology [82,93,88,80,106,42].

5.2.1. *K-Corrections*

As SN are observed at larger and larger redshifts, their light is shifted to longer wavelengths. Since astronomical observations are normally made in fixed bandpasses on Earth, corrections need to be made to account for the differences caused by the spectrum of a SN Ia shifting within these bandpasses. These corrections take the form of integrating the spectrum of a SN Ia as observed with the relevant bandpasses, and shifting the SN spectra to the correct redshifts, and re-integrating. Kim *et al.* showed that these effects can be minimized if one does not stick with a single bandpass, but rather if one chooses the closest bandpass to the redshifted rest-frame bandpass[44]. They showed the interband K-correction is given by

$$K_{ij}(z) = 2.5 \log \left[(1+z) \frac{\int F(\lambda) S_i(\lambda) d\lambda}{\int F(\lambda/(1+z)) S_j(\lambda) d\lambda)} \frac{\int Z(\lambda) S_j(\lambda) d\lambda}{\int Z(\lambda) S_i(\lambda) d\lambda} \right], \quad (12)$$

where $K_{ij}(z)$ is the correction to go from filter i to filter j, and $Z(\lambda)$ is the spectrum corresponding to zero magnitude of the filters.

The brightness of an object expressed in magnitudes, as a function of z is

$$m_i(z) = 5 \log\left(\frac{D_L(z)}{\mathrm{Mpc}}\right) + 25 + M_j + K_{ij}(z), \quad (13)$$

where $D_L(z)$ is given by equation 4, M_j is the absolute magnitude of object in filter j, and K_{ij} is given by equation 12. For example, for $H_0 = 70$ km/s/Mpc, $D_L = 2835$ Mpc ($\Omega_M = 0.3, \Omega_\Lambda = 0.7$); at maximum light a SN Ia has $M_B = -19.5$ mag and a $K_{BR} = -0.7$ mag; We therefore expect a SN Ia at $z = 0.5$ to peak at $m_R \sim 22.1$ for this set of cosmological parameters.

K-correction errors depend critically on several separate uncertainties:

(1) Accuracy of spectrophotometry of SN. To calculate the K-correction, the spectra of supernovae are integrated in equation 12. These integrals are insensitive to a grey shift in the flux calibration of the spectra, but any wavelength dependent flux calibration error, will translate into incorrect K-corrections.
(2) Accuracy of the absolute calibration of the fundamental astronomical standard systems. Equation 12 shows that the K-corrections are sensitive to the shape of the astronomical bandpasses, and the zero points of these bandpasses.
(3) Using spectrophotometry for appropriate objects to calculate the corrections. Although a relatively homogenous class, there are variations in the spectra of SN Ia. If a particular objects has, for example, a stronger Calcium triplet than average SN Ia, the K-corrections will be error, unless a subset of appropriate SN Ia spectra are used in the calculations.

Error (1) should not be an issue if correct observational procedures are used on an instrument that has no fundamental problems. Error (2) is currently small (0.01

mag), and to be improved requires a careful experiment to accurately calibrate a star such as Vega or Sirius, and to carefully infer the standard bandpass that defines the photometric system in use at all telescopes being used. The final error requires a large database to be available to match as closely as possible a SN with the spectrophotometry used to calculate the K-corrections. Nugent *et al.* have shown that by correcting the SN spectra to match the photometry of a SN needing K-corrections, it is possible to largely eliminate errors (1) and (3)[66]. The scatter in the measured K-corrections from a variety of telescopes and objects allow us to estimate the combined size of the effect for the first and last error; these appear to be of order 0.01 mag for redshifts where the high-z and low-z filters have a large region of overlap (e.g. $R \Rightarrow B$ at $z = 0.5$). The size of the second error is estimated to be approximately 0.01 mag based on the consistency of spectrophotometry and broadband photometry of the fundamental standards, Sirius and Vega[3].

5.2.2. *Extinction*

In the nearby Universe we see SN Ia in a variety of environments, and about 10% have significant extinction[26]. Since we can correct for extinction by observing two or more wavelengths, it is possible to remove any first order effects caused by the average extinction properties of SN Ia changing as a function of z. However, second order effects, such as the evolution of the average properties of intervening dust could still introduce systematic errors. This problem can also be addressed by observing distant SN Ia over a decade or so of wavelength, in order to measure the extinction law to individual objects, but this is observationally expensive. Current observations limit the total systematic effect to less than 0.06 mag, as most of our current data is based on two colour observations.

An additional problem is the existence of a thin veil of dust around the Milky Way. Measurements from the COBE satellite have measured the relative amount of dust around the Galaxy accurately[95], but there is an uncertainty in the absolute amount of extinction of about 2% or 3%[9]. This uncertainty is not normally a problem; it affects everything in the sky more or less equally. However, as we observe SN at higher and higher redshifts, the light from the objects is shifted to the red, and is less affected by the galactic dust. A systematic error as large as 0.06 mag is attributable to this uncertainty with our present knowledge.

5.2.3. *Selection Effects*

As we discover SN, we are subject to a variety of selection effects, both in our nearby and distant searches. The most significant effect is Malmquist Bias — a selection effect which leads magnitude limited searches finding brighter than average objects near their distance limit; brighter objects can be seen in a larger volume relative to their fainter counterparts. Malmquist bias errors are proportional to the square of the intrinsic dispersion of the distance method, and because SN Ia are such accurate distance indicators, these errors are quite small — approximately 0.04 mag. Monte Carlo simulations can be used to estimate these effects, and to remove them from our data sets[93,80]. The total uncertainty from selection effects is approximately 0.01 mag, and interestingly, maybe worse for lower redshift, where they are, up to now, more poorly quantified.

There are many misconceptions held about selection effects and SN Ia. It is often quoted "that our search went 1.5 magnitudes fainter than the peak magnitude of a SN Ia at $z = 0.5$ and therefore our search is not subject to selection effects for $z = 0.5$ SN Ia". This statement is wrong. It is not possible to eliminate this effect by simply going deep. Although such a search would have smaller selection effects on the $z = 0.5$ objects than one a magnitude brighter, such a search would still miss $z = 0.5$ objects due to, in decreasing importance, their age (early objects missed), extinction (heavily reddened objects missed), and the total luminosity range of SN Ia (faintest SN Ia missed). Because the sample is not complete, such a search would still find brighter than average objects, and is biased (at the $\sim 2\%$ level).

5.2.4. *Gravitational Lensing*

Several authors have pointed out that the radiation from any object, as it traverses the large scale structure between where it was emitted, and where it is detected, will be weakly lensed as it encounters fluctuations in the gravitational potential[47,114,34]. Generally, most light paths go through under-dense regions, and objects appear de-magnified. Occasionally the photons from a distant object encounter dense regions, and these lines of sight become magnified. The distribution of observed fluxes for sources is skewed by this process, such that the vast majority of objects appear slightly fainter than the canonical luminosity distance, with the few highly magnified events making the mean of all paths unbiased. Unfortunately, since we do not observe enough objects to capture the entire distribution, unless we know and include the skewed shape of the lensing, a bias will occur. At $z = 0.5$, this lensing is not a significant problem: if the Universe is flat in normal matter, the large scale structure can induce a shift of the mode of the distribution by a few percent. However, the effect scales roughly as z^2, and by $z = 1.5$, the effect can be as large as 25%[35]. While corrections can be derived by measuring the distortion on background galaxies in the line-of-sight region around each SN, at $z > 1$, this problem may be one which ultimately limits the accuracy of luminosity distance measurements, unless a large enough set of SN at each redshift can be used to characterise the lensing distribution and average out the effect. For the $z \sim 0.5$ sample it is less than 0.02 mag problem, but is of significant concern for SN at $z > 1$ such as SN 1997ff[2], especially if observed in small numbers.

5.2.5. *Evolution*

SN Ia are seen to evolve in the nearby Universe. Hamuy *et al.* plotted the shape of the SN lightcurves against the type of host galaxy[29]. Early hosts (ones without recent star formation), consistently show lightcurves which rise and fade more quickly than those objects which occur in late-type hosts (objects with on-going star formation). However, once corrected for lightcurve shape, the corrected luminosity shows no bias as a function of host type. This empirical investigation provides confidence in using SN Ia over a variety of stellar population ages. It is possible, of course, to devise scenarios where some of the more distant supernovae do not have nearby analogues; therefore, at increasingly higher redshifts it can become important to obtain sufficiently detailed spectroscopic and photometric observations of each distant supernova to recognize and reject such examples that have no nearby analogues. Recent theoretical work suggests the SN type correlation with host galaxy is due to the metallicity of the host galaxy, with white dwarfs from metal rich systems (such as ellipticals) having significant amount of ^{22}Ne, which poisons the production of ^{56}Ni during the SN explosion[103]. Theoretical work such as this should help to better pin down the likely types of evolution SN Ia will be subject to at higher and higher redshifts.

In principle, it could be possible to use the differences in the spectra and lightcurves between nearby and distant samples to correct any differences in absolute magnitude. Unfortunately theoretical investigations are not yet advanced enough to precisely quantify the effect of these differences on the absolute magnitude. A different empirical approach to handle SN evolution is to divide the supernovae into subsamples of very closely matched events, based on the details of the object's lightcurve, spectral time series, host galaxy properties, etc. A separate Hubble diagram can then be constructed for each subsample of supernovae, and each will yield an independent measurement of the cosmological parameters[5]. The agreement (or disagreement) between the results from the separate subsamples is an indicator of the total effect of evolution. A simple, first attempt at this kind of test has been performed comparing the results for supernovae found in elliptical host galaxies to supernovae found in late spirals or irregular hosts; the cosmological results from these subsamples were found to agree well[97].

Finally, it is possible to move to higher redshift and see if the SN deviate from the predictions of equation 4. At a gross level, we expect an accelerating Universe to be decelerating in the past because the matter density of the Universe increases with redshift, whereas the density of any dark energy leading to acceleration will increase at a slower rate than this (or not at all in the case of a cosmological constant). If the observed acceleration is caused by some sort of systematic effect, it is likely to continue to increase (or at least remain steady) with look-back time, rather than disappear like the effects of dark energy. A first comparison has been made with SN 1997ff[23] at $z \sim 1.7$, and it seems consistent with a decelerating Universe at this epoch[86]. More objects are necessary for a definitive answer, and these should be provided by a large program using the Hubble Space Telescope in 2002–3 by Riess and collaborators.

5.3. *High Redshift SN Ia Observations*

The SCP in 1997 announced their first results with 7 objects at a redshift around $z = 0.4$[82]. These objects hinted at a decelerating Universe with a measurement of $\Omega_M = 0.88^{+0.69}_{-0.60}$, but were not definitive. Soon after, a $z \sim 0.8$ object observed with HST[81], and the first five objects of the HZSNS[93,20] ruled out a $\Omega_M = 1$ universe with greater than 95% significance. These results were again superseded dramatically when both the HZSNS[88] and the SCP[80] announced results that showed not only were the SN observations incompatible with a $\Omega_M = 1$ universe, they were also incompatible with a Universe containing only normal matter. Both samples show that SN are, on average, fainter than what would be expected for even an empty Universe, indicating that the Universe is accelerating. The agreement between the two teams' experimental results is spectacular, especially considering the two programs have worked in near complete isolation.

The easiest solution to explain the observed acceleration is to include an additional component of matter with an equation-of-state parameter more negative than $w < -1/3$; the most familiar being the cosmological constant ($w = -1$). If we assume the universe is composed only of normal matter and a cosmological constant, then with greater than 99.9% confidence, the Universe has a cosmological constant.

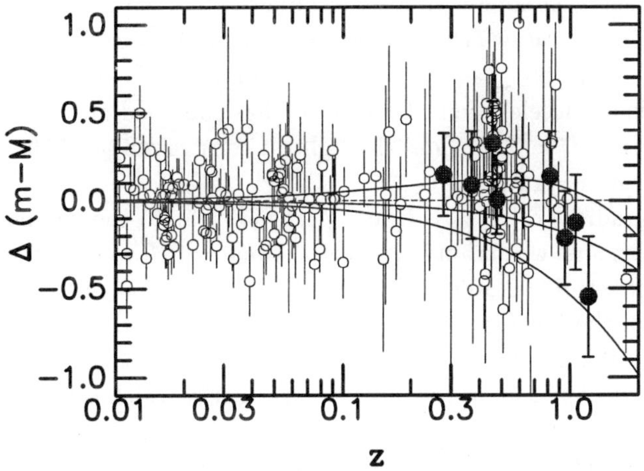

Figure 6. Data as summarised in Tonry 2003 with points shown in a residual Hubble diagram with respect to an empty universe. In this plot the highlighted points correspond to median values in six redshift bins. From top to bottom the curves show $\Omega_M, \Omega_\Lambda = 0.3, 0.7$, $\Omega_M, \Omega_\Lambda = 0.3, 0.0$, and $\Omega_M, \Omega_\Lambda = 1.0, 0.0$.

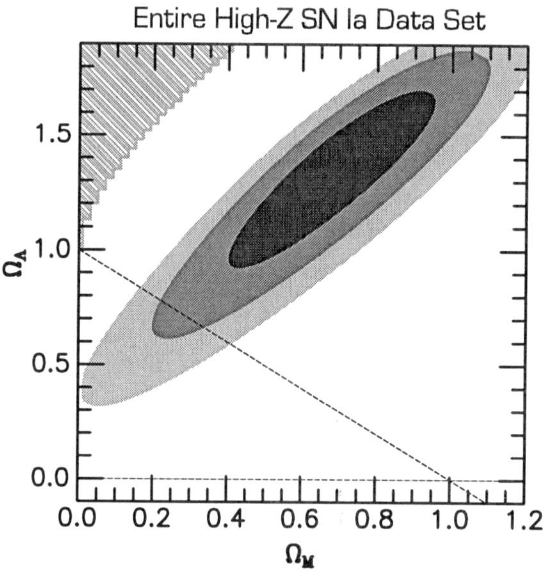

Figure 7. The joint confidence contours for Ω_M, Ω_Λ using the Tonry *et al.* compilation of objects

Since 1998, many new objects have been added and these can be used to further test past conclusions. Tonry *et al.* has compiled current data (Figure 6), and used only the new data to re-measure Ω_M, Ω_Λ, and find a more constrained, but perfectly compatible set of values with the SCP and High-Z 1998/99 results[106]. A similar study has been done with a set of objects observed using the Hubble Space Telescope by Knop *et al.* which also find concordance between the old data and new observations[42]. The 1998 results were not a statistical fluke, these independent sets of SN Ia still show acceleration. Tonry *et al.* has compiled all useful data from all sources (both teams) and provides the tightest constraints of SN Ia data so far[106]. These are shown in Figure 7.

Since the gradient of $H_0\, t_0$ is nearly perpendicular to the narrow dimension of the Ω_Λ–Ω_M contours, we obtain a a precise estimate of $H_0\, t_0$ from the SN distances. For the current set of 203 objects, we find $H_0\, t_0 = 0.96 \pm 0.04$[106], which is in good agreement with the far less precise determination of the ages of globular clusters using an $H_0 \sim 70$ km/s/Mpc.

Of course, we do not know the form of dark energy which is leading to the acceleration, and it is worthwhile investigating what other forms of energy are possible second components[21,80]. Figure 8 shows the joint confidence contours for Ω_M and w_x (the equation of state of the unknown component causing the acceleration) using the current compiled data set[106]. Because this introduces an extra parameter, we apply the additional constraint that $\Omega_M + \Omega_x = 1$, as indicated by the Cosmic Microwave Background Experiments[13,96]. The cosmological constant is preferred, but anything with a $w < -0.73$ is acceptable.

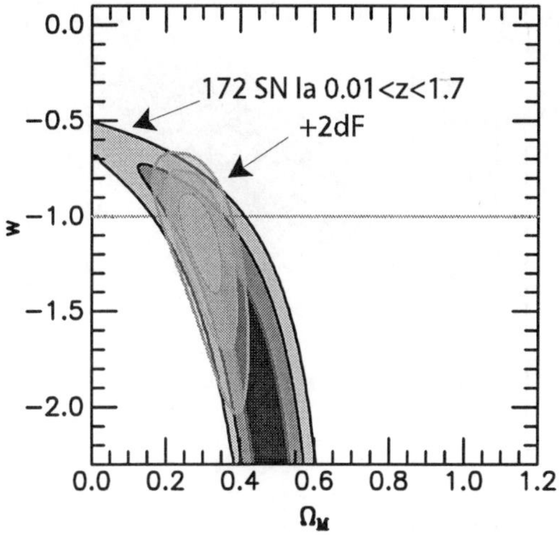

Figure 8. Contours of Ω_M versus w_x from current observational data (where $\Omega_M + \Omega_x = 1$ has been used as a prior), both with and without the additional constraint provided by the current value of Ω_M from the 2dF Galaxy Redshift Survey.

Additionally, we can add information about the value of Ω_M, as supplied by recent 2dF redshift survey results[112], as shown in the 2nd panel, where the constraint strengthens to $w < -0.73$ at 95% confidence. As a further test, if we assume a flat Λ universe, and derive Ω_M, independent of other methods, the SN Ia data give $\Omega_M = 0.28 \pm 0.05$, in perfect accord with the 2dF results. These results are essentially identical, both in value and in size of uncertainty, to those obtained from the recent WMAP experiment[96] when they combine their experiment with the 2dF results. Taken in whole, we have three cosmological experiments — SN Ia, Large Scale Structure, and the Cosmic Microwave Background, each probing parameter space in a slightly different way, and each agreeing with each other. Figure 9 shows that in order for the accelerating Universe to go away, two of these three experiments must both have severe systematic errors, and have these errors conspire in a way to overlap with each other to give a coherent story.

6. The Future

How far can we push the SN measurements? Finding more and more SN allows us to beat down statistical errors to arbitrarily small amounts, but ultimately systematic effects will limit the precision by which SN Ia distances can be applied to measure distances. A careful inspection of figure 7 shows the best fitting SN Ia cosmology

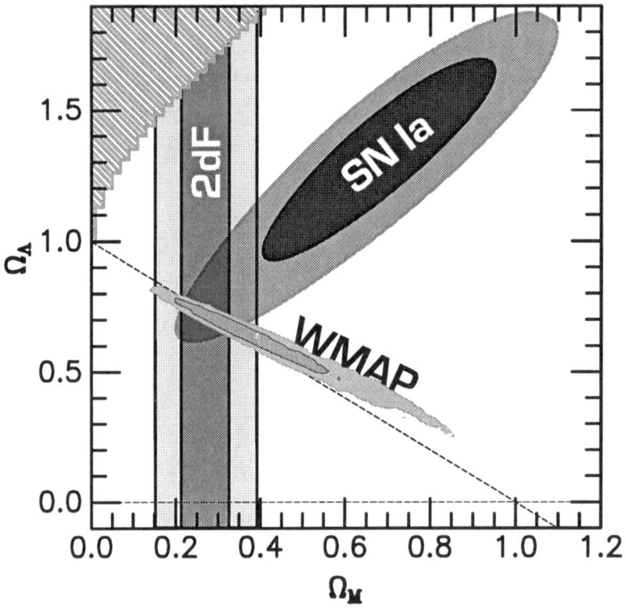

Figure 9. Contours of Ω_M versus Ω_Λ from three current observational experiments; High-Z SN Ia (Tonry *et al.* 2003), WMAP (Spergel *et al.* (2003), and the 2dF Galaxy Redshift Survey (Verde *et al.* 2002)

does not lie on the $\Omega_{tot} = 1$ line, but rather at higher Ω_M, and Ω_Λ. This is because, at a statistical significance of 1.5σ, the SN data show the onset and departure of deceleration (centred around $z = 0.5$) occurs faster than the flat model allows. The total size of the effect is roughly 0.04 mag, which is within the current allowable systematic uncertainties that this data set allows. So while this may be a real effect, it could equally plausibly be a systematic error, or just a statistical fluke.

Our best estimate is that it is possible to control systematic effects from a ground based experiment to a level of 0.03 mag. A carefully controlled ground based experiment of 200 SN will reach this statistical uncertainty in $z = 0.1$ redshift bins, and is achievable in a five year time frame. The Essence project and CFHT Legacy survey are such experiments, and should provide answers over the coming years.

The Supernova/Acceleration Probe (SNAP) collaboration has proposed launching a dedicated Cosmology satellite — the ultimate SN Ia experiment. This device will, if funded, scan many square degrees of sky, discovering a thousand SN Ia in a year, and obtain spectra and lightcurves of objects out to $z = 1.8$. Besides the large numbers of objects and their extended redshift range, space also provides the opportunity to control many systematic effects better than from the ground.

With rapidly improving CMB data from interferometers, the satellites MAP and Planck, and balloon based instrumentation planned for the next several years, CMB measurements promise dramatic improvements in precision on many of the cosmological parameters. However, the CMB measurements are relatively insensitive to the dark energy and the epoch of cosmic acceleration. SN Ia are currently the only way to directly study this acceleration epoch with sufficient precision (and control on systematic uncertainties) that we can investigate the properties of the dark energy, and any time-dependence in these properties. This ambitious goal will require complementary and cross-checking measurements of, for example, Ω_M from CMB, weak lensing, and large scale structure. The supernova measurements will also provide a test of the cosmological results independently of these other techniques (since CMB and weak lensing measurements are, of course, not themselves immune to systematic effects). By moving forward simultaneously on these experimental fronts, we have the plausible and exciting possibility of achieving a comprehensive measurement of the fundamental properties of our Universe.

References

1. Baum, W. A. Astronom. J. **62**, 6 1957
2. Benitez, N., Riess, A., Nugent, P., Dickinson, M., Chornock, R., & Filippenko, A. Astrophys. J. Lett. **577**, L1 2002
3. Bessell, M. Publ. Astro. Soc. Pac. **102**, 1181 1998
4. Branch, D., Fisher, A. Baron, E. & Nugent, P. Astrophys. J. Lett. **470,**, L7 1996
5. Branch, D., Perlmutter, S., Baron, E. & Nugent, P., in *Resource Book on Dark Energy*, ed. E.V. Linder, from Snowmass 2001 (astro-ph/0109070), 2001.
6. Branch, D., Fisher, A. & Nugent, P. Astronom. J. **106**, 2383 1993
7. Branch, D., in *Encyclopedia of Astronomy and Astrophysics*, p. 733, San Diego: Academic, 1989.
8. Branch, D. & Tammann, G.A., Annu. Rev. Astron. Astrophys., **30**, 359, 1992.
9. Burstein, D. 2003, Astronom. J. **126**, 1849 2003
10. Cadonau, R., PhD thesis, Univ. Basel, 1987.
11. Cappellaro, E. *et al.* Astron. & Astrophys. **322**, 431 1997
12. Coles, P. & Lucchin, F. 1995, Cosmology (Chicester: Wiley), 31
13. de Bernardis, P. *et al.* Nature **404**, 955 2000
14. Eastman, R. G., Kirshmer, R. P. Astrophys. J. **347**, 771 1989
15. Eastman, R. G., Schmidt, B. P., Kirshner, R. Astrophys. J. **466**, 911 1996
16. Fisher, A., Branch, D., Hoeflich, P. & Khokhlov, A. Astrophys. J. Lett. **447**, L73 1995
17. Filippenko, A.V., in *SN 1987A and Other Supernoave*, ed. I.J. Danziger, K. Kjar, p.343, Garching: ESO, 1991.
18. Fillipenko, A. V. *et al.* Astrophys. J. Lett. **384**, L15 1992
19. Freedman, W. L. *et al.* Astrophys. J. **553**, 47 2001
20. Garnavich, P. *et al.* Astrophys. J. Lett. **493**, L53 1998
21. Garnavich, P. *et al.* Astrophys. J. **509**, 74 1998
22. Germany, L. G., Riess, Schmidt, B. P. & Suntzeff, N. B . (A&A in press) 2003
23. Gilliland, R, L., Nugent, P. E., & Phillips, M. M. Astrophys. J. **521**, 30 1999
24. Goobar, A. & Perlmutter, S. 1995 Astrophys. J. **450**, 14 1995
25. Hamuy, M. *et al.* Astrophys. J. **558**, 615 2001
26. Hamuy, M. & Pinto, P. A. Astronom. J. **117**, 1185 1999
27. Hamuy, M., Phillips, M. M., Maza, J., Suntzeff, N. B., Schommer, R. A., & Aviles, R. Astronom. J. **109**, 1 1995

28. Hamuy, M., *et al.* , Astronom. J. **106**, 2392 1993
29. Hamuy, M., *et al.* 1996, Astronom. J. **112**, 2391 1996
30. Hamuy, M., *et al.* 1996, Astronom. J. **112**, 2408 1996
31. Hamuy, M., *et al.* Astronom. J. **102**, 208 1991
32. Hansen, L., Jorgensen, H. E., Norgaard-Nielsen, H. U., Ellis, R. S. & Couch, W. J., Astronomy and Astrophysics **211**, L9, 1989.
33. Harkness, R.P. & Wheeler, J.C., In *Supernovae* , ed. A.G. Petschek, p. 1, New York: Springer-Verlag, 1990.
34. Holz, D. E. & Wald, R. M., Phys. Rev **D58**, 063501 1998
35. Holz, D. E., Astrophys. J. **506**, 1 1998
36. Hubble, E. 1929, Proc.Nat.Acad.Sci 15, 168
37. Hoskin, M. A., 1976, J.Hist.Astron, 7, 169
38. Humason, M. L., Mayall, N. U., & Sandage, A. R. Astrophys. J. **61**, 97 1956
39. Jacoby, G. H.*et al.* Publ. Astro. Soc. Pac. **104**, 599 1992
40. Jensen, J. B., Tonry, J. L., Thompson, R. I., Ajhar, E. A., Lauer, T. R., Rieke, M. J., Postman, M., & Liu, M. C. Astrophys. J. **550**, 503 2001
41. Jha, S. PhD thesis, Harvard University, 2002.
42. Knop, R. A. *et al.* 2003, Astrophys. J. **598**, 102 2003
43. Kowal, C. T. Astronom. J. **272**, 1021 1968
44. Kim, A., Goobar, A. & Perlmutter, S. Publ. Astro. Soc. Pac. **108**, 190 1996
45. Kirshner, R. P., Kwan, J. Astrophys. J. **193**, 27 1974
46. Kochanek, C. S. & Schechter, P. L. 2003 in Carnegie Observatories Astrophysics Series, Vol 2: Measuring and Modeling the Universe, ed W. L. Freedman (Cambridge: Cambridge University Press)
47. Kantowski, R., Vaughan, T., & Branch, D. 1995, Astrophys. J. **447**, 35 1995
48. Kundic, T.*et al.* , Astrophys. J. **482**, 75 1997
49. Lauer, T & Postman, M Astrophys. J. Lett. **400**, L47 1992
50. Lauer, T & Postman, M Astrophys. J. Lett. **425**, L418 1994
51. Leavitt, H. S. 1908 Annals of HCO 60,4
52. Leavitt, H. S. 1912 HCO Circular 173
53. Leibundgut, B., PhD thesis, Univ. Basel, 1988.
54. Leibundgut, B., Tammann, G.A., Cadonau, R. & Cerrito, D., Astron. & Astrophys. Supp. **89**, 537 1991
55. Leibundgut, B. & Tammann, G.A., Astron. & Astrophys. **230**, 81 1990
56. Leonard, D. C., Filippenko, A. V., Ardila, D. R. & Brotherton, M. S. Astrophys. J. **553**, 861 2001
57. Leonard, D. C. *et al.* Publ. Astro. Soc. Pac. **114**, 35 2002
58. Leonard, D. C., Kanbur, S. M., Ngeow, C. C., & Tanvir, N. R. Astrophys. J. **594**, 247 2003
59. Leibundgut, B. *et al.* Astronom. J. **105**, 301 1993
60. Lemaitre, G. 1927 Ann.Soc.Sci Bruxelles, A47, 49
61. Lynden-Bell, D., Faber, S. M., Burstein, D., Davies, R. L., Dressler, A., Terlevich, R. J., & Wegner, G. Astrophys. J. **326**, 19 1988
62. Maanen, A. van 1916 ApJ, 44, 210
63. Miller, D.L. & Branch, D. Astronom. J. **100**, 530 1990
64. Muller, R.A., Newberg, H.J.M., Pennypacker, C.R.; Perlmutter, S., Sasseen, T.P. & Smith, C.K. Astrophys. J. Lett. **384**, L9 1992
65. Norgaard-Nielsen, H. U., Hansen, L., Jorgensen, H.E., Aragon Salamanca, A. & Ellis, R. S. Nature **339**, 523 1989
66. Nugent, P., Kim, A. & Perlmutter, S. Publ. Astro. Soc. Pac. **114**, 803 2002

67. Opik, E. 1922 Astrophys. J. **55**, 406 1922
68. Panagia, N., In *Supernovae as Distance Indicators*, ed. N. Bartel, p. 14, Berlin: Springer-Verlag, 1985.
69. Pain, R. *et al.* , Astrophys. J. **473**, 356 1996
70. Pain, R. *et al.* , Ap.J., in press, 2002.
71. Pearce, G., Patchett, B., Allington-Smith, J. & Parry, I, Astrophys. Space. Sci., **150**, 267, 1988.
72. Phillips, M. M. *et al.* Publ. Astro. Soc. Pac. **99**, 592 1987
73. Phillips, M. M. Astrophys. J. Lett. **413**, L105 1993
74. Phillips, M. M., Lira, P., Suntzeff, N. B., Schommer, R. A., Hamuy, M., Maza, J. Astronom. J. **118**, 1766 1999
75. Peebles, P.J.E. 1993, Principles of Physical Cosmology (Princeton: Princeton Univ. Press)
76. Perlmutter, S., Muller, R.A., Newberg, H.J.M., Pennypacker, C.R.; Sasseen, T.P. & Smith, C.K., in *Robotic telescopes in the 1990s*, ed. A. Filippenko, p. 67, 1992.
77. Perlmutter, S. *et al.* , Astrophys. J. Lett. **440**, LL41 1995
78. Perlmutter, S. *et al.* , IAU circulars 5956 (1994), 6263, & 6270, (1995).
79. Perlmutter, S. *et al.* , in *Thermonuclear Supernovae* (Aiguablava, June 1995), NATO ASI, eds. P. Ruiz-Lapuente, R. Canal, and J. Isern, 1997.
80. Perlmutter, S. *et al.* Astrophys. J. **517**, 565 1999
81. Perlmutter, S. *et al.* Nature **391**, 51 1998
82. Perlmutter, S. *et al.* , Astrophys. J. **483**, 565 1997
83. Phillips, M. M. *et al.* Astronom. J. **103**, 1632 1992
84. Permutter, S., Turner, M. & White, M.
85. Refsdal, S. Mon. Not. Roy. Astr. Soc. **128**, 307 1964
86. Riess, A. G. *et al.* Astrophys. J. **560**, 49 2001
87. Riess, A. G. *et al.* Astronom. J. **117**, 707 1999
88. Riess, A. G. *et al.* Astronom. J. **116**, 1009 1998
89. Riess, A. G., Filippenko, A. V., Li, W., Schmidt, B. P. Astronom. J. **118**, 2668 1999
90. Riess, A. G., Press, W. H., & Kirshner, R. P. Astrophys. J. **473**, 88 1996
91. Robertson, H. P., 1928, Phil. Mag. 5, 835
92. Schmidt, B. P., Kirshner, R. P., Eastman, R. G., Phillips, M. M., Suntzeff, N. B., Hamuy, N. B., Maza, J. & Aviles, R. Astrophys. J. **432**, 42 1994
93. Schmidt, B. *et al.* Astrophys. J. **507**,, 46 1998
94. Saha, A., Sandage, A., Tammann, G. A., Dolphin, A. E., Christensen, J., Panagia, N. & Macchetto, F. D. Astrophys. J. **562**, 313 2001
95. Schlegel, D. J., Finkbeiner, D. P., & Davis, M. Astrophys. J. Supp. **500**, 525 1998
96. Spergel, D. *et al.* Astrophys. J. Supp. **148**, 175 2003
97. Sullivan, M. *et al.* Mon. Not. Roy. Astr. Soc. **340**, 1057 2003
98. Sunyaev, R. A. & Zeldovich, Y. B. 1970, Comments on Astrophysics, 2, 66
99. Sunyaev, R. A. & Zeldovich, Y. B. 1972, Comments on Astrophysics, 4, 173
100. Sandage, A., & Tammann, G. A. Astrophys. J. **415**, 1 1993
101. Tammann, G. A., & Leibundgut, B. Astron. & Astrophys. **236**, 9 1990
102. Tammann, G. A., & Sandage, A. Astrophys. J. **452**, 16 1995
103. Timmes, F. X., Brown, E. F., & Truran, J. W. Astrophys. J. Lett. **590**, L83 2003
104. Tinsley, B Astrophys. J. **173**, 93 1972
105. Tonry, J. L., Dressler, A., Blakeslee, J. P., Ajhar, E. A., Fletcher, A. B., Luppino, G. A., Metzger, M. R., & Moore, C. B. Astrophys. J. **546**, 681 2001
106. Tonry, J. L.*et al.* Astrophys. J. **594**, 1 2003
107. Tully, R. B. & Fisher, J. R. Astron. & Astrophys. **54**, 661 1977

108. Uomoto, A. & Kirshner, R.P., Astronomy and Astrophysics **149**, L7, 1985.
109. van den Bergh, S. Astrophys. J. Lett. **453**, L55 1995
110. van den Bergh, S., & Pazder, J Astrophys. J. **390**, 34 1992
111. Vaughan, T.E., Branch, D., Miller, D.L. & Perlmutter, S. Astrophys. J. **439**, 558 1995
112. Verde, L. *et al.* Mon. Not. Roy. Astr. Soc. **335**, 432 2002
113. Wagoner, R. V. Astrophys. J. Lett. **250**, L65 1981
114. Wambsgabss, J., Cen, R., Guohong, X., & Ostriker, J. Astrophys. J. Lett. **475**, L81 1997
115. Wang, L, Howell, A. D., Hoeflich, P., & Wheeler, J. C. Astrophys. J. **550**, 1030 2001
116. Wheeler, J.C. & Levreault, R. Astrophys. J. Lett. **294**, L17 1985
117. Wittman, D.M., *et al.* Proc. SPIE, **3355**, 626, 1998.

INFLATION AND THE COSMIC MICROWAVE BACKGROUND

CHARLES H. LINEWEAVER
School of Physics, University of New South Wales, Sydney, Australia
email: charley@bat.phys.unsw.edu.au

I present a pedagogical review of inflation and the cosmic microwave background. I describe how a short period of accelerated expansion can replace the special initial conditions of the standard big bang model. I also describe the development of CMBology: the study of the cosmic microwave background. This cool (3 K) new cosmological tool is an increasingly precise rival and complement to many other methods in the race to determine the parameters of the Universe: its age, size, composition and detailed evolution.

1. A New Cosmology

"The history of cosmology shows that in every age devout people believe that they have at last discovered the true nature of the Universe."
- E. R. Harrison (1981)

1.1. *Progress*

Cosmology is the scientific attempt to answer fundamental questions of mythical proportion: How did the Universe come to be? How did it evolve? How will it end? If humanity goes extinct it will be of some solace to know that just before we went, incredible progress was made in our understanding of the Universe. "The effort to understand the Universe is one of the very few things that lifts human life a little above the level of farce, and gives it some of the grace of tragedy." (Weinberg 1977).

A few decades ago cosmology was laughed at for being the only science with no data. Cosmology was theory-rich but data-poor. It attracted armchair enthusiasts spouting speculations without data to test them. The night sky was calculated to be as bright as the Sun, the Universe was younger than the Galaxy and initial conditions, like animistic gods, were invoked to explain everything. Times have changed. We have entered a new era of precision cosmology. Cosmologists are being flooded with high quality measurements from an army of new instruments. We are observing the Universe at new frequencies, with higher sensitivity, higher spectral resolution and higher spatial resolution. We have so much new data that state-of-the-art computers process and store them with difficulty. Cosmology papers now include error bars — often asymmetric and sometimes even with a distinction made between statistical and systematic error bars. This is progress.

Cosmological observations such as measurements of the cosmic microwave background, and the inflationary ideas used to interpret them, are at the heart of what we know about the origin of the Universe and everything in it. Over the past century cosmological observations have produced the standard hot big bang model describing the evolution of the Universe in sharp mathematical detail. This model provides a consistent framework into which all relevant cosmological data seem to fit, and is the dominant paradigm against which all new ideas are tested. It became the dominant paradigm in 1965 with the discovery of the cosmic microwave. In the 1980s the big bang model was interpretationally upgraded to include an early short period of rapid expansion and a critical density of non-baryonic cold dark matter.

For the past 20 years many astronomers have assumed that 95% of the Universe was clumpy non-baryonic cold dark matter. They also assumed that the cosmological constant, Ω_Λ, was Einstein's biggest blunder and could be ignored. However, recent measurements of the cosmic microwave background combined with supernovae and other cosmological observations have given us a new inventory. We now find that 73% of the Universe is made of vacuum energy, while only 23% is made of non-baryonic cold dark matter. Normal baryonic matter, the stuff this paper is made of, makes up about 4% of the Universe. Our new inventory has identified a previously unknown 73% of the Universe! This has forced us to abandon the standard CDM ($\Omega_M = 1$) model and replace it with a new hard-to-fathom Λ-dominated CDM model.

1.2. *Big Bang: Guilty of Not Having an Explanation*

"...the standard big bang theory says nothing about what banged, why it banged, or what happened before it banged. The inflationary universe is a theory of the "bang" of the big bang." — Alan Guth (1997).

Although the standard big bang model can explain much about the evolution of the Universe, there are a few things it cannot explain:

- The Universe is clumpy. Astronomers, stars, galaxies, clusters of galaxies and even larger structures are sprinkled about. The standard big bang model cannot explain where this hierarchy of clumps came from — it cannot explain the origin of structure. We call this the structure problem.
- In opposite sides of the sky, the most distant regions of the Universe are at almost the same temperature. But in the standard big bang model they have never been in causal contact — they are outside each other's causal horizons. Thus, the standard model cannot explain why such remote regions have the same temperature. We call this the horizon problem.
- As far as we can tell, the geometry of the Universe is flat — the interior angles of large triangles add up to 180°. If the Universe had started out with a tiny deviation from flatness, the standard big bang model would have quickly generated a measurable degree of non-flatness. The standard

big bang model cannot explain why the Universe started out so flat. We call this the flatness problem.

- Distant galaxies are redshifted. The Universe is expanding. Why is it expanding? The standard big bang model cannot explain the expansion. We call this the expansion problem.

Thus the big bang model is guilty of not having explanations for structure, homogeneous temperatures, flatness or expansion. It tries — but its explanations are really only wimpy excuses called initial conditions. These initial conditions are

- the Universe started out with small seeds of structure
- the Universe started out with the same temperature everywhere
- the Universe started out with a perfectly flat geometry
- the Universe started out expanding

Until inflation was invented in the early 1980s, these initial conditions were tacked onto the front end of the big bang. With these initial conditions, the evolution of the Universe proceeds according to general relativity and can produce the Universe we see around us today. Is there anything wrong with invoking these initial conditions? How else should the Universe have started? The central question of cosmology is: How did the Universe get to be the way it is? Scientists have made a niche in the world by not answering this question with "That's just the way it is." And yet, that was the nature of the explanations offered by the big bang model without inflation.

"The horizon problem is not a failure of the standard big bang theory in the strict sense, since it is neither an internal contradiction nor an inconsistency between observation and theory. The uniformity of the observed universe is built into the theory by postulating that the Universe began in a state of uniformity. As long as the uniformity is present at the start, the evolution of the Universe will preserve it. The problem, instead, is one of predictive power. One of the most salient features of the observed universe — its large scale uniformity — cannot be explained by the standard big bang theory; instead it must be assumed as an initial condition."
— Alan Guth (1997)

The big bang model without inflation has special initial conditions tacked on to it in the first picosecond. With inflation, the big bang doesn't need special initial conditions. It can do with inflationary expansion and a new unspecial (and more remote) arbitrary set of initial conditions — sometimes called chaotic initial conditions — sometimes less articulately described as 'anything'. The question that still haunts inflation (and science in general) is: Are arbitrary initial conditions a more realistic ansatz? Are theories that can use them as inputs more predictive? Quantum cosmology seems to suggest that they are. We discuss this issue more in Section 6.

2. Tunnel Vision: the Inflationary Solution

Inflation can be described simply as any period of the Universe's evolution in which the size of the Universe is accelerating. This surprisingly simple type of expansion leads to our observed universe without invoking special initial conditions. The active ingredient of the inflationary remedy to the structure, horizon and flatness problems is rapid exponential expansion sometime within the first picosecond (= trillionth of a second $= 10^{-12}$ s) after the big bang. If the structure, flatness and horizon problems are so easily solved, it is important to understand how this quick cure works. It is important to understand the details of expansion and cosmic horizons. Also, since our Universe is becoming more Λ-dominated every day (Fig. 3), we need to prepare for the future. Our descendants will, of necessity, become more and more familiar with inflation, whether they like it or not. Our Universe is surrounded by inflation at both ends of time.

2.1. *Friedmann-Robertson-Walker metric $\longrightarrow$ Hubble's law and Cosmic Event Horizons*

The general relativistic description of a homogeneous, isotropic universe is based upon the Friedmann-Robertson-Walker (FRW) metric for which the spacetime interval ds, between two events, is given by

$$ds^2 = -c^2dt^2 + R(t)^2[d\chi^2 + S_k^2(\chi)d\psi^2], \tag{1}$$

where c is the speed of light, dt is the time separation, $d\chi$ is the comoving coordinate separation and $d\psi^2 = d\theta^2 + sin^2\theta d\phi^2$, where θ and ϕ are the polar and azimuthal angles in spherical coordinates. The scale factor R has dimensions of distance. The function $S_k(\chi) = \sin\chi$, χ or $\sinh\chi$ for closed (positive k), flat ($k = 0$) or open (negative k) universes respectively (see e.g. Peacock 1999 p. 69).

In an expanding universe, the proper distance D between an observer at the origin and a distant galaxy is defined to be along a surface of constant time ($dt = 0$). We are interested in the radial distance so $d\psi = 0$. The FRW metric then reduces to $ds = Rd\chi$ which, upon integration, becomes,

$$D(t) = R(t)\chi. \tag{2}$$

Taking the time derivative and assuming that we are dealing with a comoving galaxy ($\dot{\chi} = 0$) we have,

$$v(t) = \dot{R}(t)\chi, \tag{3}$$

$$v(t) = \frac{\dot{R}(t)}{R}\, R\,\chi, \tag{4}$$

$$\text{Hubble's Law} \quad v(t) = H(t)D, \tag{5}$$

$$\text{Hubble Sphere} \quad D_H = c/H(t). \tag{6}$$

The Hubble sphere is the distance at which the recession velocity v is equal to the speed of light. Photons have a peculiar velocity of $c = \dot{\chi}R$, or equivalently photons move through comoving space with a velocity $\dot{\chi} = c/R$. The comoving distance travelled by a photon is $\chi = \int \dot{\chi} dt$, which we can use to define the comoving coordinates of some fundamental concepts:

$$\textit{Particle Horizon} \quad \chi_{ph}(t) = c \int_0^t dt/R(t), \tag{7}$$

$$\textit{Event Horizon} \quad \chi_{eh}(t) = c \int_t^\infty dt/R(t), \tag{8}$$

$$\textit{Past Light Cone} \quad \chi_{lc}(t) = c \int_t^{t_o} dt/R(t). \tag{9}$$

Only the limits of the integrals are different. The horizons, cones and spheres of Eqs. 6 – 9 are plotted in Fig. 1.

2.2. *Inflationary Expansion: The Magic of a Shrinking Comoving Event Horizon*

Inflation doesn't make the observable universe big. The observable universe is as big as it is. What inflation does is make the region from which the Universe emerged, very small. How small? is unknown — (hence the question mark in Fig. 2), but small enough to allow the points in opposite sides of the sky (**A** and **B** in Fig. 4) to be in causal contact.

The exponential expansion of inflation produces an event horizon at a constant proper distance which is equivalent to a shrinking comoving horizon. A shrinking comoving horizon is the key to the inflationary solutions of the structure, horizon and flatness problems. So let's look at these concepts carefully in Fig. 1.

The new Λ-CDM cosmology has an event horizon and it is this cosmology that is plotted in Fig. 1 (the old standard CDM cosmology did not have an event horizon). To have an event horizon means that there will be events in the Universe that we will never be able to see no matter how long we wait. This is equivalent to the statement that the expansion of the Universe is so fast that it prevents some distant light rays, that are propagating toward us, from ever reaching us. In the top panel, one can see the rapid expansion of objects away from the central observer. As time goes by, Λ dominates and the event horizon approaches a constant physical distance from an observer. Galaxies do not remain at constant distances in an expanding universe. Therefore distant galaxies keep leaving the horizon, *i.e.*, with time, they move upward and outward along the lines labelled with redshift '1' or '3' or '10'. As time passes, fewer and fewer objects are left within the event horizon. The ones that are left, started out very close to the central observer. Mathematically, the $R(t)$ in the denominator of Eq. 8 increases so fast that the integral converges. As time goes by, the lower limit t of the integral gets bigger, making the integral converge on a smaller number — hence the comoving event horizon shrinks. The middle panel shows clearly that in the future, as Λ increasingly dominates the

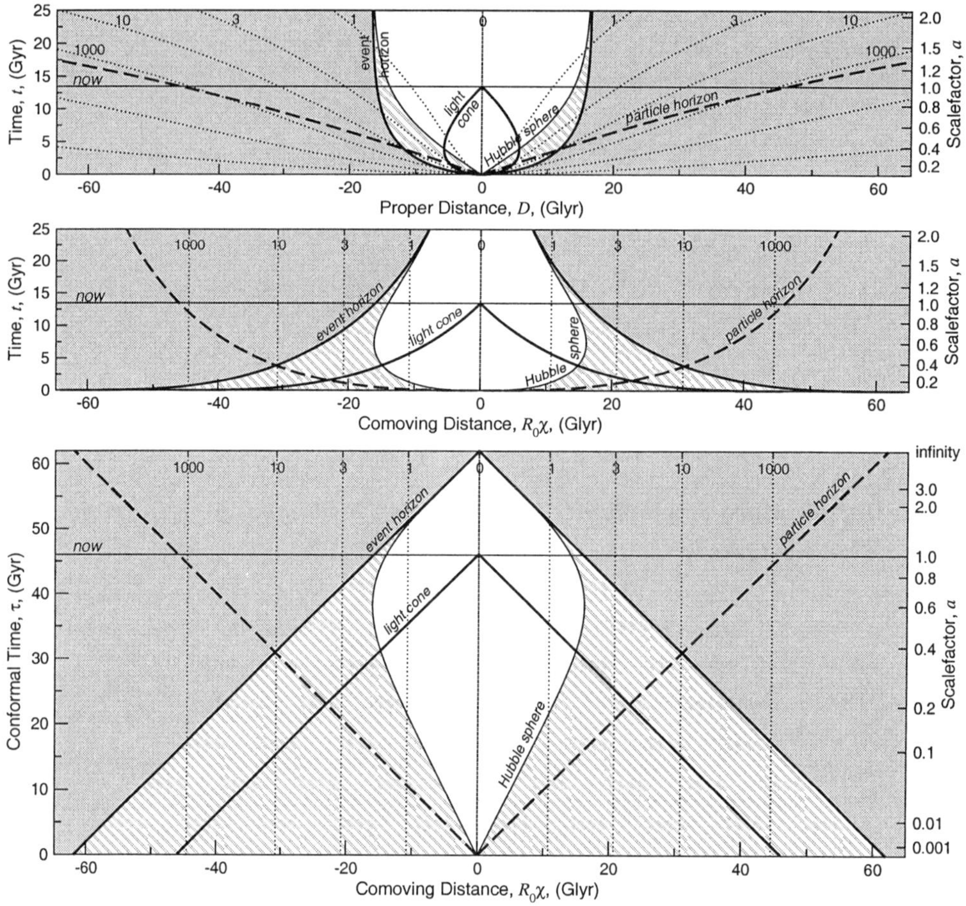

Figure 1. **Expansion of the Universe.** We live on the central vertical worldline. The dotted lines are the worldlines of galaxies being expanded away from us as the Universe expands. They are labelled by the redshift of their light that is reaching us today, at the apex of our past light cone. Top: In the immediate past our past light cone is shaped like a cone. But as we follow it further into the past it curves in and makes a teardrop shape. This is a fundamental feature of the expanding universe; the furthest light that we can see now was receding from us for the first few billion years of its voyage. The Hubble sphere, particle horizon, event horizon and past light cone are also shown (Eqs. 6 – 9). Middle: We remove the expansion of the Universe from the top panel by plotting comoving distance on the x axis rather than proper distance. Our teardrop-shaped light cone then becomes a flattened cone and the constant proper distance of the event horizon becomes a shrinking comoving event horizon — the active ingredient of inflation (Section 2.2). Bottom: the radius of the current observable Universe (the particle horizon) is 47 billion light years (Glyr), i.e., the most distant galaxies that we can see on our past light cone are now 47 billion light years away. The top panel is long and skinny because the Universe is that way — the Universe is larger than it is old — the particle horizon is 47 Glyr while the age is only 13.5 Gyr — thus producing the $3 : 1 (\approx 47 : 13.5)$ aspect ratio. In the bottom panel, space and time are on the same footing in conformal/comoving coordinates and this produces the $1 : 1$ aspect ratio. For details see Davis & Lineweaver (2003).

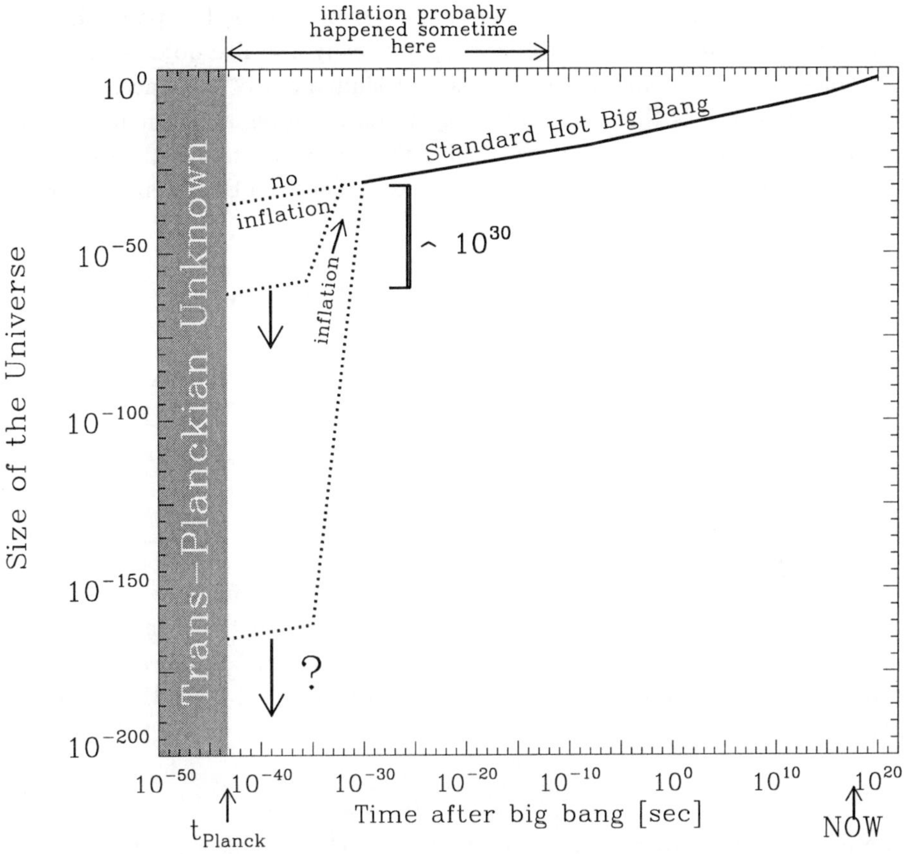

Figure 2. Inflation is a short period of accelerated expansion that probably happened sometime within the first picosecond (10^{-12} seconds) — during which the size of the Universe grows by more than a factor of $\sim 10^{30}$. The size of the Universe coming out of the 'Trans-Planckian Unknown' is unknown. Compared to its size today, maybe it was 10^{-60} as shown in one model... or maybe it was 10^{-165} as shown in the other model... or maybe even smaller (hence the question mark). In the two models shown, inflation starts near the GUT scale, ($\sim 10^{16}$ GeV or $\sim 10^{-35}$ seconds) and ends at about 10^{-30} seconds after the bang.

dynamics of the Universe, the comoving event horizon will shrink. This shrinkage is happening slowly now but during inflation it happened quickly. The shrinking comoving horizon in the middle panel of Fig. 1 is a slow and drawn out version of what happened during inflation — so we can use what is going on now to understand how inflation worked in the early universe. In the middle panel galaxies move on vertical lines upward, while the comoving event horizon shrinks. As time goes by we are able to see a smaller and smaller region of comoving space. Like using a zoom lens, or doing a PhD, we are able to see only a tiny patch of the Universe, but in amazing detail. Inflation gives us tunnel vision. The middle panel shows the narrowing of the tunnel. Galaxies move up vertically and like objects falling into black holes, from our point of view they are redshifted out of existence.

The bottom line is that accelerated expansion produces an event horizon at a given physical size and that any particular size scale, including quantum scales, expands with the Universe and quickly becomes larger than the given physical size of the event horizon.

3. Friedmann Oscillations: The Rise and Fall of Dominant Components

Friedmann's Equation can be derived from Einstein's 4x4 matrix equation of general relativity (see for example Landau & Lifshitz 1975, Kolb and Turner 1992 or Liddle & Lyth 2000):

$$R_{\mu\nu} - \frac{1}{2} g_{\mu\nu} \mathcal{R} = 8\pi G \, T_{\mu\nu} + \Lambda g_{\mu\nu} \tag{10}$$

where $R_{\mu\nu}$ is the Ricci tensor, $\mathcal{R}$ is the Ricci scalar, $g_{\mu\nu}$ is the metric tensor describing the local curvature of space (intervals of spacetime are described by $ds^2 = g_{\mu\nu} dx^\mu dx^\nu$), $T_{\mu\nu}$ is the stress-energy tensor and Λ is the cosmological constant. Taking the $(\mu, \nu) = (0, 0)$ terms of Eq. 10 and making the identifications of the metric tensor with the terms in the FRW metric of Eq. 1, yields the Friedmann Equation:

$$H^2 = \frac{8\pi G \rho}{3} - \frac{k}{R^2} + \frac{\Lambda}{3} \tag{11}$$

where R is the scale factor of the Universe, $H = \dot{R}/R$ is Hubble's constant, ρ is the density of the Universe in relativistic or non-relativistic matter, k is the constant from Eq. 1 and Λ is the cosmological constant. In words: the expansion (H) is controlled by the density (ρ), the geometry (k) and the cosmological constant (Λ). Dividing through by H^2 yields

$$1 = \frac{\rho}{\rho_c} - \frac{k}{H^2 R^2} + \frac{\Lambda}{3H^2} \tag{12}$$

where the critical density $\rho_c = \frac{3H^2}{8\pi G}$. Defining $\Omega_\rho = \frac{\rho}{\rho_c}$ and $\Omega_\Lambda = \frac{\Lambda}{3H^2}$ and using $\Omega = \Omega_\rho + \Omega_\Lambda$ we get,

$$1 - \Omega = \frac{-k}{H^2 R^2} \tag{13}$$

or equivalently,

$$(1 - \Omega) H^2 R^2 = constant. \tag{14}$$

If we are interested in only post-inflationary expansion in the radiation- or matter-dominated epochs we can ignore the Λ term and multiply Eq. 11 by $\frac{3}{8\pi G \rho}$ to get

$$\frac{3H^2}{8\pi G \rho} = 1 - \frac{3k}{8\pi G \rho R^2} \tag{15}$$

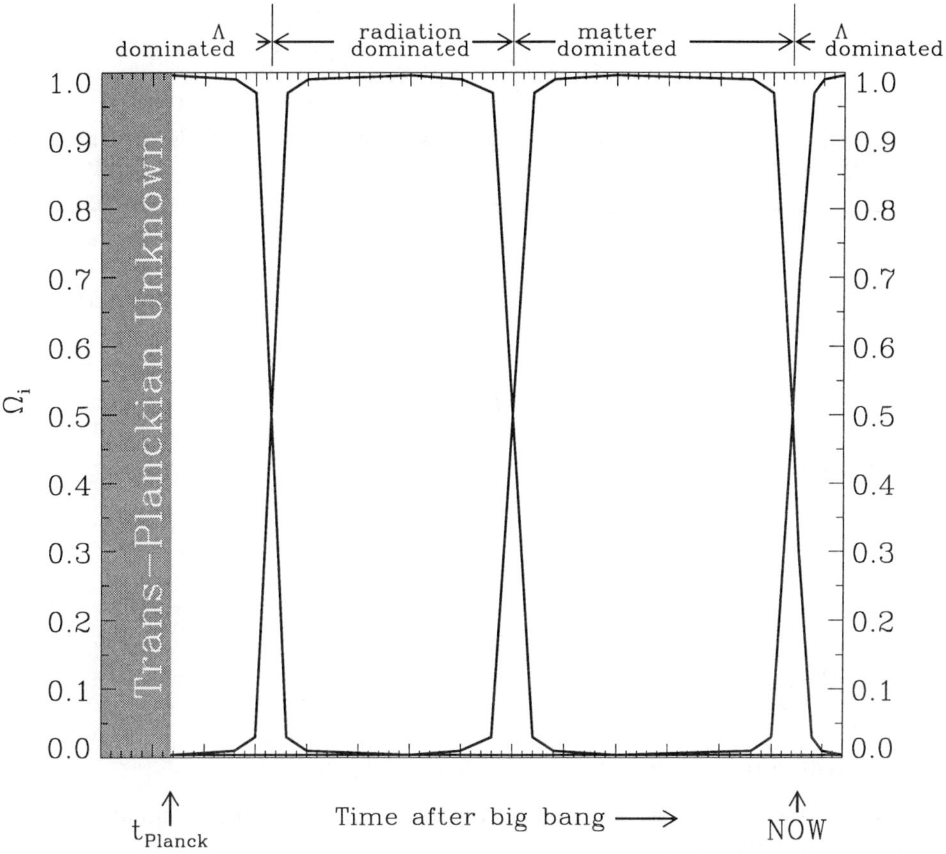

Figure 3. **Friedmann Oscillations: The rise and fall of the dominant components of the Universe.** The inflationary period can be described by a universe dominated by a large cosmological constant (energy density of a scalar field). During inflation and reheating the potential of the scalar field is turned into massive particles which quickly decay into relativistic particles and the Universe becomes radiation-dominated. Since $\rho_{rel} \propto R^{-4}$ and $\rho_{matter} \propto R^{-3}$, as the Universe expands a radiation-dominated epoch gives way to a matter-dominated epoch at $z \approx 3230$. And then, since $\rho_\Lambda \propto R^o$, the matter-dominated epoch gives way to a Λ-dominated epoch at $z \approx 0.5$. Why the initial Λ-dominated epoch became a radiation-dominated epoch is not as easy to understand as these subsequent oscillations governed by the Friedmann Equation (Eq. 11). Given the current values $(h, \Omega_m, \Omega_\Lambda, \Omega_{rel}) = (0.72, 0.27, 0.73, 0.0)$ the Friedmann Equation enables us to trace back through time the oscillations in the quantities Ω_m, Ω_Λ and Ω_{rel}.

which can be rearranged to give

$$\left(\Omega^{-1} - 1\right) \rho R^2 = constant \tag{16}$$

A more heuristic Newtonian analysis can also be used to derive Eqs. 14 & 16 (*e.g.*Wright 2003). Consider a spherical shell of radius R expanding at a velocity

$v = HR$, in a universe of density ρ. Energy conservation requires

$$2E = v^2 - \frac{2GM}{R} = H^2R^2 - \frac{8\pi G R^2 \rho}{3}. \tag{17}$$

By setting the total energy equal to zero we obtain a critical density at which $v = HR$ is the escape velocity,

$$\rho_c = \frac{3H^2}{8\pi G} = 1.879\, h^2 \times 10^{-29} g\ cm^{-3} \sim 20\ protons\ m^{-3}. \tag{18}$$

However, by requiring only energy conservation ($2E = constant$ not necessarily $E = 0$) in Eq. 17, we find,

$$constant = H^2R^2 - \frac{8\pi G R^2 \rho}{3}. \tag{19}$$

Dividing Eq. 19 by H^2R^2 we get

$$(1 - \Omega)H^2R^2 = constant, \tag{20}$$

which is the same as Eq. 14. Multiplying Eq. 19 by $\frac{3}{8\pi G\rho R^2}$ we get

$$(\Omega^{-1} - 1)\rho R^2 = constant \tag{21}$$

which is the same as Eq. 16.

3.1. *Friedmann's Equation $\longrightarrow$ Exponential Expansion*

One way to describe inflation is that during inflation, a $\Lambda_{\rm inf}$ term dominates Eq. 11. Thus, during inflation we have,

$$H^2 = \frac{\Lambda_{\rm inf}}{3} \tag{22}$$

$$\frac{dR}{dt} = R\sqrt{\frac{\Lambda_{\rm inf}}{3}} \tag{23}$$

$$\int_{R_i}^{R} \frac{dR}{R} = \int_{t_i}^{t} dt\sqrt{\frac{\Lambda_{\rm inf}}{3}} \tag{24}$$

$$ln\frac{R}{R_i} = \sqrt{\frac{\Lambda_{\rm inf}}{3}}\ (t - t_i) \tag{25}$$

$$R \approx R_i e^{Ht} \tag{26}$$

where t_i and R_i are the time and scale factor at the beginning of inflation. To get Eq. 26 we have assumed $0 \approx t_i << t < t_e$ (where t_e is the end of inflation) and we have used Eq. 22. Equation 26 is the exponential expansion of the Universe during inflation. The e-folding time is $1/H$. The doubling time is $(ln2)/H$. That is, during every interval $\Delta t = 1/H$, the size of the Universe increases by a factor of $e = 2.718281828...$ and during every interval $\Delta t = (ln2)/H$ the size of the Universe doubles.

4. Inflationary Solutions to the Flatness and Horizon Problems

4.1. *What is the Flatness Problem?*

First I will describe the flatness problem and then the inflationary solution to it. Recent measurements of the total density of the Universe find $0.95 < \Omega_o < 1.05$ (e.g. Table 1). This near flatness is a problem because the Friedmann Equation tells us that $\Omega \sim 1$ is a very unstable condition — like a pencil balancing on its point. It is a very special condition that won't stay there long. Here is an example of how special it is. Equation 16 shows us that $(\Omega^{-1} - 1)\rho R^2 = constant$. Therefore, we can write,

$$(\Omega^{-1} - 1)\rho R^2 = (\Omega_o^{-1} - 1)\rho_o R_o^2 \tag{27}$$

where the right hand side is today and the left hand side is at any arbitrary time. We then have,

$$(\Omega^{-1} - 1) = (\Omega_o^{-1} - 1)\frac{\rho_o}{\rho}\left(\frac{R_o}{R}\right)^2. \tag{28}$$

Redshift is related to the scale factor by $R = R_o/(1+z)$. Consider the evolution during matter-domination where $\rho = \rho_o(1+z)^3$. Inserting these we get,

$$(\Omega^{-1} - 1) = \frac{(\Omega_o^{-1} - 1)}{1+z}. \tag{29}$$

Inserting the current limits on the density of the Universe, $0.95 < \Omega_o < 1.05$ (for which $-0.05 < (\Omega_o^{-1} - 1) < 0.05$), we get a constraint on the possible values that Ω could have had at redshift z,

$$\frac{1}{1 + \frac{0.05}{1+z}} < \Omega < \frac{1}{1 - \frac{0.05}{1+z}}. \tag{30}$$

At recombination (when the first hydrogen atoms were formed) $z \approx 10^3$ and the constraint on Ω yields,

$$0.99995 < \Omega < 1.00005 \tag{31}$$

So the observation that $0.95 < \Omega_o < 1.05$ today, means that at a redshift of $z \sim 10^3$ we must have had $0.99995 < \Omega < 1.000005$. This range is small...special. However, Ω had to be even more special earlier on. We know that the standard big bang successfully predicts the relative abundances of the light nuclei during nucleosynthesis between ~ 1 minute and ~ 3 minutes after the big bang, so let's consider the slightly earlier time, 1 second after the big bang which is about the beginning of the epoch in which we are confident that the Friedmann Equation holds. The redshift was $z \sim 10^{11}$ and the resulting constraint on the density at that time was,

$$0.9999999999995 < \Omega < 1.0000000000005 \tag{32}$$

This range is even smaller and more special (although I have assumed matter domination for this calculation, at redshifts higher than $z_{eq} \sim 3000$, we have radiation domination and $\rho = \rho_o(1+z)^4$. This makes the $1+z$ in Eq. 30 a $(1+z)^2$ and requires that early values of Ω be even closer to 1 than calculated here).

To summarize:

$$0.95 < \Omega_o(z=0) < 1.05 \tag{33}$$
$$0.99995 < \Omega(z=10^3) < 1.000005 \tag{34}$$
$$0.9999999999995 < \Omega(z=10^{11}) < 1.0000000000005 \tag{35}$$

If the Friedmann Equation is valid at even higher redshifts, Ω must have been even closer to one. These limits are the mathematical quantification behind our previous statement that: 'If the Universe had started out with a tiny deviation from flatness, the standard big bang model would have quickly generated a measurable degree of non-flatness.' *If* we assume that Ω could have started out with any value, then we have a compelling question: Why should Ω have been so fine-tuned to 1?

Observing $\Omega_o \approx 1$ today can be compared to a pencil standing on its point. If you walk into a room and find a pencil standing on its point you think: pencils don't usually stand on their points. If a pencil is that way then some mechanism must have recently set it up because pencils won't stay that way long. Similarly, if you wake up in a universe that you know would quickly evolve away from $\Omega = 1$ and yet you find that $\Omega_o = 1$ then some mechanism must have balanced it very exactly at $\Omega = 1$.

Another way to state this flatness problem is as an oldness problem. If $\Omega_o \approx 1$ today, then the Universe cannot have gone through many e-folds of expansion which would have driven it away from $\Omega_o = 1$. It cannot be very old. If the pencil is standing on its end, then the mechanism to push it up must have just finished. But we see that the Universe *is* old in the sense that it *has* gone through many e-foldings of expansion (even without inflation).

If early values of Ω had exceeded 1 by a tiny amount then this closed Universe would have recollapsed on itself almost immediately. How did the Universe get to be so old? If early values of Ω were less than 1 by a tiny amount then this open Universe would have expanded so quickly that no stars or galaxies would have formed. How did our galaxy get to be so old? The tiniest deviation from $\Omega = 1$ grows quickly into a collapsing universe or one that expands so quickly that clumps have no time to form.

4.2. *Solving the Flatness Problem*

How does inflation solve this flatness problem? How does inflation set up the condition of $\Omega = 1$? Consider Eq. 14: $(1-\Omega)H^2R^2 = constant$. During inflation $H = \sqrt{\Lambda_{\rm inf}/3} = constant$ (Eq. 22) and the scale factor R increases by many orders of magnitude, $\gtrsim 10^{30}$. One can then see from Eq. 14 that the large increase in the scale factor R during inflation, with H constant, drives $\Omega \to 1$. This is what is meant when we say that inflation makes the Universe spatially flat. In a vacuum-dominated expanding universe, $\Omega = 1$ is a stable fixed point. During inflation H is constant and R increases exponentially. Thus, no matter how far Ω is from 1 before inflation, the exponential increase of R during inflation quickly drives it to 1 and this is equivalent to flattening the Universe. Once driven to $\Omega = 1$ by inflation, the Universe will naturally evolve away from $\Omega = 1$ in the absence of inflation as we showed in the previous section.

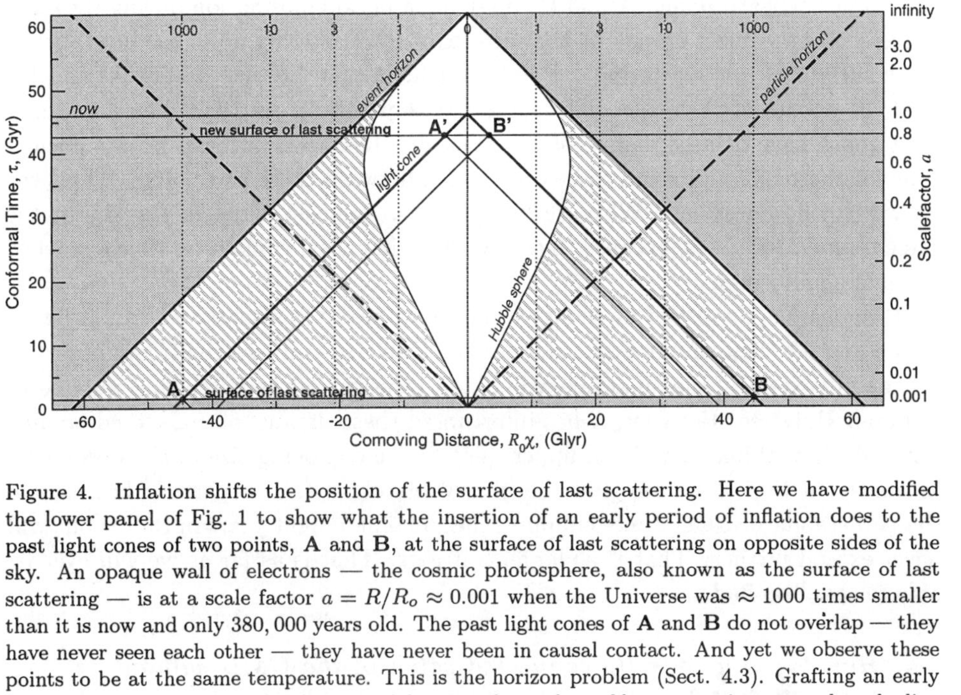

Figure 4. Inflation shifts the position of the surface of last scattering. Here we have modified the lower panel of Fig. 1 to show what the insertion of an early period of inflation does to the past light cones of two points, **A** and **B**, at the surface of last scattering on opposite sides of the sky. An opaque wall of electrons — the cosmic photosphere, also known as the surface of last scattering — is at a scale factor $a = R/R_o \approx 0.001$ when the Universe was ≈ 1000 times smaller than it is now and only 380,000 years old. The past light cones of **A** and **B** do not overlap — they have never seen each other — they have never been in causal contact. And yet we observe these points to be at the same temperature. This is the horizon problem (Sect. 4.3). Grafting an early epoch of inflation onto the big bang model moves the surface of last scattering upward to the line labelled "new surface of last scattering". Points **A** and **B** move upward to **A′** and **B′**. Their new past light cones overlap substantially. They have been in causal contact for a long time. Without inflation there is no overlap. With inflation there is. That is how inflation solves the problem of identical temperatures in 'different' horizons. The y axis shows all of time. That is, the range in conformal time [0, 62] Gyr corresponds to the cosmic time range $[0, \infty]$ (conformal time τ is defined by $d\tau = dt/R$). Consequently, there is an upper limit to the size of the observable universe. The isosceles triangle of events within the event horizon are the only events in the Universe that we will ever be able to see — probably a very small fraction of the entire universe. That is, the x axis may extend arbitrarily far in both directions. Like this ↓.

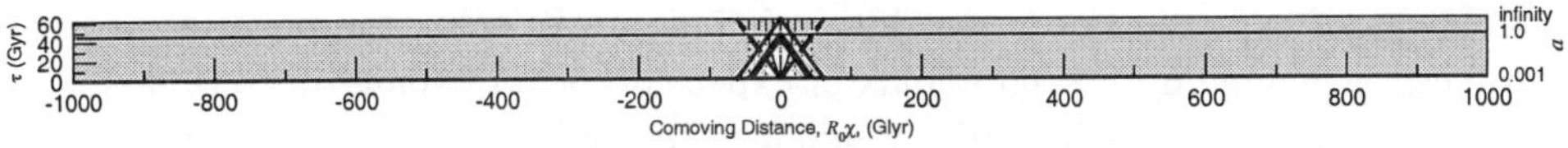

Figure 5.

4.3. *Horizon problem*

What should our assumptions be about regions of the Universe that have never been in causal contact? If we look as far away as we can in one direction and as far away as we can in the other direction we can ask the question, have those two points (points **A** and **B** in Fig. 4) been able to see each other. In the standard big bang model without inflation the answer is no. Their past light cones are the little cones beneath points **A** and **B**. Inserting a period of inflation during the early universe has the effect of moving the surface of last scattering up to the line labelled "new surface of last scattering". Points **A** and **B** then become points $\mathbf{A}'$ and $\mathbf{B}'$. And the apexes of their past light cones are at points $\mathbf{A}'$ and $\mathbf{B}'$. These two new light cones have a large degree of intersection. There would have been sufficient time for thermal equilibrium to be established between these two points. Thus, the answer to the question: "Why are two points in opposite sides of the sky at the same temperature?" is, because they have been in causal contact and have reached thermal equilibrium.

Five years ago most of us thought that as we waited patiently we would be rewarded with a view of more and more of the Universe and eventually, we hoped to see the full extent of the inflationary bubble — the size of the patch that inflated to form our Universe. However, Λ has interrupted these dreams of unfettered empiricism. We now think there is an upper limit to the comoving size of the observable universe. In Fig. 4 we see that the observable universe (= particle horizon) in the new standard Λ-CDM model approaches 62 billion light years in radius but will never extend further. That is as large as it gets. That is as far as we will ever be able to see. Too bad.

4.4. *How big is a causally connected patch of the CMB without and with inflation?*

From Fig. 4 we can read off the x axis that the comoving radius of the base of the small light cone under points **A** or **B** is $r = R_o\chi \sim$ billion light years. This is the current size of the patch that was causally connected at last scattering. The physical size D of the particle horizon today is $D \approx 47$ billion light years (Fig. 4). The fraction f of the sky occupied by one causally connected patch is $f = \pi r^2/4\pi D^2 \approx 1/9000$. The area of the full sky is about 40,000 square degrees (4π steradians). The area of a causally connected patch is $(area\ of\ the\ sky) \times f = 40,000/9,000 \approx 4$ square degrees.

With inflation, the size of the causally connected patch depends on how many e-foldings of expansion occurred during inflation. To solve the horizon problem we need a minimum of $\sim$ 60 e-folds of expansion or an expansion by a factor of $\sim 10^{30}$. But since this is only a minimum, the full size of a causally connected patch, although bigger than the observable universe, will never be known unless it happens to be between 47 Glyr (our current particle horizon) and 62 Glyr (the comoving size of our particle horizon at the end of time).

The constraint on the lower limit to the number of e-foldings ~ 60 (or $\sim 10^{30}$) comes from the requirement to solve the horizon problem. What about the upper limit to the number of e-folds? How big is our inflationary bubble? How big the inflationary patch is depends sensitively on when inflation happened, the height of the inflation potential and how long inflation lasted (t_i, t_e and $\Lambda_{\rm inf}$ at Eq. 26) — which in turn depends on the decay rate of the false vacuum. Without a proper GUT, these numbers cannot be approximated with any confidence. It is certainly reasonable to expect homogeneity to continue for some distance beyond our observable universe but there does not seem to be any reason why it should go on forever. In eternal inflation models, the homogeneity definitely does not go on forever (Liddle & Lyth 2000).

When could inflation have occurred? The earliest time is the Planck time at 10^{19} GeV or 10^{-43} seconds. The latest is at the electroweak symmetry breaking at 10^{2} GeV or 10^{-12} seconds. The GUT scale is a favourite time at 10^{16} GeV or 10^{-35} seconds. "Beyond these limits very little can be said for certain about inflation. So most papers about inflationary models are more like historical novels than real history, and they describe possible interactions that would be interesting instead of interactions that have to occur. As a result, inflation is usually described as the inflationary scenario instead of a theory or hypothesis. However, it seems quite likely that the inflation did occur, even though we don't know when, or what the potential was." — Wright (2003).

5. How Does Inflation Produce All the Structure in the Universe?

In our Universe quantum fluctuations have been expanded into the largest structures we observe and clouds of hydrogen have collapsed to form kangaroos. The larger end of this hierarchical range of structure — the range controlled by gravity, not chemistry, is what inflation is supposed to explain.

Inflation produces structure because quantum mechanics, not classical mechanics, describes the Universe in which we live. The seeds of structure, quantum fluctuations, do not exist in a classical world. If the world were classical, there would be no clumps or balls to populate classical mechanics textbooks. Inflation dilutes everything — all preexisting structure. It empties the Universe of anything that may have existed before, except quantum fluctuations. These it can't dilute. These then become the seeds of who we are.

One of the most important questions in cosmology is: What is the origin of all the galaxies, clusters, great walls, filaments and voids we see around us? The inflationary scenario provides the most popular explanation for the origin of these structures: they used to be quantum fluctuations. During the metamorphosis of quantum fluctuations into CMB anisotropies and then into galaxies, primordial quantum fluctuations of a scalar field get amplified and evolve to become classical seed perturbations and eventually large scale structure. *Primordial* quantum fluctuations are initial conditions. Like radioactive decay or quantum tunnelling, they

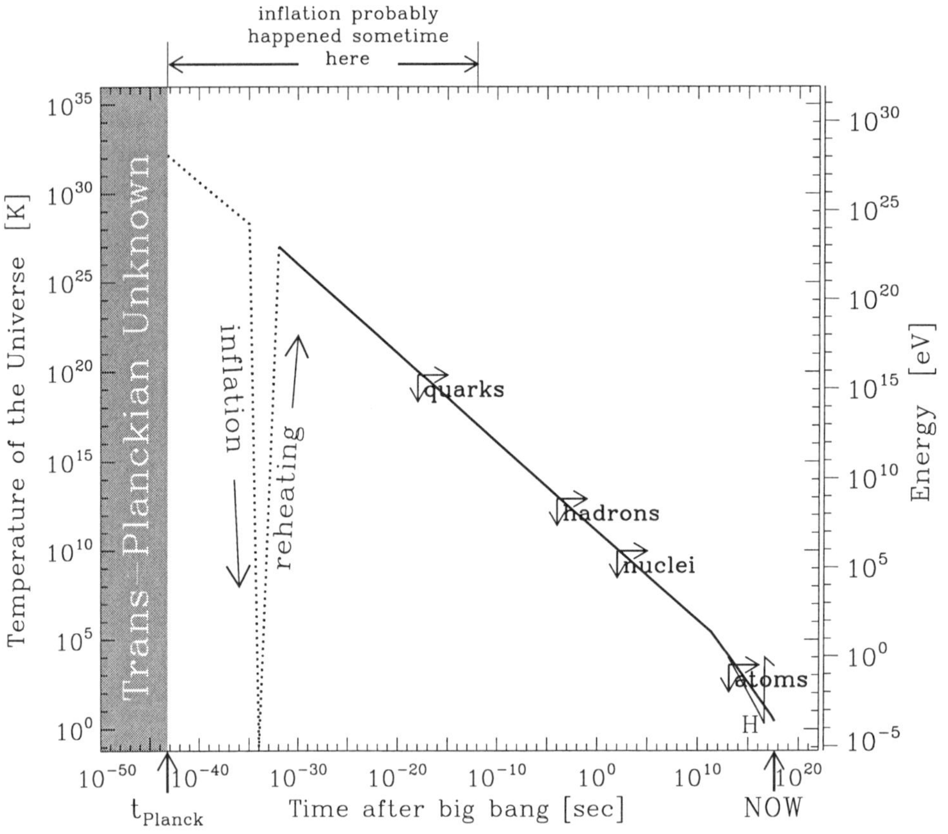

Figure 6. **Temperature of the Universe.** The temperature and composition history of the standard big bang model with an epoch of inflation and reheating inserted between 10^{-35} and 10^{-32} seconds after the big bang. Inflation increases the size of the Universe, decreases the temperature and dilutes any structure. Reheating then creates matter which decays and raises the temperature again. This plot is also an overview of the energy scales at which the various components of our Universe froze out and became permanent features. Quarks froze into protons and neutrons ($\sim$ GeV), protons and neutrons froze into light nuclei ($\sim$ MeV), and these light nuclei froze into neutral atoms ($\sim$ eV) which cooled into molecules and then gravitationally collapsed into stars. And now, huddled around these warm stars, we are living in the ice ages of the Universe with the CMB at 3K or $\sim 10^{-3}$ eV.

are not caused by any preceding event.

"Although introduced to resolve problems associated with the initial conditions needed for the Big Bang cosmology, inflation's lasting prominence is owed to a property discovered soon after its introduction: It provides a possible explanation for the initial inhomogeneities in the Universe that are believed to have led to all the structures we see, from the earliest objects formed to the clustering of galaxies to

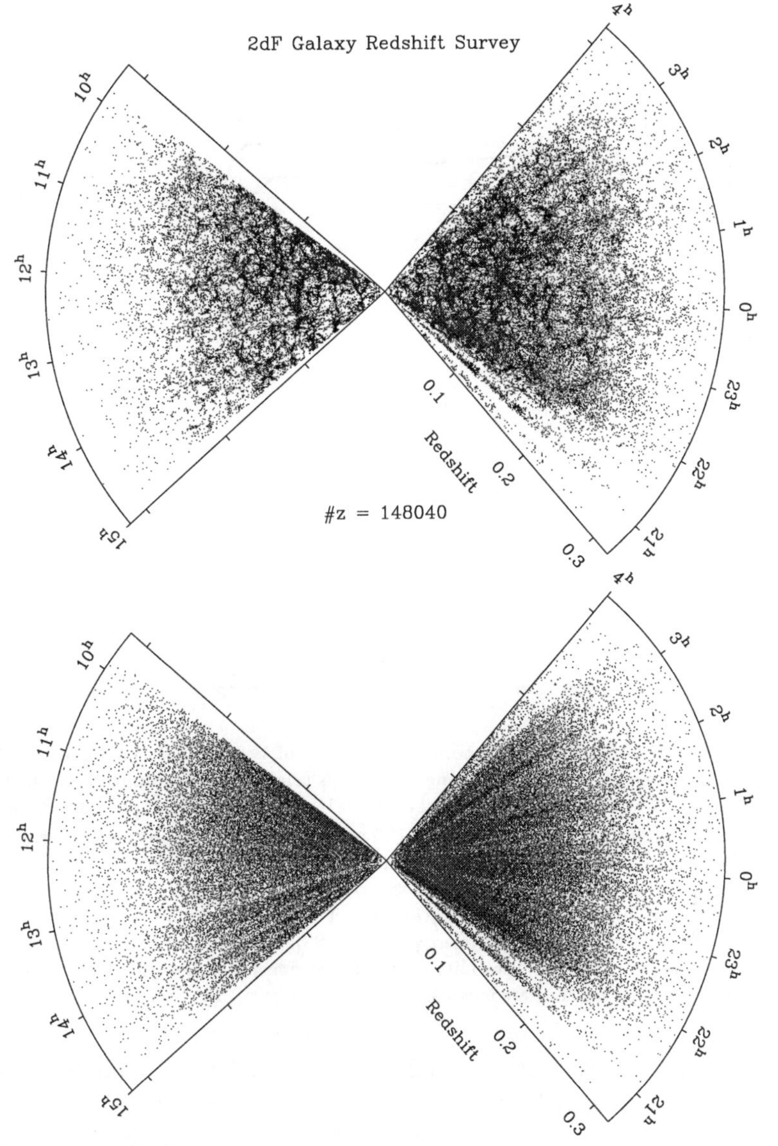

Figure 7. **Real Structure (top) is not Random (bottom).** If galaxies were distributed randomly in the Universe with no large scale structure, the 2dF galaxy redshift survey of the Local Universe would have produced the lower map. The upper map it did produce shows galaxies clumped into clusters radially smeared by the fingers of God, and empty voids surrounded by great walls of galaxies. The same number of galaxies is shown in each panel. Since all the large scale structure in the Universe has its origin in inflation, we should be able to look at the details of this structure to constrain inflationary models. A minimalistic set of parameters to describe all this structure is the amplitude and the scale dependence of the density perturbations.

the observed irregularities in the microwave background." — Liddle & Lyth (2000)

In early versions of inflation, it was hoped that the GUT scale Higgs potential

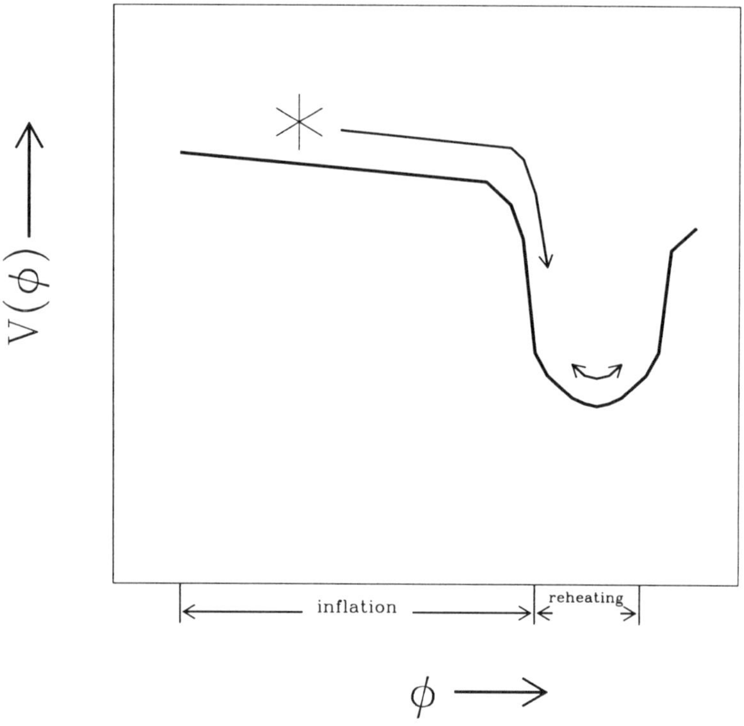

Figure 8. **Model of the Inflaton Potential.** A potential V of a scalar field ϕ with a flat part and a valley. The rate of expansion H during inflation is related to the amplitude of the potential during inflation. In the slow roll approximation $H^2 = V(\phi)/m_{pl}^2$ (where m_{pl} is the Planck mass). Thus, from Eq. 22 we have $\Lambda_{\rm inf} = 3\ V(\phi)/m_{pl}^2$. Thus, the height of the potential during inflation determines the rate of expansion during inflation. And the rate at which the ball rolls (the star rolls in this case) is determined by how steep the slope is: $\dot{\phi} = V'/3H$. In modern physics, the vacuum is the state of lowest possible energy density. The non-zero value of $V(\phi)$ is false vacuum — a temporary state of lowest possible energy density. The only difference between false vacuum and the cosmological constant is the stability of the energy density — how slow the roll is. Inflation lasts for $\sim 10^{-35}$ seconds while the cosmological constant lasts $\gtrsim 10^{17}$ seconds.

could be used to inflate. But the GUT theories had 1st order phase transitions. All the energy was dumped into the bubble walls and the observed structure in the Universe was supposed to come from bubble wall collisions. But the energy had to be spread out evenly. Percolation was a problem and so too was a graceful exit from inflation. New Inflation involves second order phase transitions (slow roll approximations). The whole universe is one bubble and structure cannot come from collisions. It comes from quantum fluctuations of the fields. There is one bubble rather than billions and the energy gets dumped everywhere, not just at the bubble wall.

One way to understand how quantum fluctuations become real fluctuations is this. Quantum fluctuations, *i.e.*virtual particle pairs of borrowed energy ΔE, get separated during the interval $\Delta t \lesssim \hbar/\Delta E$. The Δx in $\Delta x \lesssim \hbar/\Delta p$ is a measure of their separation. If during Δt the physical size Δx leaves the event horizon, the virtual particles cannot reconnect, they become real and the energy debt must be paid by the driver of inflation, the energy of the false vacuum — the $\Lambda_{\rm inf}$ associated with the inflaton potential $V(\phi)$ (see Fig. 8).

What kind of choices does the false vacuum have when it decays? If there are many pocket universes, what are they like? Do they have the same value for the speed of light? Are their true vacua the same as ours? Do the Higgs fields give the particles and forces the same values that reign in our Universe? Is the baryon asymmetry the same as in our Universe?

6. The Status of Inflation

Down to Earth astronomers are not convinced that inflation is a useful model. For them, inflation is a cute idea that takes a geometric flatness problem and replaces it with an inflaton potential flatness problem. It moves the problem to earlier times, it does not solve it. Inflation doesn't solve the fine-tuning problem. It moves the problem from "Why is the Universe so flat?" to "Why is the inflaton potential so flat?". When asked, "Why is the Universe so flat?", Mr Inflation responds, "Because my inflaton potential is so flat." "But why is your inflaton potential so flat?" "I don't know. It's just an initial condition." This may or may not be progress. If we are content to believe that spatial flatness is less fundamental than inflaton potential flatness then we have made progress.

6.1. *Inflationary Observables*

Models of inflation usually consist of choosing a form for the potential $V(\phi)$. A simple model of the potential is $V(\phi) = m^2\phi^2/2$ where the derivative with respect to ϕ is $V' = m^2\phi$ and $V'' = m^2$. This leads to a prediction for the observable spectral index of the CMB power spectrum: $n_s = 1 - 8m_p/\phi^2$ (*e.g.*Liddle & Lyth 2000). Estimates of the slope of the CMB power spectrum n_s and its derivative $\frac{dn_s}{dk}$ have begun to constrain models of the inflaton potential (Table 1 and Spergel 2003).

The observational scorecard of inflation is mixed. Based on inflation, many theorists became convinced that the Universe was spatially flat despite many measurements to the contrary. The Universe has now been measured to be flat to high precision — score one for inflation. Based on vanilla inflation, most theorists thought that the flatness would be without Λ — score one for the observers. Guth wanted to use the Georgi-Glashow GUT model as the potential to form structure. It didn't work — score one against inflation. But other plausible inflaton potentials can work. Inflation seems to be the only show in town as far as producing the seeds of structure — score one for inflation. Inflation predicts the spectral index of CMB fluctuations to be $n_s \approx 1$ — score one for inflation. But we knew that $n_s \approx 1$ before inflation (minus 1/2 point for cheating). So far most of inflation's predictions have been retrodictions — explaining things that it was designed to explain.

Inflationary models and the new ekpyrotic models make different predictions about the slope n_T of the tensor mode contribution to the CMB power spectrum. Inflation has higher amplitudes at large angular scales while ekpyrotic models have the opposite. However, since the amplitude T is unknown, finding the ratio of the amplitude of tensor to scalar modes, $r = T/S \sim 0$, does not really distinguish the two models. Finding a value $r > 0$ would however be interpreted as favouring inflation over ekpyrosis. Recent WMAP measurements of the CMB power spectrum yield $r < 0.71$ at the 95% confidence level.

Measurements of CMB polarization over the next five years will add more diagnostic power to CMB parameter estimation and may be able to usefully constrain the slope and amplitude of tensor modes if they exist at a detectable level.

One can be sceptical about the status of the problems that inflation claims to have solved. After all, the electron mass is the same everywhere. The constants of nature are the same everywhere. The laws of physics seem to be the same everywhere. If these uniformities need no explanation then why should the uniform temperatures, flat geometry and seeds of structure need an explanation. Is this first group more fundamental than the second?

The general principle seems to be that if we can't imagine plausible alternatives then no explanation seems necessary. Thus, dreaming up imaginary alternatives creates imaginary problems, to which imaginary solutions can be devised, whose explanatory power depends on whether the Universe could have been other than what it is. However, it is not easy to judge the reality of counterfactuals. Yes, inflation can cure the initial condition ills of the standard big bang model, but is inflation a panacea or a placebo?

Inflation is not a theory of everything. It is not based on M-theory or any candidate for a theory of everything. It is based on a scalar field. The inflation may not be due to a scalar field ϕ and its potential $V(\phi)$. Maybe it has more to do with extra-dimensions?

7. CMB

7.1. *History*

By 1930, the redshift measurements of Hubble and others had convinced many scientists that the Universe was expanding. This suggested that in the distant past the Universe was smaller and hotter. In the 1940s an ingenious nuclear physicist George Gamow, began to take the idea of a very hot early universe seriously, and with Alpher and Herman, began using the hot big bang model to try to explain the relative abundances of all the elements. Newly available nuclear cross-sections made the calculations precise. Newly available computers made the calculations doable. In 1948 Alpher and Herman published an article predicting that the temperature of the bath of photons left from the early universe would be 5 K. They were told by colleagues that the detection of such a cold ubiquitous signal would be impossible.

In the early 1960s, Arno Penzias and Robert Wilson discovered excess antenna noise in a horn antenna at Crawford Hill, Holmdel, New Jersey. They didn't know what to make of it. Maybe the white dielectric material left by pigeons had something to do with it? During a plane ride, Penzias explained his excess noise problem to a fellow radio astronomer Bernie Burke. Later, Burke heard about a talk by a young Princeton post-doc named Peebles, describing how Robert Dicke's Princeton group was gearing up to measure radiation left over from an earlier hotter phase of the Universe. Peebles had even computed the temperature to be about 10 K (Peebles 1965). Burke told the Princeton group about Penzias and Wilson's noise and Dicke gave Penzias a call.

Dicke did not like the idea that all the matter in the Universe had been created in the big bang. He liked the oscillating universe. He knew however that the first stars had fewer heavy elements. Where were the heavy elements that had been produced by earlier oscillations? — these elements must have been destroyed by the heat of the last contraction. Thus there must be a remnant of that heat and Dicke had decided to look for it. Dicke had a theory but no observation to support it. Penzias had noise but no theory. After the phone call Penzias' noise had become Dicke's observational support.

Until 1965 there were two competing paradigms to describe the early universe: the big bang model and the steady state model. The discovery of the CMB removed the steady state model as a serious contender. The big bang model had predicted the CMB; the steady state model had not.

7.2. *What is the CMB?*

The observable universe is expanding and cooling. Therefore in the past it was hotter and smaller. The cosmic microwave background (CMB) is the after glow of thermal radiation left over from this hot early epoch in the evolution of the Universe. It is the redshifted relic of the hot big bang. The CMB is a bath of photons coming from every direction. These are the oldest photons one can observe and they contain information about the Universe at redshifts much larger than the redshifts of galaxies and quasars ($z \approx 1000 >> z \approx$ few).

Their long journey toward us has lasted more than 99.99% of the age of the Universe and began when the Universe was one thousand times smaller than it is today. The CMB was emitted by the hot plasma of the Universe long before there were planets, stars or galaxies. The CMB is thus a unique tool for probing the early universe.

One of the most recent and most important advances in astronomy has been the discovery of hot and cold spots in the CMB based on data from the COBE satellite (Smoot *et al.*1992). This discovery has been hailed as "Proof of the Big Bang" and the "Holy Grail of Cosmology" and elicited comments like: "If you're religious it's like looking at the face of God" (George Smoot) and "It's the greatest discovery of the century, if not of all time" (Stephen Hawking). As a graduate student analysing COBE data at the time, I knew we had discovered something fundamental but its full import didn't sink in until one night after a telephone interview for BBC radio. I asked the interviewer for a copy of the interview, and he told me that would be possible if I sent a request to the *religious affairs* department.

The CMB comes from the surface of last scattering of the Universe. When you look into a fog, you are looking at a surface of last scattering. It is a surface defined by all the molecules of water which scattered a photon into your eye. On a foggy day you can see 100 meters, on really foggy days you can see 10 meters. If the fog is so dense you cannot see your hand then the surface of last scattering is less than an arm's length away. Similarly, when you look at the surface of the Sun you are seeing photons last scattered by the hot plasma of the photosphere. The early universe is as hot as the Sun and similarly the early universe has a photosphere (the surface of last scattering) beyond which (in time and space) we cannot see. As its name implies, the surface of last scattering is where the CMB photons were scattered for the last time before arriving in our detectors. The 'surface of last screaming' presented in Fig. 9 is a pedagogical analog.

7.3. *Spectrum*

The big bang model predicts that the cosmic background radiation will be thermalized — it will have a blackbody spectrum. The measurements of the antenna temperature of the radiation at various frequencies between 1965 and 1990 had shown that the spectrum was approximately blackbody but there were some measurements at high frequencies that seemed to indicate an infrared excess — a bump in the spectrum that was not easily explained. In 1989, NASA launched the COBE (Cosmic Background Explorer) satellite to investigate the cosmic microwave and infrared background radiation. There were three instruments on board. After one year of observations the FIRAS instrument had measured the spectrum of the CMB and found it to be a blackbody spectrum. The most recent analysis of the FIRAS data gives a temperature of 2.725 ± 0.002 K (Mather *et al.*1999).

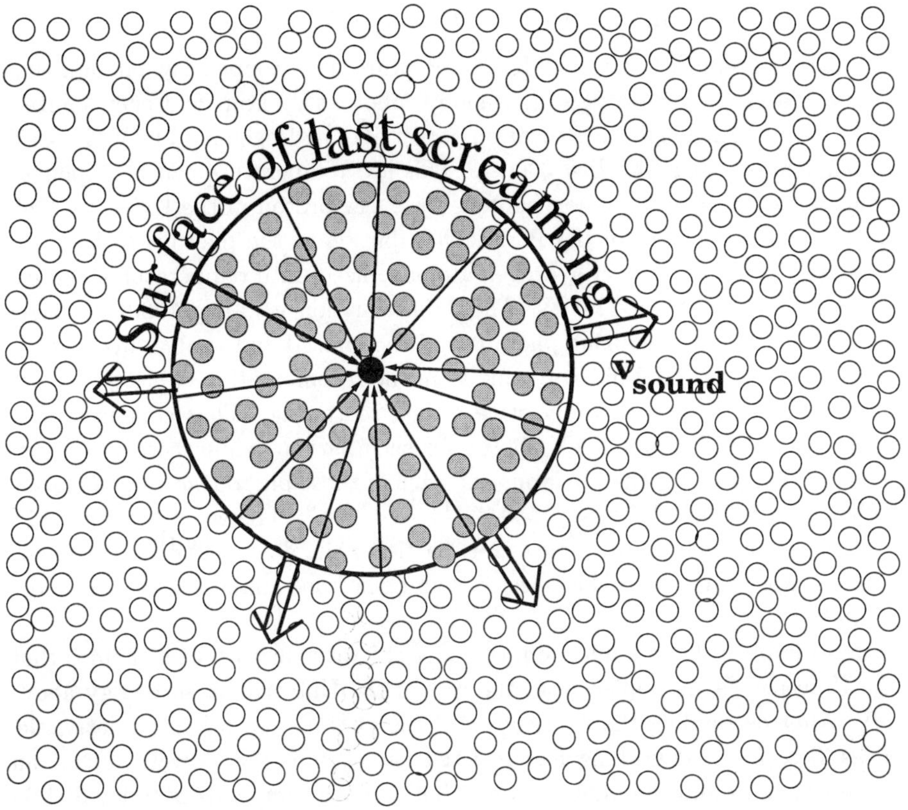

Figure 9. **The Surface of Last Screaming.** Consider an infinite field full of people screaming. The circles are their heads. You are screaming too. (Your head is the black dot.) Now suppose everyone stops screaming at the same time. What will you hear? Sound travels at 330 m/s. One second after everyone stops screaming you will be able to hear the screams from a 'surface of last screaming' 330 meters away from you in all directions. After 3 seconds the faint screaming will be coming from 1 km away...etc. No matter how long you wait, faint screaming will always be coming from the surface of last screaming — a surface that is receding from you at the speed of sound ('v_{sound}'). The same can be said of any observer — each is the centre of a surface of last screaming. In particular, observers on your surface of last screaming are currently hearing you scream since you are on their surface of last screaming. The screams from the people closer to you than the surface of last screaming have passed you by — you hear nothing from them (gray heads). When we observe the CMB in every direction we are seeing photons from the surface of last scattering. We are seeing back to a time soon after the big bang when the entire universe was opaque (screaming).

A CMB of cosmic origin (rather than one generated by starlight processed by iron needles in the intergalactic medium) is expected to have a blackbody spectrum and to be extremely isotropic. COBE FIRAS observations show that the CMB is very well approximated by an isotropic blackbody.

7.4. *Where did the energy of the CMB come from?*

Recombination occurs when the CMB temperature has dropped low enough such that there are no longer enough high energy photons to keep hydrogen ionized; $\gamma + H \leftrightarrow e^- + p^+$. Although the ionization potential of hydrogen is 13.6 eV ($T \sim 10^5$ K), recombination occurs at $T \approx 3000$ K. This low temperature can be explained by the fact that there are a billion photons for every proton in the Universe. This allows the high energy tail of the Planck distribution of the photons to keep the comparatively small number of hydrogen atoms ionized until temperatures and energies are much lower than 13.6 eV. The Saha equation *e.g.*Lang 1980) describes this balance between the ionizing photons and the ionized and neutral hydrogen.

The energy in the CMB did not come from the recombination of electrons with protons to form hydrogen at the surface of last scattering. That contribution is negligible — only about one 10 eV photon for each baryon, while there are $\sim 10^{10}$ times more CMB photons than baryons and each of those photons *at recombination* had an energy of ~ 0.3 eV: $\frac{\Delta E_{rec}}{E_{CMB}} = \frac{10eV \times 10^{-10}}{0.3eV} \sim 10^{-9}$. The energy in the CMB came from the annihilation of particle/anti-particle pairs during a very early epoch called baryogenesis and later when electrons and positrons annihilated at an energy of ~ 1 MeV.

As an example of energy injection, consider the thermal bath of neutrinos that fills the Universe. It decoupled from the rest of the Universe at an energy above an MeV. After decoupling the neutrinos and the photons, both being relativistic, cooled as $T \propto R^{-1}$. If nothing had injected energy into the Universe below an MeV, the neutrinos and the photons would both have a temperature today of 1.95 K. However the photons have a temperature of 2.725 K. Where did this extra energy come from? It came from the annihilation of electrons and positrons when the temperature of the Universe fell below an MeV. This process injected energy into the Universe by heating up the residual electrons, which in turn heated up the CMB photons. The relationship between the CMB and neutrino temperatures is $T_{CMB} = (11/4)^{1/3}\, T_\nu$. Derivation of this result using entropy conservation during electron/positron annihilation can be found in Wright (2003) or Peacock (2000). The bottom line: $T_{CMB} = 2.7$ K $> T_\nu = 1.9$ K because the photons were heated up by $e^\pm$ annihilation while the neutrinos were not. This temperature for the neutrino background has not yet been confirmed observationally.

7.5. *Dipole*

To a very good approximation the CMB is a flat featureless blackbody; there are no anisotropies and the temperature is a constant $T_o = 2.725$ K in every direction. When we remove this mean value, the next largest feature visible at 1000 times smaller amplitude is the kinetic dipole. Just as the 17 satellites of the Global Positioning System (GPS) provide a reference frame to establish positions and velocities on the Earth, the CMB gives all the inhabitants of the Universe a special common rest frame with respect to which all velocities can be measured — the comoving frame in which the observers see no CMB dipole. People who enjoy special relativity but not general relativity often baulk at this concept. A profound question that may make sense is: Where did the rest frame of the CMB come from? How was it chosen? Was there a mechanism for a choice of frame, analogous to the choice of vacuum during spontaneous symmetry breaking?

7.6. *Anisotropies*

Since the COBE discovery of hot and cold spots in the CMB, anisotropy detections have been reported by more than two dozen groups with various instruments, at various frequencies and in various patches and swathes of the microwave sky. Figure 10 is a compilation of the world's measurements (including the recent WMAP results). Measurements on the left (low ℓs) are at large angular scales while most recent measurements are trying to constrain power at small angular scales. The dominant peak at $\ell \sim 200$ and the smaller amplitude peaks at smaller angular scales are due to acoustic oscillations in the photon-baryon fluid in cold dark matter gravitational potential wells and hills. The detailed features of these peaks in the power spectrum are dependent on a large number of cosmological parameters.

7.7. *What are the oldest fossils we have from the early universe?*

It is sometimes said that the CMB gives us a glimpse of the Universe when it was $\sim 300,000$ years old. This is true but it also gives us a glimpse of the Universe when it was less than a trillionth of a second old. The acoustic peaks in the power spectrum (the spots of size less than about 1 degree) come from sound waves in the photon-baryon plasma at $\sim 300,000$ years after the big bang but there is much structure in the CMB on angular scales greater than 1 degree. When we look at this structure we are looking at the Universe when it was less than a trillionth of a second old. The large scale structure on angular scales greater than ~ 1 degree is the oldest fossil we have and dates back to the time of inflation. In the standard big bang model, structure on these acausal scales can only be explained with initial conditions.

The large scale features in the CMB, *i.e.*, all the features in the top map of Fig. 13 but none of the features in the lower map, are the largest and most distant objects ever seen. And yet they are probably also the smallest for they are quantum fluctuations zoomed in on by the microscope called inflation and hung up in the sky. So this map belongs in two different sections of the Guinness book of world records.

The small scale structure on angular scales less than ~ 1 degree (lower map) results from oscillations in the photon-baryon fluid between the redshift of equality and recombination. Figure 11 describes these oscillations in more detail.

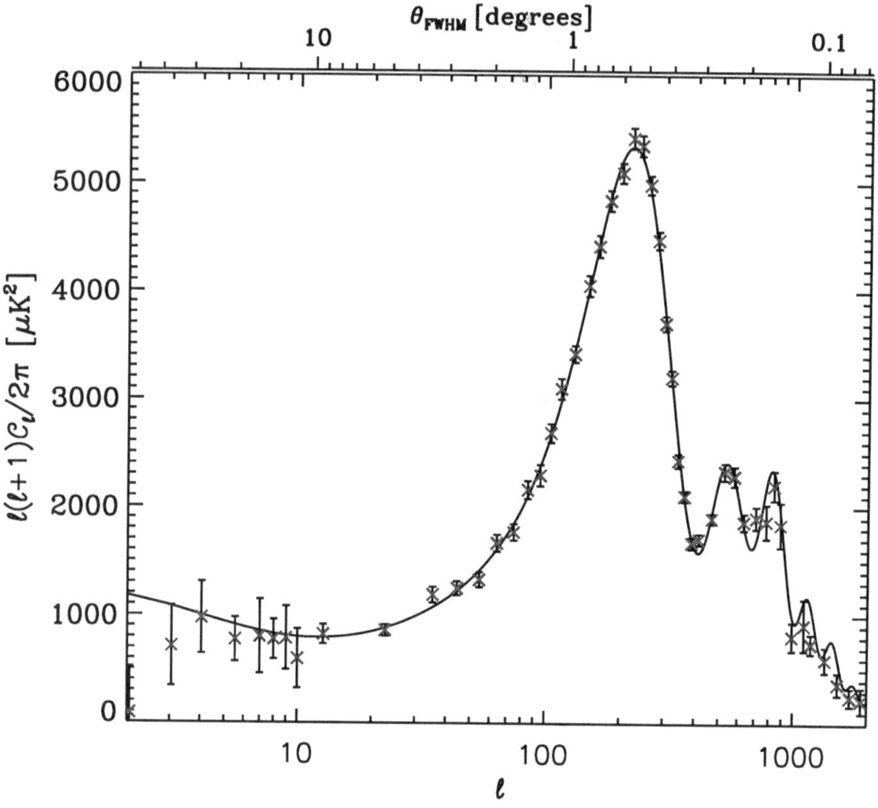

Figure 10. **Measurements of the CMB power spectrum.** CMB power spectrum from the world's combined data, including the recent WMAP satellite results (Hinshaw *et al.*2003). The amplitudes of the hot and cold spots in the CMB depend on their angular size. Angular size is noted in degrees on the top x axis. The y axis is the power in the temperature fluctuations. No CMB experiment is sensitive to this entire range of angular scale. When the measurements at various angular scales are put together they form the CMB power spectrum. At large angular scales ($\ell \lesssim 100$), the temperature fluctuations are on scales so large that they are 'non-causal', *i.e.*, they have physical sizes larger than the distance light could have travelled between the big bang (without inflation) and their age at the time we see them (300,000 years after the big bang). They are either the initial conditions of the Universe or were laid down during an epoch of inflation $\sim 10^{-35}$ seconds after the big bang. New data are being added to these points every few months. The concordance model shown has the following cosmological parameters: $\Omega_\Lambda = 0.743$, $\Omega_{CDM} = 0.213$, $\Omega_{baryon} = 0.0436$, $h = 0.72$, $n = 0.96$, $\tau = 0.12$ and no hot dark matter (neutrinos) (τ is the optical depth to the surface of last scattering). χ^2 fits of this data to such model curves yield the estimates in Table 1. The physics of the acoustic peaks is briefly described in Fig. 11.

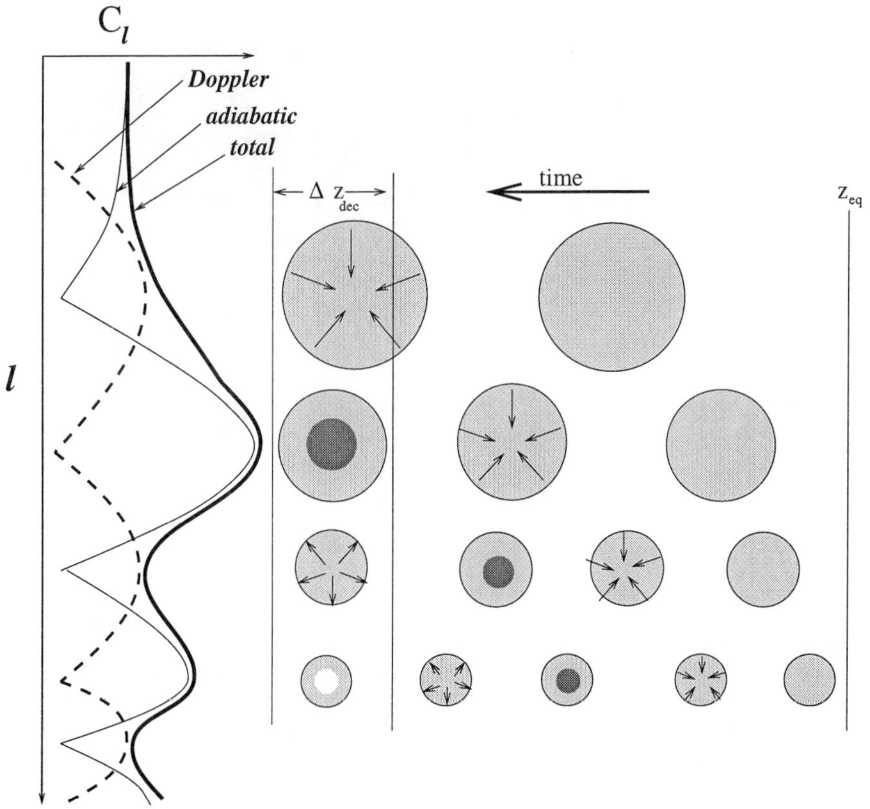

Figure 11. The dominant acoustic peaks in the CMB power spectra are caused by the collapse of dark matter over-densities and the oscillation of the photon-baryon fluid into and out of these over-densities. After matter becomes the dominant component of the Universe, at $z_{eq} \approx 3233$ (see Table 1), cold dark matter potential wells (grey spots) initiate in-fall and then oscillation of the photon-baryon fluid. The phase of this in-fall and oscillation at z_{dec} (when photon pressure disappears) determines the amplitude of the power as a function of angular scale. The bulk motion of the photon-baryon fluid produces 'Doppler' power out of phase with the adiabatic power. The power spectrum (or C_ℓs) is shown here rotated by 90° compared to Fig. 10. Oscillations in fluids are also known as sound. Adiabatic compressions and rarefactions become visible in the radiation when the baryons decouple from the photons during the interval marked Δz_{dec} ($\approx 195 \pm 2$, Table 1). The resulting bumps in the power spectrum are analogous to the standing waves of a plucked string. This very old music, when converted into the audible range, produces an interesting roar (Whittle 2003). Although the effect of over-densities is shown, we are in the linear regime so under-densities contribute an equal amount. That is, each acoustic peak in the power spectrum is made of equal contributions from hot and cold spots in the CMB maps (Fig. 12). Anisotropies on scales smaller than about $8'$ are suppressed because they are superimposed on each other over the finite path length of the photon through the surface Δz_{dec}.

7.8. Observational Constraints from the CMB

Our general relativistic description of the Universe can be divided into two parts, those parameters like Ω_i and H which describe the global properties of the model

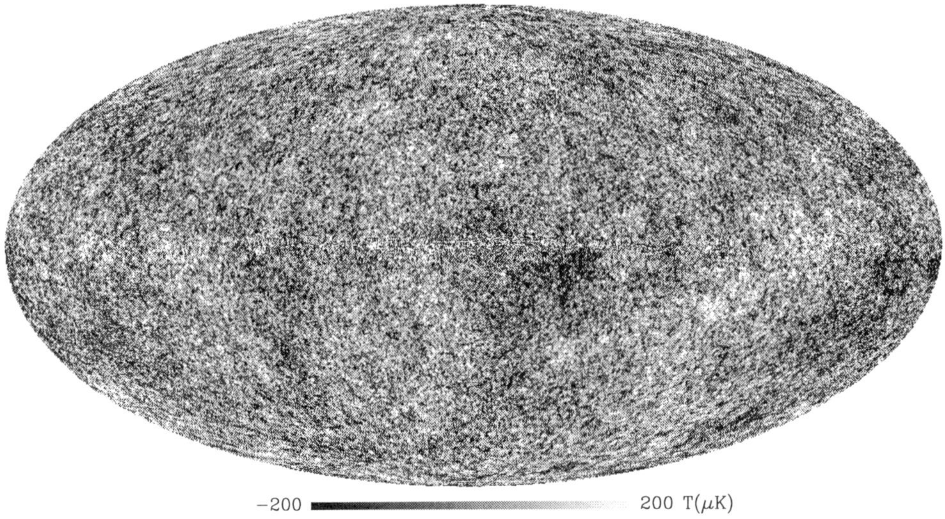

Figure 12. Full sky temperature map of the cosmic microwave background derived from the WMAP satellite (Bennett *et al.*2003, Tegmark *et al.*2003). The disk of the Milky Way runs horizontally through the centre of the image but has been almost completely removed from this image. The angular resolution of this map is about 20 times better than its predecessor, the COBE-DMR map in which the hot and cool spots shown here were detected for the first time. The large and small scale power of this map is shown separately in the next figure.

and those parameters like n_s and A which describe the perturbations to the global properties and hence describe the large scale structure (Table 1).

In the context of general relativity and the hot big bang model, cosmological parameters are the numbers that, when inserted into the Friedmann equation, best describe our particular observable universe. These include Hubble's constant H (or $h = H/100\ km\ s^{-1}Mpc^{-1}$), the cosmological constant $\Omega_\Lambda = \Lambda/3H^2$, geometry $\Omega_k = -k/H^2R^2$, the density of matter, $\Omega_M = \Omega_{CDM} + \Omega_{baryon} = \rho_{CDM}/\rho_c + \rho_{baryon}/\rho_c$ and the density of relativistic matter $\Omega_{rel} = \Omega_\gamma + \Omega_\nu$. Estimates for these have been derived from hundreds of observations and analyses. Various methods to extract cosmological parameters from cosmic microwave background (CMB) and non-CMB observations are forming an ever-tightening network of interlocking constraints. CMB observations now tightly constrain Ω_k, while type Ia supernovae observations tightly constrain the deceleration parameter q_o. Since lines of constant Ω_k and constant q_o are nearly orthogonal in the $\Omega_M - \Omega_\Lambda$ plane, combining these measurements optimally constrains our Universe to a small region of parameter space.

The upper limit on the energy density of neutrinos comes from the shape of the small scale power spectrum. If neutrinos make a significant contribution to the

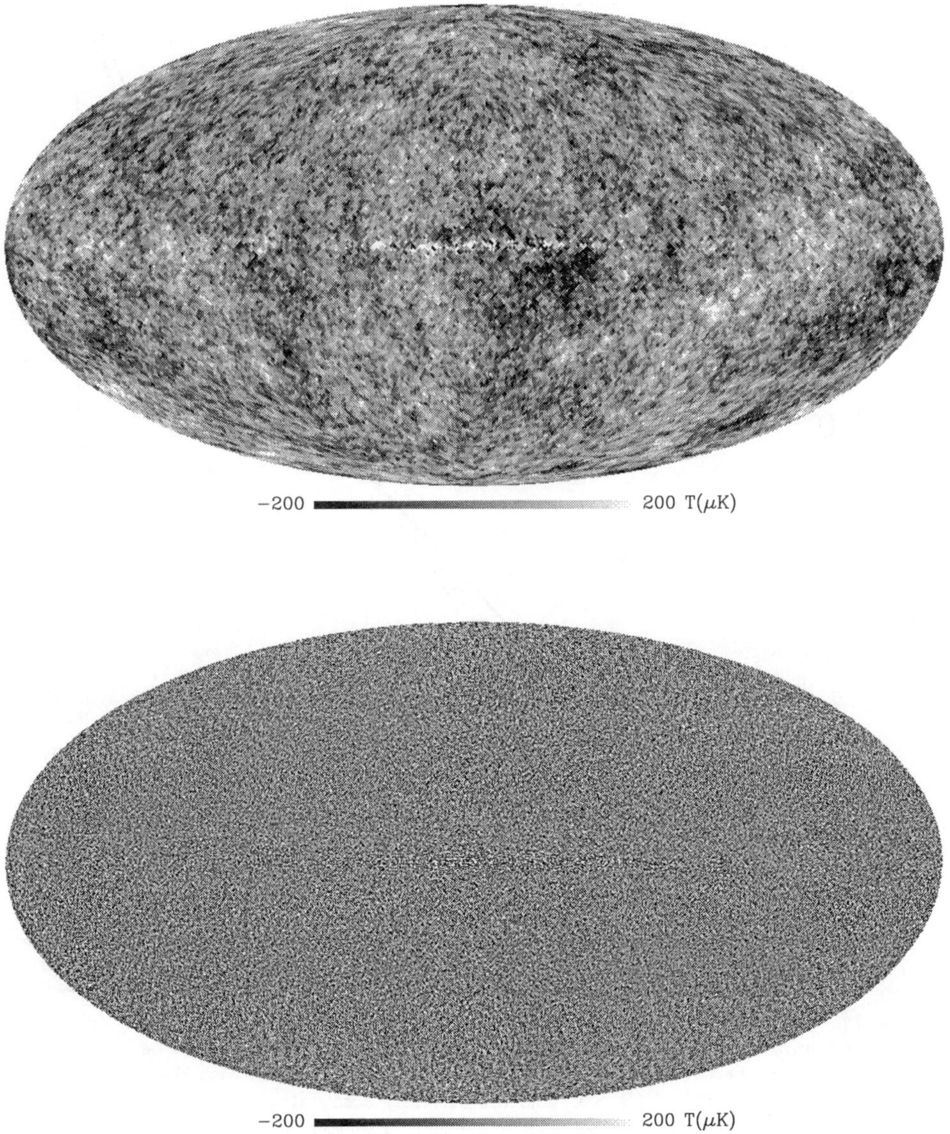

Figure 13. Two basic ingredients: old quantum fluctuations (top) and new sound (bottom). These two maps were constructed from Fig. 12. The top map is a smoothed version of Fig. 12 and shows only power at angular scales greater than $\sim 1\,\text{deg}$ ($\ell \lesssim 100$, see Fig. 10). This footprint of the inflationary epoch was made in the first picosecond after the big bang. In the standard big bang without inflation, all the structure here has to be attributed to initial conditions. The lower map was made by subtracting the top map from Fig. 12. That is, all the large scale power was subtracted from the CMB leaving only the small scale power in the acoustic peaks ($\ell > 100$, see Fig. 10) — these are the crests of the sound waves generated after radiation/matter equality (Fig. 11). Thus, the top map shows quantum fluctuations imprinted when the age of the Universe was in the range $[10^{-43}, 10^{-12}]$ seconds old, while the bottom map shows foreground contamination from sound generated when the Universe was $\sim 10^{13}$ seconds old.

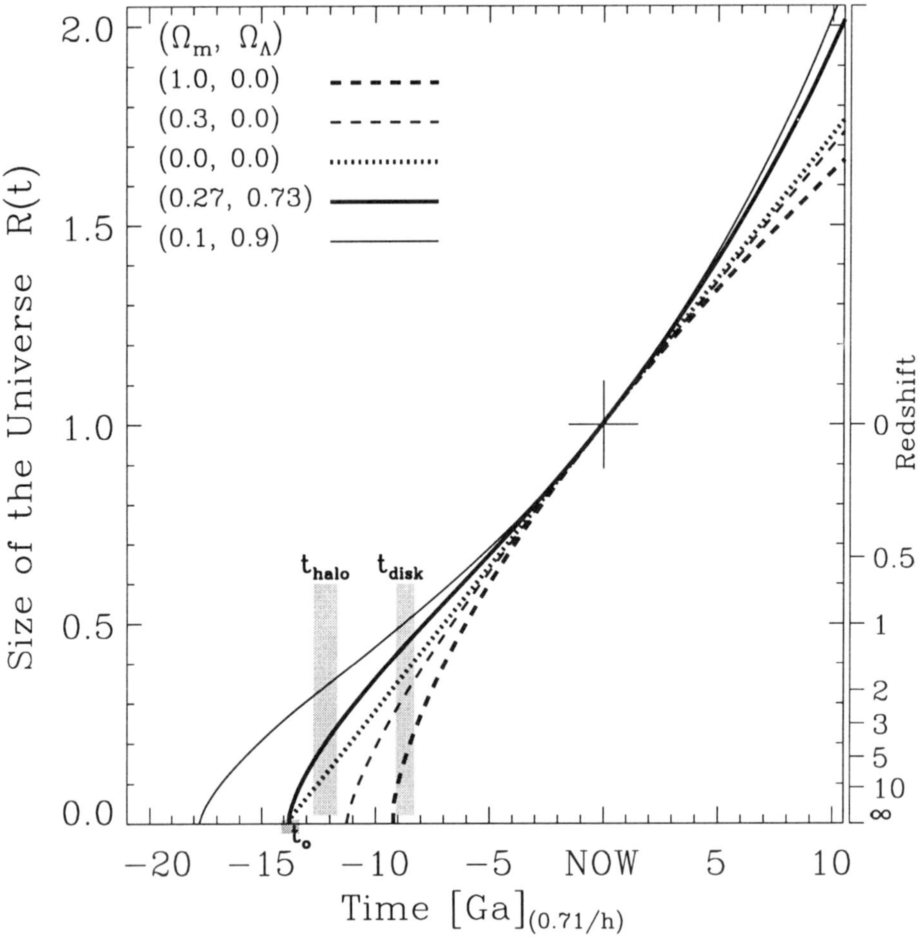

Figure 14. **Size and Destiny of the Universe.** This plot shows the size of the Universe, in units of its current size, as a function of time. The age of the five models can be read from the x axis as the time between 'NOW' and the intersection of the model with the x axis. Models containing Ω_Λ curve upward ($\ddot{R} > 0$) and are currently accelerating. The empty universe has $\ddot{R} = 0$ (dotted line) and is 'coasting'. The expansion of matter-dominated universes is slowing down ($\ddot{R} < 0$). The $(\Omega_\Lambda, \Omega_M) \approx (0.27, 0.73)$ model is favoured by the data. Over the past few billion years and on into the future, the rate of expansion of this model increases. This acceleration means that we are in a period of slow inflation — a new period of inflation is starting to grab the Universe. Knowing the values of h, Ω_M and Ω_Λ yields a precise relation between age, redshift and size of the Universe allowing us to convert the ages of local objects (such as the disk and halo of our galaxy) into redshifts. We can then examine objects at those redshifts to see if disks are forming at a redshift of ~ 1 and halos are forming at $z \sim 4$. This is an example of the tightening network of constraints produced by precision cosmology.

density, they suppress the growth of small scale structure by free-streaming out of over-densities. The CMB power spectrum is not sensitive to such small scale power or its suppression, and is not a good way to constrain Ω_ν. And yet the best limits on Ω_ν come from the WMAP normalization of the CMB power spectrum used to normalize the power spectrum of galaxies from the 2dF redshift survey (Bennett *et al.*2003).

The parameters in Table 1 are not independent of each other. For example, the age of the Universe, $t_o = h^{-1}f(\Omega_M, \Omega_\Lambda)$. If $\Omega_m = 1$ as had been assumed by most theorists until about 1998, then the age of the Universe would be simple:

$$t_o(h) = \frac{2}{3}H_o^{-1} = 6.52\ h^{-1}Gyr. \tag{36}$$

However, current best estimates of the matter and vacuum energy densities are $(\Omega_M, \Omega_\Lambda) = (0.27, 0.73)$. For such flat universes ($\Omega = \Omega_M + \Omega_\Lambda = 1$) we have (Carroll *et al.*1992):

$$t_o(h, \Omega_M, \Omega_\Lambda) = \frac{1}{\sqrt{\Omega_\Lambda}}\left(\frac{1+\sqrt{\Omega_\Lambda}}{\sqrt{\Omega_M}}\right) 6.52\ h^{-1}Gyr. \tag{37}$$

for $t_o(h = 0.71, \Omega_M = 0.27, \Omega_\Lambda = 0.73) = 13.7$ Gyr.

If the Universe is to make sense, independent determinations of Ω_Λ, Ω_M and h and the minimum age of the Universe must be consistent with each other. This is now the case (Lineweaver 1999). Presumably we live in a universe which corresponds to a single point in multidimensional parameter space. Estimates of h from HST Cepheids and the CMB overlap. Deuterium and CMB determinations of $\Omega_{baryon}h^2$ are consistent. Regions of the $\Omega_M - \Omega_\Lambda$ plane favoured by supernovae and CMB overlap with each other and with other independent constraints (*e.g.*Lineweaver 1998). The geometry of the Universe does not seem to be like the surface of a ball ($\Omega_k < 0$) nor like a saddle ($\Omega_k > 0$) but seems to be flat ($\Omega_k \approx 0$) to the precision of our current observations.

There has been some speculation recently that the evidence for Ω_Λ is really evidence for some form of stranger dark energy (dubbed 'quintessence') that we have been incorrectly interpreting as Ω_Λ. The evidence so far indicates that the cosmological constant interpretation fits the data as well as or better than an explanation based on quintessence.

7.9. *Background and the Bumps on it and the Evolution of those Bumps*

Equation 11 is our hot big bang description of the unperturbed Friedmann-Robertson-Walker universe. There are no bumps in it, no over-densities, no inhomogeneities, no anisotropies and no structure. The parameters in it are the background parameters. It describes the evolution of a perfectly homogeneous universe.

However, bumps are important. If there had been no bumps in the CMB thirteen billion years ago, no structure would exist today. The density bumps seen

Composition of Universe[a]		
Total density	Ω_o	1.02 ± 0.02
Vacuum energy density	Ω_Λ	0.73 ± 0.04
Cold Dark Matter density	Ω_{CDM}	0.23 ± 0.04
Baryon density	Ω_b	0.044 ± 0.004
Neutrino density	Ω_ν	< 0.0147 95% CL
Photon density	Ω_γ	$4.8 \pm 0.014 \times 10^{-5}$
Fluctuations		
Spectrum normalization[b]	A	$0.833^{+0.086}_{-0.083}$
Scalar spectral index[b]	n_s	0.93 ± 0.03
Running index slope[b]	$dn_s/d\ln k$	$-0.031^{+0.016}_{-0.018}$
Tensor-to-scalar ratio[c]	$r = T/S$	< 0.71 95% CL
Evolution		
Hubble constant	h	$0.71^{+0.04}_{-0.03}$
Age of Universe (Gyr)	t_0	13.7 ± 0.2
Redshift of matter-energy equality	z_{eq}	3233^{+194}_{-210}
Decoupling Redshift	z_{dec}	1089 ± 1
Decoupling epoch (kyr)	t_{dec}	379^{+8}_{-7}
Decoupling Surface Thickness (FWHM)	Δz_{dec}	195 ± 2
Decoupling duration (kyr)	Δt_{dec}	118^{+3}_{-2}
Reionization epoch (Myr, 95% CL))	t_r	180^{+220}_{-80}
Reionization Redshift (95% CL)	z_r	20^{+10}_{-9}
Reionization optical depth	τ	0.17 ± 0.04

[a] $\Omega_i = \rho_i/\rho_c$ where $\rho_c = 3H^2/8\pi G$
[b] at a scale corresponding to wavenumber $k_0 = 0.05$ Mpc^{-1}
[c] at a scale corresponding to wavenumber $k_0 = 0.002$ Mpc^{-1}

as the hot and cold spots in the CMB map have grown into gravitationally enhanced light-emitting over-densities known as galaxies (Fig. 7). Their gravitational growth depends on the cosmological parameters — much as tree growth depends on soil quality (see Efstathiou 1990 for the equations of evolution of the bumps). We measure the evolution of the bumps and from them we infer the background. Specifically, matching the power spectrum of the CMB (the C_ℓs which sample the $z \sim 1000$ universe) to the power spectrum of local galaxies (the $P(k)$ which sample the $z \sim 0$ universe) we can constrain cosmological parameters. The limit on Ω_ν is an example.

7.10. *The End of Cosmology?*

When the WMAP results came out at the end of this school I was asked "So is this the end of cosmology? We know all the cosmological parameters...what is there left to do? To what precision does one really want to know the value of Ω_m?" In his talk, Brian Schmidt asked the rhetorical question: "We know Hubble's parameter to about 10%, is that good enough?" Well, now we know it to about 5%. Is that good enough? Obviously the more precision on any one parameter the better, but we are talking about constraining an entire model of the universe defined by a network of parameters. As we determine 5 parameters to less than 10%, it enables us to turn a former upper limit on another parameter into a detection. For example we still have only upper limits on the tensor to scalar ratio r and this limits our ability to test inflation. We only have an upper limit on the density of neutrinos Ω_ν and this limits our ability to go beyond the standard model of particle physics. And we have only a tenuous detection of the running of the scalar spectral index $dn/dlnk \neq 0$, and this limits our ability to constrain inflaton potential model builders.

We still know next to nothing about $\Omega_\Lambda \sim 0.7$, most of the Universe. ΛCDM is an observational result that has yet to be theoretically confirmed. From a quantum field theoretic point of view $\Omega_\Lambda \sim 0.7$ presents a huge problem. It is a quantum term in a classical equation. But the last time such a quantum term appeared in a classical equation, Hawking radiation was discovered. A similar revelation may be in the offing. The Friedmann equation will eventually be seen as a low energy approximation to a more complete quantum model in much the same way that $\frac{1}{2}mv^2$ is a low energy approximation to pc.

Inflation solves the origin of structure problem with quantum fluctuations, and this is just the beginning of quantum contributions to cosmology. Quantum cosmology is opening up many new doors. Varying coupling constants are expected at high energy (Wilczek 1999) and c variation, G variation, α (fine structure constant) variation, and Ω_Λ variation (quintessence) are being discussed. We may be in an ekpyrotic universe or a cyclic one (Steinhardt & Turok 2002). The topology of the Universe is also alluringly fundamental (Levin 2002). Just as we were getting precise estimates of the parameters of classical cosmology, whole new sets of quantum cosmological parameters are being proposed. The next high profile goal of cosmology may be trying to figure out if we are living in a multiverse. And what, pray tell, is the connection between inflation and dark matter?

7.11. *Tell me More*

For a well-written historical (non-mathematical) review of inflation see Guth (1997). For a detailed mathematical description of inflation see Liddle and Lyth (2000). For a concise mathematical summary of cosmology for graduate students see Wright (2003). Three authoritative texts on cosmology that include inflation and the CMB are 'Cosmology' by P. Coles and F. Lucchin, 'Physical Cosmology' by P. J. E. Peebles and 'Cosmological Physics' by J. Peacock.

Acknowledgments

I thank Matthew Colless for inviting me to give these five lectures to such an appreciative audience. I thank John Ellis for useful discussions as we bushwhacked in the gloaming. I thank Tamara Davis for Figs. 1, 4 & 5. I thank Roberto dePropris for preparing Fig. 7. I thank Louise Griffiths for producing Fig. 10 and Patrick Leung for producing Figs. 12 & 13. The HEALPix package (Górski, Hivon and Wandelt 1999) was used to prepare these maps. I acknowledge a Research Fellowship from the Australian Research Council.

References

1. Alpher, R.A. and Herman, R. 1948 Nature, 162, 774–775
2. Bennett, C.L. *et al.*2003, Astrophys. J. Suppl. 148, 97
3. Carroll, S.M., Press, W.H., Turner, E.L. 1992, Ann. Rev. Astron. Astrophy. 30, 499
4. Coles, P. & Lucchin, F. 1995 "Cosmology: The Origin and Evolution of Cosmic Structure" Wiley: NY
5. Davis, T.M. & Lineweaver, C.H. 2004, "Expanding Confusion: common misconceptions of horizons and the superluminal expansion of the universe" PASA 2(1) 97
6. Dicke, R.H., Peebles, P.J.E., Roll, P.G. and Wilkinson, D.T. 1965, Astrophys. J 142, 414
7. Efstathiou, G. 1990, in *Physics of the Early Universe*, 36th Scottish Universities Summer School in Physics, ed J.A. Peacock, A.F. Heavens, A.T. Davies, Adam Hilger, p. 361
8. Górski, K.M., Hivon, E. and Wandelt, B.D. 1999, in Proceedings of the MPA/ESO Cosmology Conference *Evolution of Large Scale Structure* eds. A.J. Banday, R.S. Sheth and L. DaCosta, PrintPartners Ipskamp, NL, pp. 37–42, astro-ph/9812350.
9. Guth, A.H. 1997 *The Inflationary Universe: The Quest for a New Theory of Cosmic Origins*, Random House, London, quotes cited are from pp. xiii and 184
10. Harrison, E.R. 1981, *Cosmology: Science of the Universe*, Cambridge University Press
11. Hinshaw, G. *et al.*2003, Astrophys. J. submitted astro-ph/0302217
12. Kolb, E.R. and Turner, M.S. 1990 *The Early Universe* Addison-Wesley, Redwood City
13. Kragh, H. 1996 *Cosmology and Controversy*, Princeton Univ. Press
14. Landau, L.D., Lifshitz, E.M. 1975, *The Classical Theory of Fields* Fourth Revised Edition, Course of Theoretical Physics, Vol 2., Pergamon Press, Oxford
15. Lang, K.R. 1980 *Astrophysical Formulae*, 2nd Edition Springer-Verlag, Berlin
16. Levin, J. 2002 Phys. Rept. 365, 251-333, gr-qc/0108043
17. Liddle, A.R. and Lyth, D.H. 2000 *Cosmological Inflation and Large-Scale Structure* (Cambridge Univ. Press, Cambridge) quote from page 1.
18. Lineweaver, C.H. 1998, Astrophys. J. 505, L69–73
19. Lineweaver, C.H. Science 1999, 284, 1503–1507 astro-ph/9901234
20. Mather, J. et al. 1999, Astrophys. J. 512, 511
21. Peacock, J. 1999, *Cosmological Physics* Cambridge Univ. Press.
22. Peebles, P.J.E. 1965 "Cosmology, Cosmic Black Body Radiation, and the Cosmic Helium Abundance" Physical Review, submitted, unpublished.
23. Peebles, P.J.E. 1993, *Principles of Physical Cosmology* Princeton Univ. Press
24. Penzias, A.A. and Wilson, R.W. Astrophy. J., 142, pp 419–421
25. Smoot, G. F. *et al.*1992 Astrophys. J. L32.
26. Spergel, D. *et al.*2003 Astrophys. J. in press. astro-ph/0302209
27. Steinhardt, P. & Turok, N. 2002, Science, 296, 1436-1439
28. Tegmark, M., de Oliveira-Costa, A. Hamilton, A. 2003, astro-ph/0302496, available at http://www.hep.upenn.edu/~max/wmap.html
29. Weinberg, S. 1977, *The First Three Minutes* Basic Books, NY p 144
30. Whittle, M. 2003 Mark Whittle with the help of Louise Griffiths, Joe Wolfe and Alex

Tarnopolsky produced the CMB music available at http://bat.phys.unsw.edu.au/~charley/cmb.wav.
31. Wilczek, F. 1999, Nucl. Phys. Proc. Suppl. 77, 511–519, hep-ph/9809509
32. Wright, E. 2003, Astronomy 275, UCLA Graduate Course Lecture Notes, available at http://www.astro.ucla.edu/~wright/cosmolog.htm (file A275.ps).

THE LARGE-SCALE STRUCTURE OF THE UNIVERSE

MATTHEW COLLESS

Research School of Astronomy and Astrophysics, The Australian National University, Cotter Road, Weston Creek, ACT 2611, Australia

E-mail: colless@mso.anu.edu.au

These three lectures give an introduction to galaxy redshift surveys as probes of the large-scale structure in the Universe, and describe recent measurements of fundamental cosmological parameters from both the redshift surveys and observations of the cosmic microwave background. The first lecture deals with the large-scale structure (LSS) revealed by the galaxy distribution, and its interpretation in terms of cosmological parameters. The topics covered include: a descriptive review of large-scale structure; redshift surveys, cosmography and cosmology; the statistical characterization of LSS; an introduction to the theory of structure formation; the density and velocity fields; bias and the relation of light to mass; redshift-space distortions; the observed correlation function and power spectrum; and the Gaussianity and topology of the density field. The second lecture discusses the current state of the art in redshift surveys, describing the results on large-scale structure and cosmology emerging from the 2dF Galaxy Redshift Survey (2dFGRS). The third lecture discusses the important new results from observations of the cosmic microwave background (CMB) by the Wilkinson Microwave Anisotropy Probe (WMAP) satellite that were reported during the course of the Summer School.

1. Redshift Surveys, Large-scale Structure and Cosmology

1.1. *Redshift Surveys*

A redshift survey is a systematic mapping of a volume of space by measuring the cosmological redshifts of galaxies (Geller & Huchra 1989; Giovanelli & Haynes 1991; Strauss & Willick 1995). A galaxy's redshift is related to the ratio of the observed wavelengths of its spectral features to their emitted (rest-frame) values, and directly measures the relative scale factor of the Universe, $a(t)$, between the time the light was detected by the observer and the time it was emitted by the galaxy:

$$1 + z = \lambda_{obs}/\lambda_{emi} = a(t_{obs})/a(t_{emi}) \ . \tag{1}$$

Redshifts can be viewed as distance coordinates. For cosmologically small distances, the redshift is approximately linearly related to the recession velocity of the galaxy and its distance (the Hubble law; Hubble 1934),

$$cz = v_{recession} = H_0 d \quad (\text{for } z \ll 1) \ , \tag{2}$$

where H_0 is the Hubble constant, in $\mathrm{km\,s^{-1}\,Mpc^{-1}}$. Another way of stating this is that for a low-z galaxy moving with the Hubble flow, redshift distance ($s \equiv cz$) is the same as true distance ($r \equiv H_0 d$), where s and r are conveniently measured in

$\mathrm{km\,s^{-1}}$. Note that $1\,\mathrm{h^{-1}\,Mpc}$ corresponds to $100\,\mathrm{km\,s^{-1}}$ in redshift space, using the convention that $H_0 = 100h\,\mathrm{km\,s^{-1}\,Mpc^{-1}}$.

For larger distances the Hubble law breaks down, and the radial co-moving distance to an object (the measure of distance that remains constant if the object is purely moving with the Hubble expansion) is given by

$$d_C = \frac{c}{H_0}\int_0^z \frac{dz'}{\sqrt{\Omega_m(1+z')^3 + \Omega_k(1+z')^2 + \Omega_\Lambda}}\,, \tag{3}$$

where Ω_m and Ω_Λ are the densities of matter and the cosmological constant in units of the critical energy density for producing a flat Universe, and Ω_k is the curvature of space defined by $\Omega_m + \Omega_\Lambda + \Omega_k = 1$ (for a flat Universe, $\Omega_k = 0$ and so $\Omega_m + \Omega_\Lambda = 1$). Other important measures of distance, such as the transverse co-moving distance d_M (the co-moving distance between two objects at the same redshift), the luminosity distance (defined by $d_L = \sqrt{L/4\pi S}$, where L and S are an object's total luminosity and observed flux), and the angular diameter distance (defined by $d_A = D/\theta$, the ratio of an object's physical size to its angular size), are directly related to the co-moving distance (and hence to redshift). For a flat Universe, these relations are:

$$d_C = d_M = d_L/(1+z) = d_A(1+z)\,. \tag{4}$$

Taking redshifts as distance coordinates is the viewpoint in low-z surveys of spatial structure. But redshifts can also be considered as time coordinates; the look-back time to a galaxy is

$$t_L = H_0^{-1}\int_0^z \frac{dz'}{(1+z')\sqrt{\Omega_m(1+z')^3 + \Omega_k(1+z')^2 + \Omega_\Lambda}}\,; \tag{5}$$

this is the viewpoint in high-z surveys of galaxy evolution.

As well as moving with the Hubble flow, galaxies also have 'peculiar velocities' due to the net gravitational attraction of the surrounding mass field. The full relation between redshift-space and real-space coordinates is therefore

$$s = r + \vec{v}\cdot\vec{r}/r = r + v_p \quad (\text{for } s \ll c)\,, \tag{6}$$

where $\vec{v}\cdot\vec{r}$ is the galaxy's peculiar velocity along the observer's line of sight (only this radial component of the galaxy's peculiar velocity is observed, since redshifts only measure radial motions).

To summarize the above discussion, there are three (partial) views of redshift: (i) z measures the distance needed to map 3D positions and number density; (ii) z measures the look-back time needed to map histories and evolution; and (iii) $cz - H_0 d$ measures the peculiar velocity needed to map the velocity field and mass density. The three main uses of redshift surveys correspond to emphasizing one of these views. Firstly, one can map the galaxy distribution, in order to chart the large-scale structures (cosmography), to test whether structure grows through gravitational instability, and to determine the nature and density of the dark matter. Secondly, one can determine the properties of galaxies at different look-back

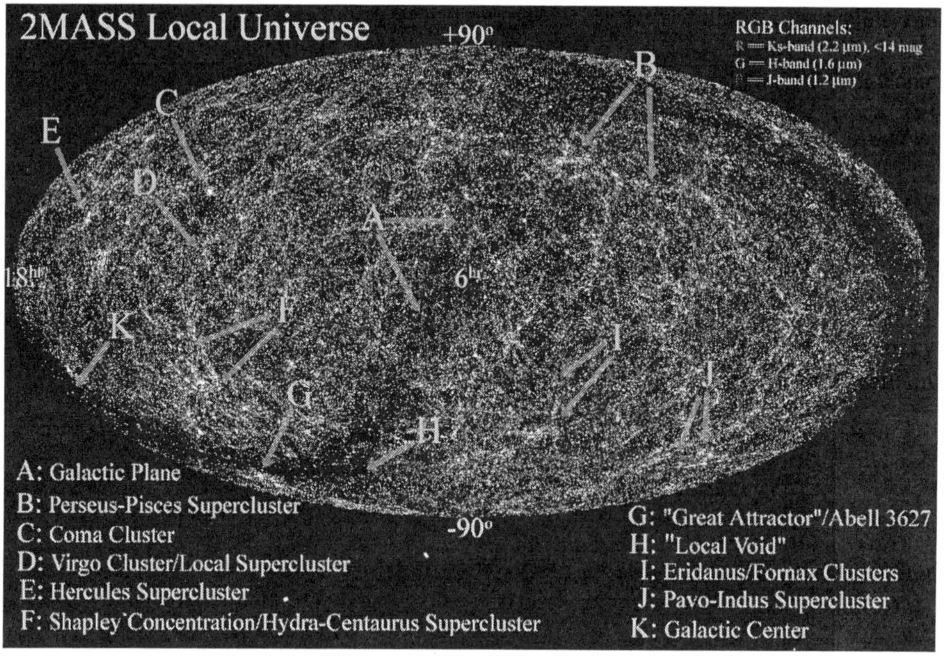

Figure 1. The large-scale structures in the local Universe as revealed by the galaxy density distribution over the whole sky from 2MASS (T.Jarrett, 2003, priv.comm.).

times, in order to characterise the galaxy population at each epoch, determine the physical mechanisms by which the population evolves, and so probe the history of galaxy formation. Thirdly, one can combine redshifts with independent distance measurements to determine peculiar velocities, mapping the velocity field and hence 'see' the underlying mass distribution through its gravitational effects.

1.2. *Cosmography*

The main structures in the local (low-redshift) galaxy distribution include (Tully & Fisher 1987; Strauss & Willick 1995):

(1) The Local Group: Milky Way, Andromeda and retinue of smaller galaxies.

(2) The Virgo cluster: the nearest significant galaxy cluster; the Local Group is falling towards Virgo.

(3) The Local Supercluster: a flattened distribution of galaxies within $cz < 3000\ \mathrm{km\,s^{-1}}$; supergalactic coordinates (X, Y, Z) are defined with X and Y in the supergalactic plane and Z perpendicular to this plane.

(4) The 'Great Attractor': a large mass concentration lying at one end of the Local Supercluster at $(X, Y, Z) = (-3400, +1500, \pm 2000)\,\mathrm{km\,s^{-1}}$, towards which both the Local Group and Virgo are falling.

(5) The Perseus-Pisces supercluster: lies at the other end of the Local Supercluster, at $(X, Y, Z) = (+4500, \pm 2000, \pm 2000)\,\mathrm{km\,s^{-1}}$.

(6) The Coma cluster: the nearest very rich cluster, at $(X, Y, Z) = (0, +7000, 0)$; a major node in the 'Great Wall' filamentary structure.

(7) The Shapley supercluster: the most massive supercluster within $z < 0.1$, at a distance of 14,000 $\mathrm{km\,s^{-1}}$ behind the Great Attractor.

(8) Voids: the Local Void, Sculptor Void, and others, lie between these mass concentrations.

Figure 1 shows these features, and other large-scale structures in the local Universe, as they appear in the galaxy density distribution mapped over the whole sky by the Two Micron All-Sky Survey (2MASS). As well as these high-contrast features, yet larger structures are seen at lower contrast on scales of $100\,h^{-1}$ Mpc and beyond (at the mean depth of this survey, $z = 0.05$–0.1, this corresponds to about 20–40°).

1.3. *Describing the Density Field*

We would like to determine the mass distribution (the mass density field), represented by the dimensionless density perturbation,

$$\delta(\vec{r}) = \rho(\vec{r})/\langle\rho\rangle - 1 \ . \tag{7}$$

The paradigm is that structures grow from 'initial' density fluctuations by gravitational instability amplification. Up to the decoupling of matter and radiation, the evolution of the density perturbations is complex and depends on the interactions of the matter and radiation fields that go into 'CMB physics' (see Lineweaver's lectures).

After decoupling, the linear growth of fluctuations is simple and depends only on the cosmology and the fluctuations in the density at the surface of last scattering (see, *e.g.*, Peebles 1980, 1993; Coles & Lucchin 1995; Peacock 1999, 2004). This is often referred to as large-scale structure in the linear regime. As the density perturbations grow the evolution becomes non-linear, and complex structures like galaxies and clusters form. In this regime additional complications also emerge, like gas dynamics and star formation. Here we concentrate mainly on structures in the linear regime, which corresponds to the largest scales and to density contrasts $\delta < 1$.

It is helpful to express the density distribution $\delta(\vec{r})$ in the Fourier domain as

$$\delta(\vec{k}) = V^{-1} \int \delta(\vec{r}) e^{i\vec{k}\cdot\vec{r}} d^3r \ . \tag{8}$$

The power spectrum is the mean squared amplitude of each Fourier mode:

$$P(k) =< |\delta(\vec{k})|^2 > \ . \tag{9}$$

Note that we have $P(k)$ not $P(\vec{k})$ because of the isotropy of the distribution (*i.e.* scales matter but directions don't). $P(k)$ gives the power in fluctuations with a scale $r = 2\pi/k$, so that $k = (1.0, 0.1, 0.01)\,h\,\mathrm{Mpc}^{-1}$ corresponds to

$r \approx (6, 60, 600)\,\mathrm{h}^{-1}\,\mathrm{Mpc}$. The power spectrum can be written in dimensionless form as the variance per unit $\ln k$:

$$\Delta^2(k) = \frac{d\langle|\delta(k)|^2\rangle}{d\ln k} = \frac{V}{2\pi^2}k^3\,P(k)\ , \tag{10}$$

so that $\Delta^2(k) = 1$ means the modes in the logarithmic bin around wavenumber k have rms density fluctuations of order unity.

The autocorrelation function of the density field (often just called the correlation function) is given by

$$\xi(r) = \langle\delta(x)\delta(x+r)\rangle\ . \tag{11}$$

The correlation function and the power spectrum are a Fourier transform pair:

$$\xi(r) = \frac{V}{(2\pi)^3}\int |\delta(k)|^2\, e^{-i\vec{k}\cdot\vec{r}}\, d^3k \tag{12}$$

$$= \frac{1}{2\pi^2}\int P(k)\frac{\sin kr}{kr}k^2dk\ . \tag{13}$$

They therefore contain precisely the same information about the density field.

When applied to galaxies rather than the density field, $\xi(r)$ is often referred to as the 'two-point correlation function', as it gives the excess probability (over the mean) of finding two galaxies in volumes dV separated by r:

$$dP = \rho_0^2[1+\xi(r)]\, d^2V \tag{14}$$

(by isotropy, only separation r matters, and not the vector $\vec{r}$). We can thus think of $\xi(r)$ as the mean over-density of galaxies at distance r from a random galaxy.

The fluctuations in the density field can also be characterised by the (filtered) variance as a function of scale. The filter (or its FT, the window function) specifies the effective volume over which the variance in the density field is determined. To obtain the variance, the correlation function is convolved with the filter in real space, or the power spectrum is multiplied by the window in Fourier space:

$$\sigma^2 = \frac{V}{(2\pi)^3}\int P(k)|W(k)|^2 d^3k\ . \tag{15}$$

For example, the variance in a uniform sphere of radius r (the 'top-hat' filter) is:

$$\sigma^2(r) = \frac{1}{2\pi^2}\int P(k)W^2(kr)k^2dk\ , \tag{16}$$

where W(x) is a spherical Bessel function

$$W(x) = 3(sin(x) - xcos(x))/x^3\ . \tag{17}$$

There are various ways of setting the normalization of the power spectrum. One is to normalize to the variance in a sphere of radius $8\,\mathrm{h}^{-1}\,\mathrm{Mpc}$ ($\sigma_8 \approx 1$). Another is to use the J_3 integral over the correlation function

$$J_3(r) = \int_0^r \xi(x)x^2dx\ , \tag{18}$$

which is observed to be $J_3(10\,\mathrm{h}^{-1}\,\mathrm{Mpc}) \approx 277\,h^{-3}\,\mathrm{Mpc}^3$. A third way is to use the volume-averaged correlation function

$$\bar{\xi}(r) = \frac{3}{4\pi r^3}\int_0^r \xi(x)4\pi x^2 dx = \frac{3}{r^3}J_3(r) \ , \tag{19}$$

which is observed to be about 0.83 on a scale of $10\,\mathrm{h}^{-1}\,\mathrm{Mpc}$. Finally, rather than normalize the power spectrum at small scales today, one can use CMB observations to normalize it at large scales and early times.

To recover the galaxy density field directly, rather than statistically through the power spectrum or correlation function, we take the observed distribution of galaxies from a redshift survey and weight inversely with the survey's selection function, $1/\phi(r)$, then smooth with a window function $W(r/r_0)$. The smoothed, weighted density is:

$$\delta(r) = \frac{1}{n}\sum_i \frac{W(|r - r_i|/r_0)}{\phi(r_i)} - 1 \tag{20}$$

where $\int W(r/r_0)d^3r = 1$ and r_0 is the smoothing radius. Common choices for W (with r_0 scaled to galaxy separation) include the spherical tophat,

$$W(x)r_0^3 = (3/4\pi) \ \ (x < 1) \ , \tag{21}$$

and the spherical Gaussian,

$$W(x)r_0^3 = (2/\pi)^{1/2} exp(-x^2/2) \ . \tag{22}$$

Errors in the density field (and their cures) include (i) shot-noise (apply Wiener or other noise-reduction filter), (ii) errors in $\phi(r)$ (use a volume-limited sample), (iii) peculiar velocities (model the velocity field), and (iv) sky coverage (interpolate over the Zone of Avoidance or other gaps).

1.4. *The Form and Evolution of the Density Field*

Most simple inflationary cosmological models predict that the initial density field emerging from the Big Bang will have Fourier modes with random phases (*i.e.* where different wavenumbers are independent). Superposing many Fourier density modes with random phases results, by the central limit theorem, in a Gaussian density field, with the property that the joint probability distribution of the density at any number of points is a multivariate Gaussian. Linear amplification of a Gaussian field leaves it Gaussian, so the large-scale galaxy distribution should also be Gaussian. A Gaussian field is fully characterized by its mean and variance (as a function of scale). Hence $\langle\rho\rangle$ and $P(k)$ provide a complete statistical description of the density field if it is Gaussian, and should provide a complete description of the galaxy (and mass) distribution in the linear regime (*i.e.* on large scales).

Unless some physical process imposes a scale, the initial power spectrum should be scale-free, *i.e.* a power-law,

$$P(k) \propto k^n \ . \tag{23}$$

The index n determines the balance between large- and small-scale power, with rms fluctuations on a mass scale M given by

$$\delta_{rms} \propto M^{-(n+3)/6} \; . \tag{24}$$

The 'natural' initial power spectrum is the power-law with $n = 1$ (called the Zel'dovich, or Harrison-Zel'dovich, spectrum). The $P(k) \propto k^1$ spectrum is also referred to as the scale-invariant spectrum, since it gives variations in the gravitational potential that are the same on all scales. Since potential governs the curvature, this means that space-time has the same amount of curvature variation on all scales (*i.e.* the metric is a fractal). In fact, inflationary models predict that the initial power spectrum of the density fluctuations will be approximately scale-invariant.

The (non-relativistic) equations governing fluid motion under gravity can be linearized to give the following equation governing the growth of linear density perturbations:

$$\ddot{\delta} + 2\frac{\dot{a}}{a}\dot{\delta} = \delta\left(4\pi G\rho_0 - \frac{c_s^2 k^2}{a^2}\right) , \tag{25}$$

where c_s is the sound speed, $c_s^2 = \partial p/\partial \rho$.

This has growing solutions for large scales (small k) and oscillating solutions for small scales (large k); the cross-over scale between the two is the Jeans length,

$$\lambda_J = c_s\sqrt{\frac{\pi}{G\rho}} \; . \tag{26}$$

For $\lambda < \lambda_J$, sound waves cross an object on the same time-scale as the gravitational collapse, so pressure can counter gravity. In an expanding Universe, λ_J varies with time; perturbations on some scales swap between growing and oscillating solutions.

The evolution of density fluctuations at different scales are independent, so

$$P(k, t_0) \propto T_k^2 P(k, t_i) \; , \tag{27}$$

where

$$T_k = D(z)^{-1}\delta_k(t_0)/\delta_k(t_i) \tag{28}$$

is the transfer function and $D(z)$ is the linear growth factor from redshift $z = z(t_i)$ to the present, $z = z(t_0) = 0$.

Pressure counters gravity for scales less than the Jeans length, which is close to the size of the horizon while the Universe is radiation-dominated, as in this epoch $c_s = c/\sqrt{3}$. It reaches a maximum at the redshift of matter-radiation equality, z_{eq}, after which the sound speed drops. For scales greater than the Jeans length at matter-radiation equality, $\lambda_J(z_{eq})$, the density grows under gravity, while for smaller scales the pressure damps the growth. The power spectrum thus becomes bent at the scale of $\lambda_J(z_{eq})$.

The world model enters through $\lambda_J(z_{eq})$, which is related to the co-moving horizon scale at matter-radiation equality:

$$R_0 r_H(z_{eq}) = 2(\sqrt{2}-1)(c/H_0)(\Omega_m z_{eq})^{-1/2} \approx 16/(\Omega_m h)\, h^{-1}\,\text{Mpc} \; , \tag{29}$$

so if k is given in $\mathrm{h}^{-1}\,\mathrm{Mpc}$, then T is only a function of $k/(\Omega_m h)$. We therefore write T_k as $T(k/\Gamma)$, where Γ encodes the world model: $\Gamma \approx \Omega_m h$, or, more precisely,

$$\Gamma = \Omega_m h \exp[-\Omega_b(1 + (2h)^{1/2}/\Omega_m)] \ . \tag{30}$$

The transfer function $T(x)$ contains the other physics, including (i) the nature of the dark matter (CDM or HDM); (ii) small-scale damping via free-streaming; and (iii) the acoustic oscillations of the baryons. For CDM the full (numerical) solution for the transfer function can be approximated by

$$T(x) \approx \frac{\log(1 + Bx)}{1 + (Ax)^2} \ , \tag{31}$$

with $A = 4.0\,\mathrm{h}^{-1}\,\mathrm{Mpc}$ and $B = 2.4\,\mathrm{h}^{-1}\,\mathrm{Mpc}$.

1.5. *Peculiar Velocities, Bias and Redshift-space Distortions*

To recap, the observed redshift is the combination of the Hubble redshift due to the expansion of the Universe and the peculiar velocity due to the gravitationally-induced motion. At low redshift,

$$cz = H_0 r + \vec{v} \cdot \hat{r} \ , \tag{32}$$

where $\vec{v}$ is the peculiar velocity. The (linearized) equation of motion is

$$\dot{\vec{v}} + 2\frac{\dot{a}}{a}\vec{v} = \frac{\vec{g}}{a} \tag{33}$$

where $\vec{g} = -\nabla\Phi/a$ is the peculiar gravitational potential. This has the solution

$$\vec{v} = \frac{2f(\Omega_m)}{3H_0\Omega_m}\vec{g} \tag{34}$$

where

$$f(\Omega_m) = d\ln\delta/d\ln a \approx \Omega_m^{0.6} \ . \tag{35}$$

Another useful relation links the divergence of the velocity field to the mass fluctuation:

$$\nabla \cdot v(\vec{r}) = -H_0 f(\Omega_m)\delta_m(r) \approx -H_0\Omega_m^{0.6}\delta_m(r) \ . \tag{36}$$

The development of gravitational instability theory above is in terms of the mass distribution, but observations are of the galaxy distribution. What is the relation between these two distributions? There is much more mass in dark matter than in baryons, and more mass in baryons than in galaxies ($\rho_m \gg \rho_b > \rho_g$), so why suppose $\delta_g = \delta_m$? A bias factor b parameterizes our ignorance: $\delta_g = b\delta_m$; *i.e.* fractional variations in the galaxy density are proportional to fractional variations in the mass density, with ratio b.

What might produce a bias? Do galaxies form only at the peaks of the mass field, due (say) to a star-formation threshold? Is there a variation in bias with scale? A scale variation is plausible at small scales (where there are many potential mechanisms), but not at large scales. Any theory for the bias must explain the observed variation with galaxy type; the ratio of the numbers of ellipticals to spirals is large in clusters ($\delta_g \gg 1$), but small in the field ($\delta_g \leq 1$).

The bias also affects the peculiar velocities, since replacing δ_m by δ_g/b gives

$$\nabla \cdot v(\vec{r}) = -H_0 \beta \delta_g(r) \ , \tag{37}$$

where $\beta = f(\Omega_m)/b \approx \Omega_m^{0.6}/b$.

Because of peculiar velocities, the redshift-space correlation function is distorted w.r.t. the real-space correlation function. In real space the contours of the correlation function are circular. But in redshift space coherent infall on large scales (in the linear regime) squashes the contours along the line of sight, while rapid motions in collapsed structures on small scales stretch the contours along the line of sight. Likewise, peculiar velocities distort the power spectrum in redshift space, $P^s(\vec{k})$ w.r.t. the power spectrum in real space, $P(k)$. Far from the observer (in the plane-parallel approximation), this distortion takes the form

$$P^s(\vec{k}) = (1 + \beta \mu_k^2)^2 P(k) \ , \tag{38}$$

where μ_k is the cosine of the angle between k and the radial line of sight (note that P^s depends on $\vec{k}$, not just k, because it is no longer isotropic). The angle-averaged z-space power spectrum becomes

$$P^s(k) = \frac{1}{4\pi} \int P^s(\vec{k}) \, d\theta_k = (1 + \frac{2}{3}\beta + \frac{1}{5}\beta^2) P(k) \ , \tag{39}$$

so the ratio of the redshift-space and real-space power spectra (in the linear regime) constrains $\beta \approx \Omega_m^{0.6}/b$ (*i.e.* the mass density, up to biasing). With a redshift survey, one is measuring $P^s(k)$ not $P(k)$. This does not affect the shape analysis, since they are proportional. But to use these distortions to measure β one also needs $P(k)$. This can be obtained by inverting the angular power spectrum $w(\theta)$, or by linearly evolving the CMB mass power spectrum. Alternatively, the degree of distortion of $P^s(k)$, and hence the value of β, can be determined by measuring the ratio of its quadruple and monopole moments:

$$\frac{P_2^s(k)}{P_0^s(k)} = \frac{4/3\beta + 4/7\beta^2}{1 + 2/3\beta + 1/5\beta^2} \ . \tag{40}$$

The estimates of β from $P(k)$ using linear redshift-space distortions depend of course on the bias parameter, which differs for different galaxy samples. Comparing β from optically-selected and infrared-selected surveys shows that the relative bias is $b_{optical}/b_{IRAS} \approx 1.5$.

Finally, the spatial correlation function $\xi(r)$ can be recovered from the redshift-space correlation function $\xi(s)$ by computing $\xi(s)$ as a function of the separations in plane of sky, σ, and the line of sight, π, to obtain $\xi_s(\sigma, \pi)$. The projection of $\xi_s(\sigma, \pi)$ onto the σ-axis is

$$w(\theta) = 2 \int \xi_s(\sigma, \pi) \, d\pi = 2 \int \xi_r(\sqrt{(\sigma^2 + y^2)}) \, dy \ . \tag{41}$$

For a power-law, $\xi(r) = (r/r_0)^{-\gamma}$, and we have

$$\Xi(\sigma) = w(\sigma)/\sigma = \xi_r(\sigma)\Gamma(1/2)\Gamma((\gamma-1)/2)/\Gamma(\gamma/2) , \quad (42)$$

where Γ is the standard gamma function.

1.6. *Gaussianity and Topology*

On large scales, all the evidence appears consistent with the initial density fluctuations having random phases (*i.e.* Gaussian fluctuations), although the evidence is not yet conclusive. On small scales, non-linear evolution of the density field occurs, resulting in 3-point and 4-point correlation functions that are non-zero (*i.e.* the density field has non-random phases). The higher-order correlation functions appear to obey hierarchical scaling relations, whereby the spherically-averaged N-point correlation function is related to the 2-point correlation function by

$$< \xi_N(V) >= S_N < \xi_2(V) >^{N-1} , \quad (43)$$

where S_N is a scaling factor, as predicted by perturbation theory for Gaussian initial conditions and gravitational instability.

Another diagnostic of non-linear clustering (or non-Gaussian initial conditions) is the topology of the large-scale structure. This can be characterized by $g(\nu)$, the topological genus of the surface described by the isodensity contour as a function of density threshold ν; the genus of a surface is g $=$ # holes $-$ # pieces $+ 1$. On small scales, the observed $g(\nu)$ undergoes a slight 'meatball' shift compared to the $g(\nu)$ for a Gaussian density field (*i.e.* the isodensity surface contains high-density 'meatballs' in a low-density 'stew'); this is as expected from non-linear evolution. On large scales, the genus provides another test of the Gaussianity of the initial density distribution.

1.7. *Open Questions*

Up until the last few years, many of the major questions regarding the large-scale structure of the Universe were still open, including:

(1) What is the shape of the power spectrum? What is the nature of the dark matter? What is the value of the power spectrum shape parameter, $\Gamma = \Omega_m h$?
(2) How are the mass and light distributions related? What is the value of the redshift-space distortion parameter, $\beta = \Omega_m^{0.6}/b$? Can we obtain β and the bias parameter, b, independently of each other? What are the relative biases of different galaxy populations, and why do they differ?
(3) Can we check the gravitational instability paradigm? Can we demonstrate that the large-scale structures we see result from gravitational amplification of small initial density perturbations? Were these initial density fluctuations random-phase (Gaussian)?

(4) What is the non-linear evolution of the galaxy and mass distributions? Can we link galaxy properties (luminosity, mass, type) to local density and/or large-scale structure? Which properties are primordial? Which are contingent on detailed evolution?

In the last couple of years, massive new redshift surveys covering 10^5–10^6 galaxies at $\langle z \rangle \approx 0.1$, such as the 2dF Galaxy Redshift Survey (see the following section and Colless *et al.* 2001) and the Sloan Digital Sky Survey (Stoughton *et al.* 2002), have vastly improved our understanding of large-scale structure and provided higher-precision estimates of the cosmological parameters. The results from the 2dFGRS are discussed in detail in following sections. In coming years, deep redshift surveys of $\sim 10^5$ galaxies out to $z \sim 1$, such as the VIRMOS-VLT survey (Le Fèvre & Vettolani 2004) and the DEEP survey (Davis *et al.* 2003), will extend our understanding of the evolution of both the large-scale structure and the galaxy population, while surveys of the local Universe, such as the 6dF Galaxy Survey (Colless *et al.* 2004) will measure both the redshifts and the distances of nearby galaxies, yielding the velocity field as well as the density field, and giving a yet more detailed picture of the large-scale structure and the relationship between the galaxies and the dark matter.

2. The 2dF Galaxy Redshift Survey

2.1. *Survey Observations*

The state-of-the-art redshift surveys of the early 1990s, such as the Las Campanas Redshift Survey (Shectman *et al.* 1996) and the *IRAS* Point Source Catalogue redshift survey (Saunders *et al.* 2000), either did not cover sufficiently large volumes to be statistically representative of the large-scale structure, or covered large volumes too sparsely to provide precise measurements. An order-of-magnitude increase in the survey volume and sample size was needed to enter the regime of 'precision cosmology'. The 2dF Galaxy Redshift Survey (2dFGRS) was specifically conceived as a massive redshift survey for precisely measuring fundamental cosmological parameters.

The source catalogue for the 2dFGRS was a revised and extended version of the APM galaxy catalogue (Maddox *et al.* 1990), which was created by scanning the photographic plates of the UK Schmidt Telescope Southern Sky Survey. The survey targets were chosen to be galaxies with extinction-corrected magnitudes brighter than $b_J = 19.45$ mag. The main survey regions were two declination strips, one in the southern Galactic hemisphere spanning $80° \times 15°$ around the South Galactic Pole (the SGP strip), and the other in the northern Galactic hemisphere spanning $75° \times 10°$ along the celestial equator (the NGP strip); in addition, there were 99 individual 2dF "random" fields spread over the southern Galactic cap (see Fig. 2). The large volume that is sparsely probed by the random fields allows the survey to measure structure on scales greater than would be permitted by the relatively narrow widths of the main survey strips. In total, the survey covers approximately 1800 deg^2, and has a median redshift depth of $z = 0.11$. Further information on the 2dF Galaxy Redshift Survey can be found in Colless *et al.* (2001) and on the WWW at http://www.mso.anu.edu.au/2dFGRS.

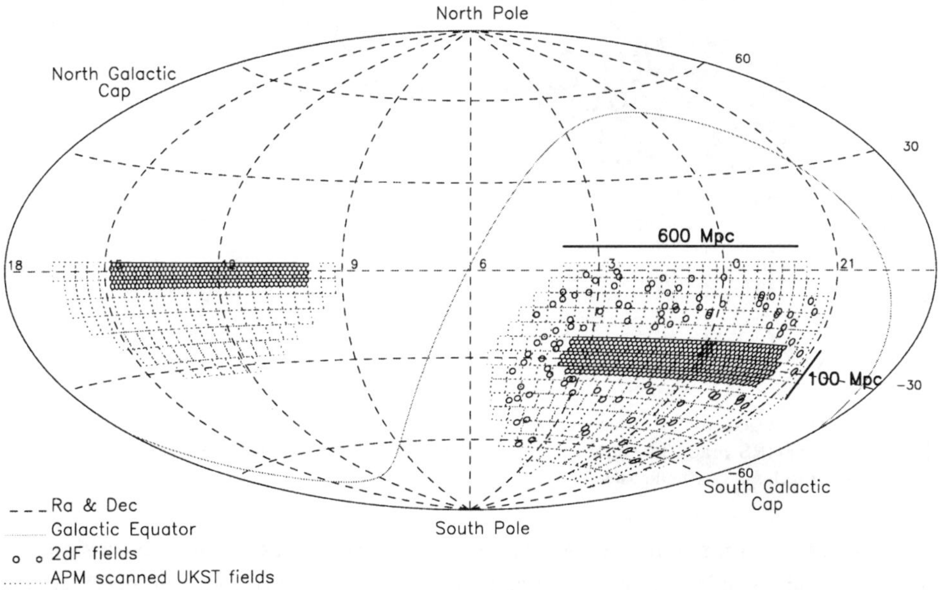

Figure 2. A map of the sky showing the locations of the two 2dFGRS survey strips (NGP strip at left, SGP strip at right) and the random fields. Each 2dF field in the survey is shown as a small circle; the sky survey plates from which the source catalogue was constructed are shown as dotted squares. The scale of the strips at the mean redshift of the survey is indicated.

Figure 3 shows a thin slice through the three-dimensional map of over 221,000 galaxies produced by the 2dFGRS. This 3°-thick slice passes through both the NGP strip (at left) and the SGP strip (at right). The decrease in the number of galaxies toward higher redshifts is an effect of the survey selection by magnitude—only intrinsically more luminous galaxies are brighter than the survey magnitude limit at higher redshifts. The clusters, filaments, sheets and voids making up the large-scale structures in the galaxy distribution are clearly resolved. The fact that there are many such structures visible in the figure is a qualitative demonstration that the survey volume comprises a representative sample of the Universe.

2.2. *The Large-scale Structure of the Galaxy Distribution*

The statistical properties of the large-scale structure of the galaxy distribution observed in redshift space are summarized in Figure 4, which shows both the correlation function and the power spectrum obtained from the 2dFGRS. The structure on very large scales (several tens to hundreds of Mpc) is best represented by the

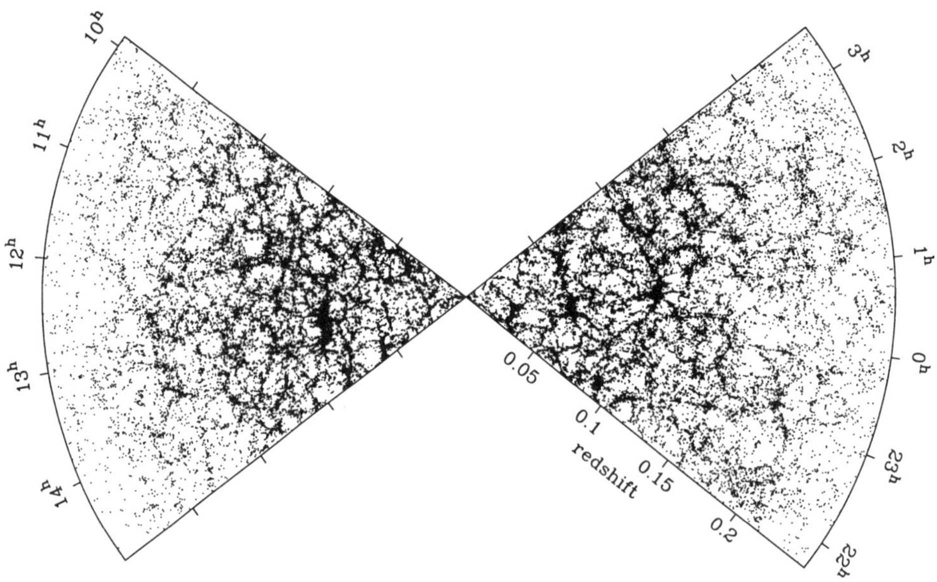

Figure 3. The large-scale structures in the galaxy distribution are shown in this 3°-thick slice through the 2dFGRS map. The slice cuts through the NGP strip (at left) and the SGP strip (at right), and contains 63,000 galaxies.

power spectrum; on smaller scales, where peculiar velocities become more significant and the shape of the power spectrum (as well as the amplitude) differs between redshift space and real space, the redshift-space structure is most clearly shown in the two-dimensional correlation function (see §2.4 below).

The power spectrum, shown in the left panel of Figure 4, is well determined from the 2dFGRS on scales less than about $400\,h^{-1}$ Mpc (wavenumbers $k > 0.015$), and its shape is little affected by nonlinear evolution of the galaxy distribution on scales greater than about $40\,h^{-1}$ Mpc ($k < 0.15$). Over this decade in scale, the power spectrum is well fitted by a cold dark matter (CDM) model having a shape parameter $\Gamma = \Omega_m h = 0.20 \pm 0.03$ (Percival *et al.* 2001). For a Hubble constant around $70\,\mathrm{km\,s^{-1}\,Mpc^{-1}}$ (*i.e.*, $h \approx 0.7$), this implies a mean mass density $\Omega_m \approx 0.3$. The power spectrum also shows some evidence for acoustic oscillations produced by baryon-photon coupling in the early Universe (see §2.5).

The right panel of Figure 4 shows the redshift-space two-point correlations as a function of the separations along and across the line of sight, and reveals two main deviations from circular symmetry due to peculiar velocity effects. On intermediate scales, for transverse separations of a few tens of Mpc, the contours of the correlation function are flattened along the line of sight due to the coherent infall of galaxies as structures form in the linear regime. The detection of this effect in the 2dFGRS is a clear confirmation that large-scale structure grows by the gravitational amplification of density fluctuations (Peacock *et al.* 2001), and allows a direct measurement of the mean mass density of the Universe (see §2.5). The other effect is the stretching

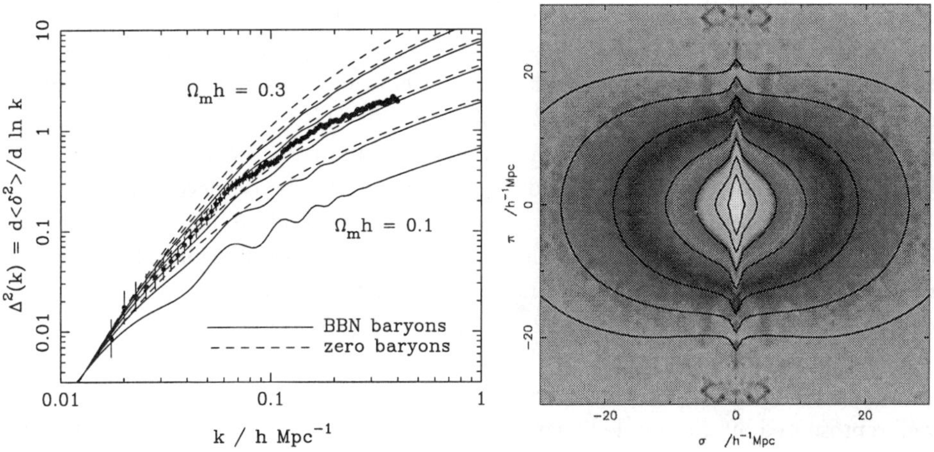

Figure 4. Large-scale structure statistics from the 2dFGRS. The left panel shows the dimensionless power spectrum $\Delta^2(k)$ (Percival *et al.* 2001; Peacock *et al.* 2004). Overlaid are the predicted linear-theory CDM power spectra with shape parameters $\Omega h = 0.1$, 0.15, 0.2, 0.25, and 0.3, with the baryon fraction predicted by Big Bang nucleosynthesis (solid curves) and with zero baryons (dashed curves). The right panel shows the two-dimensional galaxy correlation function, $\xi(\sigma, \pi)$, where σ is the separation across the line of sight and π is the separation along the line of sight (Hawkins *et al.* 2003). The greyscale image is the observed $\xi(\sigma, \pi)$, and the contours show the best-fitting model.

of the contours along the line of sight at small transverse separations. This is the finger-of-God effect due to the large peculiar velocities of collapsed structures in the non-linear regime.

2.3. *The Bias of the Galaxy Distribution*

The simplest model for galaxy biasing postulates a linear relation between fluctuations in the galaxy distribution and fluctuations in the mass distribution. In this case the galaxy power spectrum is related to the mass power spectrum by $P_g(k) = b^2 P_m(k)$. Such a relationship is expected to hold in the linear regime (up to stochastic variations). The first-order relationship between galaxies and mass can therefore be determined by comparing the measured galaxy power spectrum to the matter power spectrum based on a model fit to the cosmic microwave background (CMB) power spectrum, linearly evolved to $z = 0$ and extrapolated to the smaller scales covered by the 2dFGRS power spectrum. Applying this approach, Lahav *et al.* (2002) find that the linear bias parameter for a galaxy of characteristic luminosity L^* at zero redshift is $b(L^*, z = 0) = (0.96 \pm 0.08) \exp[-\tau + 0.5(n - 1)]$, where τ is the optical depth due to re-ionization and n is the spectral index of the primordial mass power spectrum.

An alternative way of determining the bias employs the higher-order correlations between galaxies in the intermediate, quasi-linear regime. The higher-order correlations are generated by nonlinear gravitational collapse, and so depend on the clustering of the dominant dark matter rather than the galaxies. Thus the stronger the higher-order clustering, the higher the dark matter normalization, and the lower the bias. An analysis of the bispectrum (the Fourier transform of the three-point correlation function) by Verde *et al.* (2002) yields $b(L^*, z = 0) = 0.92 \pm 0.11$, a result based solely on the 2dFGRS. Moreover, including a second-order quadratic bias term does not improve the fit of the bias model to the observed bispectrum.

For the blue, optically-selected 2dFGRS sample, it therefore seems that L^* galaxies are nearly unbiased tracers of the low-redshift mass distribution. However, this broad conclusion masks some very interesting variations of the bias parameter with galaxy luminosity and type (Fig. 5). Norberg *et al.* (2001, 2002a) show conclusively that the bias parameter varies with luminosity, ranging from $b = 1.5$ for bright galaxies to $b = 0.8$ for faint galaxies. The relation between bias and luminosity is well represented by the simple linear relation $b/b^* = 0.85 + 0.15L/L^*$. They also find that, at all luminosities, early-type galaxies have a higher bias than late-type galaxies. A detailed comparison of the clustering of passive and actively star-forming galaxies by Madgwick *et al.* (2003) shows that at small separations, the passive galaxies cluster much more strongly, and the relative bias ($b_{\rm passive}/b_{\rm active}$) is a decreasing function of scale. On the largest scales, however, the relative bias tends to a constant value of around 1.3.

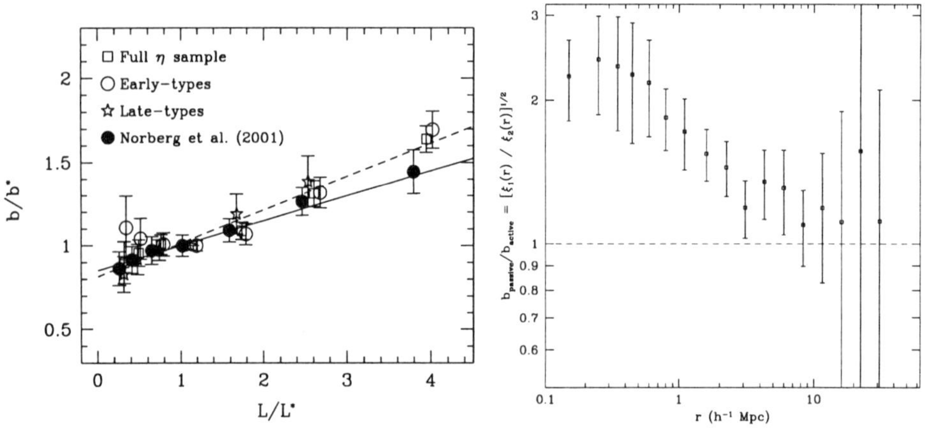

Figure 5. Variations in the bias parameter with luminosity and spectral type. The left panel shows the variation with luminosity of the galaxy bias on a scale of $\sim 5\,h^{-1}$ Mpc, relative to an L^* galaxy (Norberg *et al.* 2002a). The bias variations of the full 2dFGRS sample are compared to subsamples with early and late spectral types, and to earlier results by Norberg *et al.* (2001). The right panel shows the relative bias of passive and actively star-forming galaxies as a function of scale, over the range 0.2–20 h^{-1} Mpc (Madgwick *et al.* 2003).

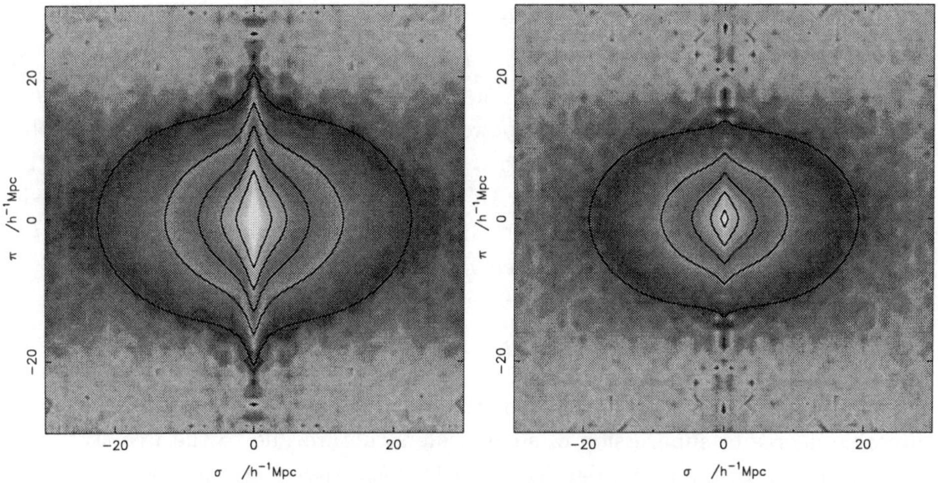

Figure 6. The two-dimensional galaxy correlation function, $\xi(\sigma, \pi)$, for passive (left) and actively star-forming (right) galaxies (Madgwick *et al.* 2003). The grayscale image is the observed $\xi(\sigma, \pi)$, and the contours show the best-fitting model.

2.4. *Redshift-Space Distortions*

The redshift-space distortion of the clustering pattern can be modelled as the combination of coherent infall on intermediate scales and random motions on small scales. The compression of structures along the line of sight due to coherent infall is quantified by the distortion parameter $\beta \simeq \Omega^{0.6}/b$ (Kaiser 1987). The random motions are modelled by an exponential distribution, $f(v) = 1/(a\sqrt{2})\exp(-\sqrt{2}|v|/a)$, where a is the pairwise peculiar velocity dispersion (also called σ_{12}).

The initial analysis of a subset of the 2dFGRS by Peacock *et al.* (2001) obtained best-fit values of $\beta(L_s, z_s) = 0.43 \pm 0.07$ and $a = 385$ km s^{-1} at an effective weighted survey luminosity $L_s = 1.9L^*$ and survey redshift $z_s = 0.17$. A more sophisticated re-analysis of the full 2dFGRS by Hawkins *et al.* (2003) obtains $\beta(L_s, z_s) = 0.49 \pm 0.09$ and $a = 506 \pm 52$ km s^{-1}, with $L_s = 1.4L^*$ and $z_s = 0.15$ (right panel of Fig. 4). These results, using different fitting methods, are consistent, although the earlier result underestimates the uncertainties by 20%. Applying corrections based on the variation in the bias parameter with luminosity and a constant galaxy clustering model (Lahav *et al.* 2002) to the Hawkins *et al.* value for the distortion parameter yields $\beta(L^*, z = 0) = 0.47 \pm 0.08$.

Madgwick *et al.* (2003) extend this analysis to a comparison of the active and passive galaxies, where the two-dimensional correlation function, $\xi(\sigma, \pi)$, reveals differences in both the bias parameter on large scales and the pairwise velocity dispersion on small scales (Fig. 6). The distortion parameter is $\beta_{\rm passive} \simeq \Omega_m^{0.6}/b_{\rm passive} = 0.46 \pm 0.13$ for passive galaxies and $\beta_{\rm active} \simeq \Omega_m^{0.6}/b_{\rm active} = 0.54 \pm 0.15$ for active galaxies; over the range 8–20 h^{-1} Mpc the effective pairwise velocity dispersions are 618 ± 50 km s^{-1} and 418 ± 50 km s^{-1} for passive and active galaxies, respectively.

2.5. *The Mass Density of the Universe*

The 2dFGRS provides a variety of ways to measure the mean mass density of the Universe, along with the relative amounts of dark matter, baryons and neutrinos.

Fitting the shape of the galaxy power spectrum in the linear regime with a model including both CDM and baryons (Percival *et al.* 2001), and assuming that the Hubble constant is $h = 0.7$ with a 10% uncertainty, yields a total mass density for the Universe of $\Omega_m = 0.29 \pm 0.07$ and a baryon fraction of $15\% \pm 7\%$ (*i.e.*, $\Omega_b = 0.044 \pm 0.021$). This analysis used 150,000 galaxies; a preliminary re-analysis of the complete final sample of 221,000 galaxies with the additional constraint that $n = 1$ yields $\Omega_m = 0.26 \pm 0.05$ and $\Omega_b = 0.044 \pm 0.016$ (Peacock *et al.* 2004; left panel of Fig. 7). Including neutrinos as a further constituent of the mass allows an upper limit to be placed on their contribution to the total density, based on the allowable degree of suppression of small-scale structure due to the free streaming of neutrinos out of the initial density perturbations (right panel of Fig. 7). Elgarøy *et al.* (2002) obtain an upper limit on the neutrino mass fraction of 13% at the 95% confidence level (*i.e.* $\Omega_\nu < 0.034$). This translates to an upper limit on the total neutrino mass (summed over all species) of $m_\nu < 1.8\,\text{eV}$.

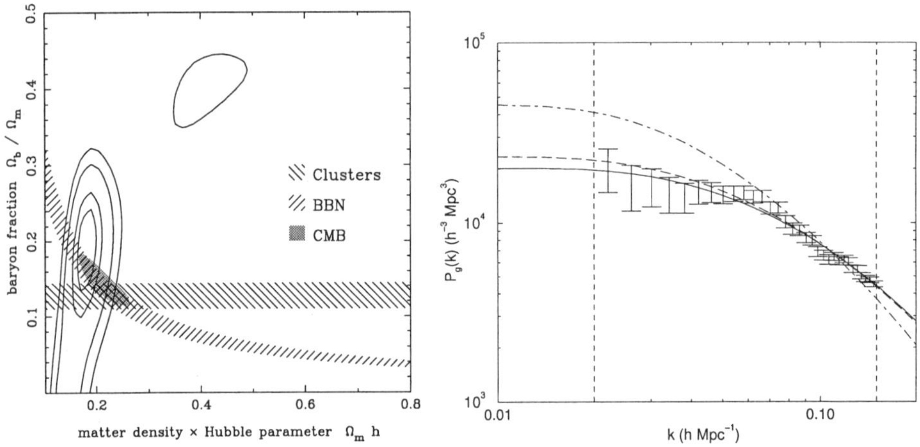

Figure 7. Determinations of the mean mass density, Ω_m, and the baryon and neutrino mass fractions. The left panel shows the likelihood surfaces obtained by fitting the full 2dFGRS power spectrum for the shape parameter, $\Omega_m h$, and the baryon fraction, Ω_b/Ω_m (Peacock *et al.* 2004; *cf.* Percival *et al.* 2001). The fit is over the well-determined linear regime ($0.02 < k < 0.15\,h\,\text{Mpc}^{-1}$) and assumes a prior on the Hubble constant of $h = 0.7 \pm 0.07$. The right panel shows the fits to the 2dFGRS power spectrum (Elgarøy *et al.* 2002), assuming $\Omega_m = 0.3$, $\Omega_\Lambda = 0.7$, and $h = 0.7$ for three different neutrino densities: $\Omega_\nu = 0$ (solid), 0.01 (dashed), and 0.05 (dot-dashed).

An alternative approach to deriving the total mass density is to use the measurements in the quasi-linear regime of the redshift-space distortion parameter $\beta \simeq \Omega_m^{0.6}/b$, in combination with estimates of the bias parameter b (Peacock *et al.*

2001; Hawkins *et al.* 2003). Using the Lahav *et al.* (2002) estimate for b gives $\Omega_m = 0.31 \pm 0.11$, while the Verde *et al.* (2002) value for b gives $\Omega_m = 0.23 \pm 0.09$.

2.6. *Joint LSS–CMB Estimates of Cosmological Parameters*

Stronger constraints on these and other fundamental cosmological parameters can be obtained by combining the power spectrum of the present-day galaxy distribution from the 2dFGRS with the power spectrum of the mass distribution at very early times derived from observations of the anisotropies in the CMB. A general analysis of the combined CMB and 2dFGRS data sets (Efstathiou *et al.* 2002) shows that, at the 95% confidence level, the Universe has a near-flat geometry ($\Omega_k \approx 0 \pm 0.05$), with a low total matter density ($\Omega_m \approx 0.25 \pm 0.08$) and a large positive cosmological constant ($\Omega_\Lambda \approx 0.75 \pm 0.10$, consistent with the independent estimates from observations of high-redshift supernovae).

Table 1. Cosmological parameters from joint fits to the CMB and 2dFGRS power spectra, assuming a flat geometry (Percival *et al.* 2002). The best-fit parameters and rms errors are obtained by marginalizing over the likelihood distribution of the remaining parameters. Results are given for scalar-only and scalar+tensor models, and for the CMB power spectrum only and the CMB and 2dFGRS power spectra jointly.

Parameter	Results: scalar only		Results: with tensor component	
	CMB	CMB + 2dFGRS	CMB	CMB + 2dFGRS
$\Omega_b h^2$	0.0205 ± 0.0022	0.0210 ± 0.0021	0.0229 ± 0.0031	0.0226 ± 0.0025
$\Omega_c h^2$	0.118 ± 0.022	0.1151 ± 0.0091	0.100 ± 0.023	0.1096 ± 0.0092
h	0.64 ± 0.10	0.665 ± 0.047	0.75 ± 0.13	0.700 ± 0.053
n_s	0.950 ± 0.044	0.963 ± 0.042	1.040 ± 0.084	1.033 ± 0.066
n_t	–	–	0.09 ± 0.16	0.09 ± 0.16
r	–	–	0.32 ± 0.23	0.32 ± 0.22
Ω_m	0.38 ± 0.18	0.313 ± 0.055	0.25 ± 0.15	0.275 ± 0.050
$\Omega_m h$	0.226 ± 0.069	0.206 ± 0.023	0.174 ± 0.063	0.190 ± 0.022
$\Omega_m h^2$	0.139 ± 0.022	0.1361 ± 0.0096	0.123 ± 0.022	0.1322 ± 0.0093
Ω_b / Ω_m	0.152 ± 0.031	0.155 ± 0.016	0.193 ± 0.048	0.172 ± 0.021

If the models are limited to those with flat geometries (Percival *et al.* 2002), then tighter constraints emerge (see Table 1). In this case the best estimate of the matter density is $\Omega_m = 0.31 \pm 0.06$, and the physical densities of CDM and baryons are $\omega_c = \Omega_c h^2 = 0.12 \pm 0.01$ and $\omega_b = \Omega_b h^2 = 0.022 \pm 0.002$; the latter agrees very well with the constraints from Big Bang nucleosynthesis. This analysis also provides an estimate of the Hubble constant ($H_0 = 67 \pm 5\,\mathrm{km\,s^{-1}\,Mpc^{-1}}$) that is independent of, but in excellent accord with, the results from the Hubble Space Telescope Key Project. Comparing the uncertainties on the various parameters in the CMB-only and CMB+2dFGRS columns of Table 1 shows the very significant improvements that are obtained by combining the CMB and 2dFGRS data sets.

Joint fits to the 2dFGRS and CMB power spectra also constrain the equation of state parameter $w = p_{\rm vac}/\rho_{\rm vac}c^2$ for the dark energy. Percival *et al.* (2002) find that in a flat Universe the joint power spectra, together with the Hubble Key Project estimate for H_0, imply an upper limit of $w < -0.52$ at the 95% confidence level.

3. Cosmological Results from WMAP

The Wilkinson Microwave Anisotropy Probe (WMAP) is a satellite that has mapped the CMB over the entire sky with higher resolution than in any previous all-sky map. The results from the first year of WMAP observations were released during this Summer School (see Bennett 2003, and references therein) and are briefly reported here.

3.1. *The WMAP Mission*

WMAP was designed to minimize systematic errors in its measurements of the CMB by exploiting to the full the advantages of differential observing techniques. The probe was placed in orbit at L2, and in 6 months maps the whole sky. These first-year results therefore contain two sets of full-sky observations. The WMAP team argue that systematic errors are well understood and controlled, based on multiple checks and detailed tests. Calibration is based on the Earth-velocity modulation of the CMB dipole, which is claimed to provide a calibration good to 0.5%. Beam patterns are measured by observing Jupiter (the uncertainties in the beam pattern affect the window function).

The famous forerunner of WMAP was the COBE satellite (Smoot *et al.* 1992). A direct comparison of WMAP's capabilities with those of COBE shows the dramatic improvement over the intervening decade: COBE had a resolution of 7° and observed in 3 spectral bands, while WMAP had a resolution of 0.2° and observed in 5 spectral bands.

The interpretation of CMB results, especially at the sensitivity and resolution of WMAP, is critically dependent on the ability to correct for Galactic emission and extragalactic point sources. The CMB is separated from the foregrounds using the spectral information in the five WMAP bands. Sky regions with bright foreground emission are masked. Low-level diffuse emission is removed by forming a map based on a maximum entropy method linear combination of the five bands—but this map has complex error properties and is not used in the analysis.

Cosmological parameters are derived from a map based on masking bright sources and subtracting foregrounds based on spectral templates for the various components (IRAS for the dust, 408MHz radio maps for the synchrotron radiation, Hα maps for the free-free ionized gas). This method leaves rms foreground contaminations of $<7\mu$K in the Q-band and $<3\mu$K in the V and W bands for Galactic latitudes $|l| > 15°$.

3.2. *The CMB Power Spectrum*

The power spectrum is a complete statistical description of the CMB anisotropies only if they are a Gaussian random field. Most inflationary models predict that the fluctuations should be Gaussian (at least at currently detectable levels). The WMAP maps are tested for non-Gaussian behaviour using Minkowski functionals and the bispectrum. These are used to determine the lowest-order non-Gaussian term in a Taylor expansion of the curvature perturbations. The non-Gaussianity is characterized in terms of a non-linear coupling parameter f_{NL} (where $f_{NL} = 0$ means the CMB anisotropies are Gaussian). Using the bispectrum (the next highest order description of the Fourier-space CMB map after the power spectrum) gives limits of $-58 < f_{NL} < 134$ (95% confidence interval); using Minkowski functionals to estimate the non-linear contribution gives $f_{NL} < 139$ (with 95% confidence). These results are with Gaussianity, but it's not clear what values of f_{NL} might reasonably be expected from non-standard models.

If the CMB anisotropies are Gaussian, then they can be described by their multipole expansion (*i.e.* their angular power spectrum). The lowest order terms of this expansion are the dipole and quadrupole. WMAP measures the dipole amplitude and direction to be 3.346 ± 0.017mK and (l, b)=(263.85°±0.10°,48.25°±0.04°), compared to the COBE dipole, 3.353±0.024mK and (l, b)=(264.26°±0.33°,48.22°±0.13°); WMAP obtains a quadrupole amplitude of $Q_{rms} = 8\ (+2, -2)$, compared to the COBE result of $Q_{rms} = 10\ (+7, -4)$. These results agree well within the errors, but the WMAP results are obviously significantly more precise.

The multipole amplitudes (power spectrum) are computed from the WMAP maps using both a quadratic estimator (QE) and a maximum likelihood (ML) technique. The QE power spectrum is used in the cosmological analysis, with the ML power spectrum used only as a cross-check. The power spectrum, shown in Figure 8 has a first peak at multipole $l = 220.1 \pm 0.8$ and a second peak at $l = 546 \pm 10$. The vast improvement in the precision with which the CMB power spectrum is known is immediately apparent from comparing the WMAP power spectrum with that deduced from all previous CMB observations.

The shape of the WMAP power spectrum is in excellent agreement with that predicted on the basis of the cosmological parameters derived from previous CMB observations and the 2dFGRS. Note, however, that the WMAP power spectrum is normalized $\sim$ 10% higher at large multipoles compared to previous CMB results. This change in normalization between the old CMB results and the WMAP power spectrum is essentially the whole difference between the old CMB +2dFGRS prediction and the new result.

3.3. *CMB Polarization and TE Cross-Correlation*

In these results from the first year of WMAP observations, the CMB polarization map is based on measurements of the Stokes I parameter alone (although maps using

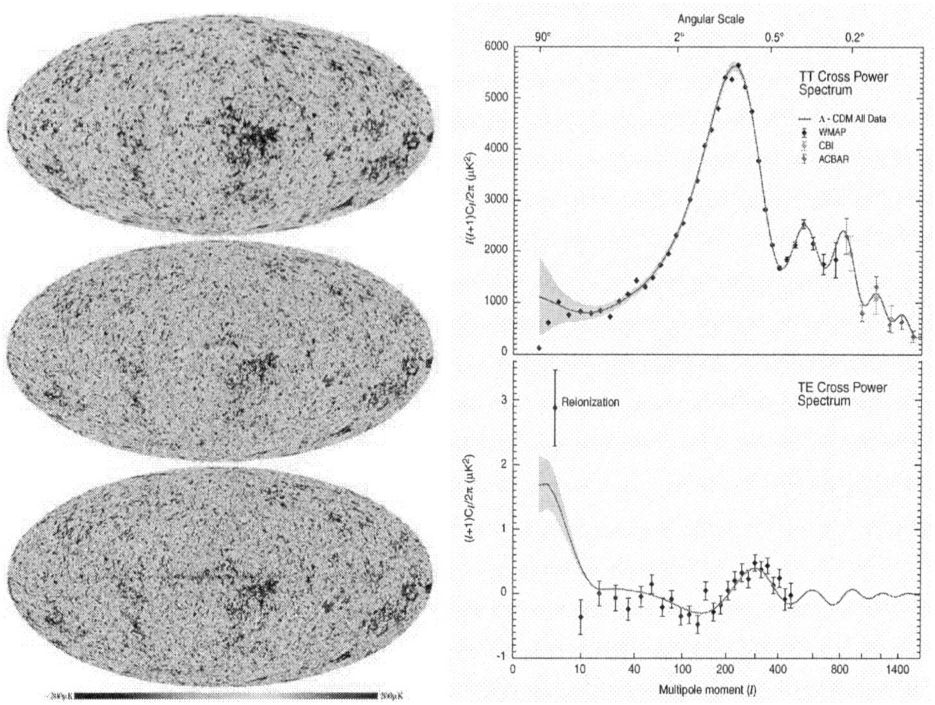

Figure 8. The WMAP CMB maps in three bands, and the TT and TE power spectra.

the Q and U parameters are expected to follow). The polarization measurements are calibrated against observations of Taurus A.

The temperature-polarization (TE) cross-power spectrum shows both correlations on large scales (low l) due to re-ionization and correlations on small scales (high l) from adiabatic fluctuations. The re-ionization feature in the TE cross-power spectrum corresponds to an integrated optical depth $\tau = 0.17 \pm 0.04$. In 'plausible' models for the re-ionization process, this optical depth implies a redshift of re-ionization of $z_r = 20(+10, -9)$ at the 95% c.l., corresponding to an epoch of re-ionization at $t_r = 100$–400 Myr. Re-ionization suppressed the acoustic peak amplitudes by $\sim$30%.

The high value of z_r obtained by WMAP is incompatible with significant amounts of warm dark matter, as WDM would suppress clustering on small scales and delay the formation of stars and QSOs, giving a later epoch of re-ionization.

The anti-correlations observed in the cross-power spectrum imply super-horizon-scale fluctuation modes, as predicted by inflationary models. But it is not clear whether this is consistent with the lack of power at low l in the power spectrum.

3.4. *Cosmological Models*

A flat Universe with a scale-invariant spectrum of adiabatic Gaussian fluctuations, with re-ionization, is an acceptable fit to the WMAP data. This is also an acceptable fit to the combination of the WMAP + ACBAR + CBI anisotropies, the 2dFGRS galaxies + Ly α forest clustering data, the HST Key Project H0, and the SN Ia data. The TE correlations and the acoustic peaks imply the initial fluctuations were primarily adiabatic (the primordial ratios of dark matter/photons and baryons/photons do not vary spatially). The initial fluctuations are consistent with a Gaussian field, as expected from most inflationary models.

The WMAP data (combined with any one of the HST H_0, the 2dFGRS Ω_m, or the SN Ia data) implies that $\Omega_{tot} = 1.02 \pm 0.02$. The dominant constituent of the Universe is dark energy, with $\Omega_\Lambda = 0.73 \pm 0.04$; cold dark matter contributes $\Omega_{CDM} = 0.23 \pm 0.04$, baryons contribute $\Omega_b = 0.044 \pm 0.004$, and neutrinos make up only $\Omega_\nu < 0.015$.

However, this simple model is not the best fit; one can do better if a scale-dependent initial spectral index is included. In this case the best fit model has an initial spectral index $n_s = 0.93$ (at $k_0 = 0.05\,\mathrm{h\,Mpc^{-1}}$, *i.e.* $120\,\mathrm{h^{-1}\,Mpc}$) and a variation with scale $dn_s/d\ln k = -0.03 \pm 0.017$ (also at k_0). This running index implies lower amplitude fluctuations on the smallest scales, altering the dark matter profiles on these scales. If correct, this might be part of the solution to the problem of the dark matter halo profiles in dwarf galaxies.

The new 'standard cosmological model' combining WMAP, 2dFGRS, SN Ia and HST Key Project results is summarized in Table 2. The cosmic timeline has the following dates: CMB last scattering surface at $t_{dec} = 379 \pm 8$ kyr ($z_{dec} = 1089 \pm 1$); epoch of re-ionization at $t_r = 100$–400 Myr; age of the Universe today, $t_0 = 13.7 \pm 0.2$ Gyr. Finally, the Hubble constant is measured to be $H_0 = 71 \pm 4\,\mathrm{km\,s^{-1}\,Mpc^{-1}}$ (*cf.* the HST Key Project value of $H_0 = 72 \pm 7\,\mathrm{km\,s^{-1}\,Mpc^{-1}}$).

3.5. *Inflation and New Physics*

WMAP provides some support for inflation and some hints of new physics. Inflation predicts that (i) the Universe is flat (WMAP finds Ω_{tot} is consistent with unity); (ii) that the initial fluctuations were a Gaussian random field (WMAP finds no evidence for non-Gaussian fluctuations); (iii) that the initial spectral index should be close to unity (WMAP finds $n_s = 0.93 \pm 0.03$); and (iv) that fluctuations should exist on super-horizon scales (WMAP sees evidence for this in the TE correlations). These generic predictions of inflationary models are therefore all supported by WMAP.

In addition, however, the WMAP data provide some intriguing hints: (i) the scalar spectral index is not exactly unity; (ii) the spectral index may change with scale ($dn_s/d\ln k = -0.03 \pm 0.017$); (iii) the tensor-to-scalar ratio is found to be <0.71 at $0.002\,\mathrm{h\,Mpc^{-1}}$; and (iv) the dark energy equation of state parameter, w, is found to be less than -0.78 at the 95% confidence level. Whether these hints are

Table 2. The best-fit cosmological parameters from WMAP, 2dFGRS, SN Ia and HST Key Project results (Bennett et al. 2003).

Description	Symbol	Value	+ Uncertainty	− Uncertainty
Total density	Ω_{tot}	1.02	0.02	0.02
Equation of state of quintessence	w	< -0.78	95% CL	...
Dark energy density	Ω_Λ	0.73	0.04	0.04
Baryon density	$\Omega_b h^2$	0.0224	0.0009	0.0009
Baryon density	Ω_b	0.044	0.004	0.004
Baryon density (cm^{-3})	n_b	2.5×10^{-7}	0.1×10^{-7}	0.1×10^{-7}
Matter density	$\Omega_m h^2$	0.135	0.008	0.009
Matter density	Ω_m	0.27	0.04	0.04
Light neutrino density	$\Omega_\nu h^2$	<0.0076	95% CL	...
CMB temperature (K)[a]	T_{CMB}	2.725	0.002	0.002
CMB photon density (cm^{-3})[b]	n_γ	410.4	0.9	0.9
Baryon-to-photon ratio	η	6.1×10^{-10}	0.3×10^{-10}	0.2×10^{-10}
Baryon-to-matter ratio	$\Omega_b \Omega_m^{-1}$	0.17	0.01	0.01
Fluctuation amplitude in 8 h^{-1} Mpc spheres	σ_8	0.84	0.04	0.04
Low-z cluster abundance scaling	$\sigma_8 \Omega_m^{0.5}$	0.44	0.04	0.05
Power spectrum normalization (at $k_0 = 0.05$ Mpc^{-1})[c]	A	0.833	0.086	0.083
Scalar spectral index (at $k_0 = 0.05$ Mpc^{-1})[c]	n_s	0.93	0.03	0.03
Running index slope (at $k_0 = 0.05$ Mpc^{-1})[c]	$dn_s/d\ln k$	-0.031	0.016	0.018
Tensor-to-scalar ratio (at $k_0 = 0.002$ Mpc^{-1})	r	<0.90	95% CL	...
Redshift of decoupling	z_{dec}	1089	1	1
Thickness of decoupling (FWHM)	Δz_{dec}	195	2	2
Hubble constant	h	0.71	0.04	0.03
Age of universe (Gyr)	t_0	13.7	0.2	0.2
Age at decoupling (kyr)	t_{dec}	379	8	7
Age at reionization (Myr, 95% CL)	t_r	180	220	80
Decoupling time interval (kyr)	Δt_{dec}	118	3	2
Redshift of matter-energy equality	z_{eq}	3233	194	210
Reionization optical depth	τ	0.17	0.04	0.04
Redshift of reionization (95% CL)	z_r	20	10	9
Sound horizon at decoupling (deg)	θ_A	0.598	0.002	0.002
Angular size distance (Gpc)	d_A	14.0	0.2	0.3
Acoustic scale[d]	ℓ_A	301	1	1
Sound horizon at decoupling (Mpc)[d]	r_s	147	2	2

reliable will emerge more clearly as the WMAP dataset grows over time.

There are a number of puzzles in these initial WMAP results, which may have an uninteresting explanation (*e.g.* remaining systematic errors), or which may lead to new insights. These problems include:

(1) The standard model predicts higher values of the correlation function for small l (large angular scales). This is best seen in the correlation function.
(2) The WMAP normalization of the CMB power spectrum is 10% higher than most previous results. (This may be an artefact of the way Wang *et al.*combined the previous CMB data.)
(3) Is the lack of power at low l in the TT power spectrum consistent with the super-horizon-scale fluctuation modes inferred from the anti-correlations observed in the TE cross-power spectrum?
(4) Is the high redshift of re-ionization ($z_r = 20$) found from WMAP compatible with the observations of $\sim$ 6 QSOs which seem to suggest a more recent epoch of re-ionization?

Acknowledgments

The results from the 2dF Galaxy Redshift Survey are the combined work of the 2dF-GRS team: Ivan K. Baldry, Carlton M. Baugh, Joss Bland-Hawthorn, Sarah Bridle, Terry Bridges, Russell Cannon, Shaun Cole, Matthew Colless, Chris Collins, Warrick Couch, Nicholas Cross, Gavin Dalton, Roberto De Propris, Simon P. Driver, George Efstathiou, Richard S. Ellis, Carlos S. Frenk, Karl Glazebrook, Edward Hawkins, Carole Jackson, Bryn Jones, Ofer Lahav, Ian Lewis, Stuart Lumsden, Steve Maddox, Darren Madgwick, Peder Norberg, John A. Peacock, Will Percival, Bruce A. Peterson, Will Sutherland, and Keith Taylor. The 2dFGRS was made possible through the dedicated efforts of the staff of the Anglo-Australian Observatory, both in creating the 2dF instrument and in supporting it on the telescope.

References

1. C.L. Bennett *et al.*, *Astrophys. J. Suppl.* **148**, 1 (2003).
2. P. Coles and F. Lucchin, *Cosmology: The Origin and Evolution of Cosmic Structure*, (John Wiley & Sons, Chichester, 1995).
3. M.M. Colless *et al.*, *Mon. Not. Roy. Astr. Soc.* **328**, 1039 (2001).
4. M.M. Colless *et al.*, in *Maps of the Cosmos*, eds M.M. Colless and L. Staveley-Smith, (ASP Conf. Series, San Francisco, 2004).
5. M. Davis *et al.*, *Proc. SPIE* **4834**, 161 (2003).
6. G. Efstathiou *et al.*, *Mon. Not. Roy. Astr. Soc.* **330**, L29 (2002).
7. Ø. Elgarøy *et al.*, *Phys. Rev. Lett.* **89**, 061301 (2002).
8. M.J. Geller and J.P. Huchra, *Science* **246**, 897 (1989).
9. R. Giovanelli and M.P. Haynes, *Ann. Rev. Astron. Astrophys* **29**, 499 (1991).
10. A.J.S. Hamilton, *Astrophys. J.* **385**, L5 (1992).
11. E. Hawkins *et al.*, *Mon. Not. Roy. Astr. Soc.* **346**, 78 (2003).
12. E.P. Hubble, *Astrophys. J.* **79**, 8 (1934).
13. N. Kaiser, *Mon. Not. Roy. Astr. Soc.* **227**, 1 (1987).
14. O. Lahav *et al.*, *Mon. Not. Roy. Astr. Soc.* **333**, 961 (2002).
15. O. Le Fèvre and G. Vettolani, in *Maps of the Cosmos*, eds M.M. Colless and L. Staveley-Smith, (ASP Conf. Series, San Francisco, 2004).
16. S.J. Maddox *et al.*, *Mon. Not. Roy. Astr. Soc.* **242**, 43P (1990).
17. D.S. Madgwick *et al.*, *Mon. Not. Roy. Astr. Soc.* **344**, 847 (2003).
18. P. Norberg *et al.*, *Mon. Not. Roy. Astr. Soc.* **328**, 64 (2001).
19. P. Norberg *et al.*, *Mon. Not. Roy. Astr. Soc.* **332**, 827 (2002).
20. J.A. Peacock, *Cosmological Physics*, (Cambridge University Press, Cambridge, 1999).
21. J.A. Peacock, in *Maps of the Cosmos*, eds M.M. Colless and L. Staveley-Smith, (ASP Conf. Series, San Francisco, 2004).
22. J.A. Peacock *et al.*, *Nature* **410**, 169 (2001).
23. P.J.E. Peebles, *The Large-Scale Structure of the Universe*, (Princeton Series in Physics, Princeton, 1980).
24. P.J.E. Peebles, *Principles of Physical Cosmology*, (Princeton Series in Physics, Princeton, 1993).
25. W.J. Percival *et al.*, *Mon. Not. Roy. Astr. Soc.* **327**, 1297 (2001).
26. W.J. Percival *et al.*, *Mon. Not. Roy. Astr. Soc.* **337**, 1068 (2002).
27. W. Saunders *et al.*, *Mon. Not. Roy. Astr. Soc.* **317**, 55 (2000).
28. S.A. Shectman *et al.*, *Astrophys. J.* **470**, 172 (1996).

29. G.F. Smoot *et al.*, *Astrophys. J.* **396**, L1 (1992).
30. C. Stoughton *et al.*, AJ **123**, 485 (2002).
31. M.A. Strauss and J.A. Willick, *Physics Reports* **261**, 271 (1995).
32. R.B. Tully and J.R. Fisher, *Atlas of Nearby Galaxies*, (Cambridge University Press, Cambridge, 1987).
33. L. Verde *et al.*, *Mon. Not. Roy. Astr. Soc.* **335**, 432 (2002).

THE FORMATION AND EVOLUTION OF GALAXIES

G. KAUFFMANN
Max Planck Institute for Astrophysics, Karl Schwarzschildstrasse 1, D-85748 Garching , Germany
E-mail: gamk@mpa-garching.mpg.de

1. Introduction

The past decade has witnessed the establishment of a "standard paradigm" for structure formation in the Universe. It is now universally accepted that the dominant matter component of the Universe is in some form of non-baryonic, weakly-interacting dark matter. Structure in the dark matter originated from inhomogeneities that were generated shortly after the Big Bang during a period of accelerated expansion, termed inflation. These early inhomogeneities were gravitationally amplified as the Universe expanded. Eventually, material contained in initially over-dense regions began to collapse. Small objects were the first to form and these later merged together to form larger and larger structures.

This picture has received spectacular confirmation from a series of experiments designed to probe anisotropies in the cosmic microwave background radiation. As a result of these experiments, cosmologists now believe they know the values of most of the basic parameters of the Universe (for example the density parameter Ω, the value of the Hubble and cosmological constants and the amplitude of the power spectrum of initial fluctuations) to better than 10%. The development of structure in the dark matter component of the Universe is also extremely well understood, thanks to a program of detailed numerical simulations that have elucidated how structures such as clusters form from the merging of smaller lumps as they stream in along filaments of dark matter.

In spite of these advances, the formation and evolution of galaxies remains poorly understood. In the standard picture, a galaxy will form when gas is able to reach high enough densities to cool, sink to the centre of a high density lump of dark matter (called a "halo") and form stars. What happens to the galaxy after that depends on the interplay between a host of complex physical processes. The most massive stars quickly run out of fuel and end their lives as supernovae. These supernovae may be responsible for reheating gas and expelling heavy elements from the galaxy, thereby altering its structure and slowing down the rate at which it can form stars. Galaxies will also merge with each other as their surrounding dark matter halos coalesce. During these mergers gas is compressed and the star formation rates in galaxies may increase by several orders of magnitude for a short period. Mergers also cause gas to lose angular momentum and sink to the centre of the galaxy. It has been speculated that the supermassive black holes that are now known to exist at the centre of almost every bright galaxy in the local Universe, may be formed in such events.

In these lecture notes, I will attempt to provide an overview of how we believe galaxies formed from the small density fluctuations present in the Early Universe and outline some of techniques that astrophysicists use in order to model the formation and evolution of galaxies from very high redshifts to the present day.

2. Methods for Calculating the Evolution of the Dark Matter

2.1. *The Linear Regime*

When the density fluctuations $\delta\rho/\rho$ are small, their evolution can followed using linear perturbation theory. A detailed derivation can be found in almost any textbook on cosmology (*e.g.*Chapter II.10 of Peebles [1]).

The analysis assumes that non-gravitational forces on the material can be neglected. Matter is treated as an ideal pressureless fluid. The three equations governing the evolution of the density and the velocity of this fluid are the continuity equation (mass conservation), the Euler equation (momentum conservation) and the Poisson equation. One then changes variables to co-moving coordinates, a "peculiar" velocity with respect to the Hubble expansion, and a dimensionless density contrast. After suitable substitutions, and keeping only terms that are first-order in the density or peculiar velocity, one arrives at a second order differential equation with a growing and a decaying mode solution.

In an Einstein de Sitter Universe, density fluctuations simply grow in proportion to the scale factor of the Universe. In low-density Universes, the density fluctuations stop growing or "freeze out" at late times.

2.2. *Spherical Collapse*

When the density fluctuations are large ($\delta\rho/\rho > 1$), linear theory no longer holds. A number of analytic approximations have been proposed to treat the collapse of the dark matter, the simplest of which is the spherical collapse model (see Chapter II.19 of Peebles [1]).

Consider a spherical region with uniform overdensity $\bar{\rho}$, physical radius R and enclosed mass M in an otherwise uniform universe. A result from General Relativity known as Birkhoff's theorem states that external matter exerts no force on the material within the sphere. Hence we can write

$$\frac{d^2R}{dt^2} = -\frac{GM}{R^2} = -\frac{4\pi G}{3}\bar{\rho}(1+\bar{\delta})R \tag{1}$$

The first integral of the evolution equation is

$$\frac{1}{2}\left(\frac{dR}{dt}\right)^2 - \frac{GM}{R} = E \tag{2}$$

For $E < 0$, the equation has a parametric solution

$$R/R_m = \frac{1}{2}(1 - \cos\eta); \quad t/t_m = (\eta - \sin\eta)/\pi \tag{3}$$

For small η, one can do a Taylor expansion and eliminate η. One then derives that the mean overdensity with respect to an Einstein-de Sitter Universe of the same age is

$$\bar{\delta} = \frac{3}{20}(6\pi t/t_m)^{2/3} \propto a, \tag{4}$$

where a is the scale factor. The spherical region reaches maximum expansion at $\eta = \pi$. After that, it starts to contract and collapses at $\eta = 2\pi$. At the time of collapse, the linearly extrapolated density contrast has value

$$\delta_{collapse} = \frac{3}{20}(12\pi)^{2/3} = 1.686. \tag{5}$$

In reality, a density perturbation is neither perfectly spherical nor homogeneous. Shell crossing occurs and collapse does not proceed to a point, but reaches virial equilibrium, with the virial radius equal to half the maximum expansion radius (this follows from $U = -2K$ and $U \propto r^{-1}$). The overdensity at virialisation is thus $18\pi^2$=178. This is why a density threshold of ~ 200 is often used to define a collapsed object in N-body simulations or in the real Universe.

2.3. *The Press-Schechter Theory*

Press and Schechter [2] proposed an analytic formula for the comoving abundance of collapsed structures in the Universe at a given redshift z.

Consider the initial density fluctuations, extrapolated to the present day by linear perturbation theory. If these fluctuations are smoothed with a spherical top-hat filter of radius R and average enclosed mass $M = 4\pi R^3\bar{\rho}/3$, the rms fluctuation amplitude $\sigma(M)$ can be estimated from the linear theory power spectrum.

If the initial fluctuations are Gaussian, then the fraction of the volume in which the smoothed linear theory density exceeds the critical density for collapse δ_c is erfc(δ_c/σ), where erfc is the integral of the Gaussian distribution from δ_c/σ to infinity. Press and Schechter suggested that the mass in such regions could be assumed to reside in collapsed objects of mass M or greater.

Differentiating the collapsed mass fraction with respect to the smoothing mass M yields the fraction of mass in objects of mass between M and $M + dM$ and this in turn yields the Press-Schechter mass function:

$$n(M)dM = -\left(\frac{2}{\pi}\right)^{1/2} \frac{\bar{\rho}}{M}\frac{\delta_c}{\sigma^2}\frac{d\sigma}{dM} \exp\left[-\frac{\delta_c^2}{2\sigma^2}\right] dM, \tag{6}$$

where σ^2 is the variance of the linear density field smoothed on mass scale M.

It has been shown that the Press-Schechter formula agrees reasonably well with the results of N-body simulations. Most recently, Sheth, Mo & Tormen [3] have derived an improved formula using an ellipsoidal collapse model and this has been shown to provide a substantially better fit to simulation data [4].

Figure 1 (taken from Mo & White [5]) illustrates a number of well known properties of the standard ΛCDM model. Haloes as massive as a rich galaxy cluster like Coma ($M \sim 10^{15} M_\odot$) have an average spacing of about $100h^{-1}$Mpc today, but their abundance drops dramatically in the relatively recent past. By $z = 1.5$ it is already down by a factor exceeding 1000, corresponding to a handful of objects in the observable Universe. The decline in the abundance of haloes with mass similar to that of the Milky Way ($M \sim 10^{12} M_\odot$) is much more gentle. By $z = 5$ the drop is only about one order of magnitude. At the smallest masses shown ($M \sim 10^7$ to $10^8 M_\odot$) there is little change in abundance over the full redshift range $0 < z < 20$ that we plot. Notice also that the abundance of such low mass haloes is actually declining slowly at low redshifts as members of these populations merge into larger systems faster than new members are formed. It is interesting that haloes of mass $10^9 M_\odot$ are as abundant at $z = 20$ as L_* galaxies are today, and haloes of $10^{10} M_\odot$ are as abundant as present-day rich galaxy clusters.

2.4. *The Extended Press-Schechter Theory*

The so-called "extended" Press-Schechter theory was first developed by Bond *et al.*[6] and independently by Bower [7]. The extended Press-Schechter theory allows one to evaluate the probability that a mass element of the Universe that has collapsed into an object of mass M_1 and time t_1, will form part of a larger object of mass M_2 at some later time t_2. Straightforward manipulation of the calculus of probabilities then allows one to derive expressions for [8]:

(1) The merger rate between objects of mass M_1 and M_2 at time t.
(2) The distribution of "formation times" of objects that have mass M at time t.
(3) The distribution of "survival" times of objects. This allows one to calculate what fraction of galaxies of given mass seen at high redshift correspond to isolated galaxies of similar mass today, and what fraction have been accreted onto larger systems.

The extended Press-Schechter formalism can also be used to generate Monte Carlo realizations of the formation history of a halo of given mass at the present day [9 10 11]. These Monte Carlo realizations of the merging process are often referred to as "merger trees". An example of such a merging tree for a cluster-mass halo ($10^{15} M_\odot$ is shown in Figure 2. At redshifts $z < 1$, the cluster grows mainly through accretion of low mass halos. However, by $z \sim 2$, mergers between halos of nearly equal mass occur very frequently. It is likely no coincidence that $z \sim 2 - 3$ corres-

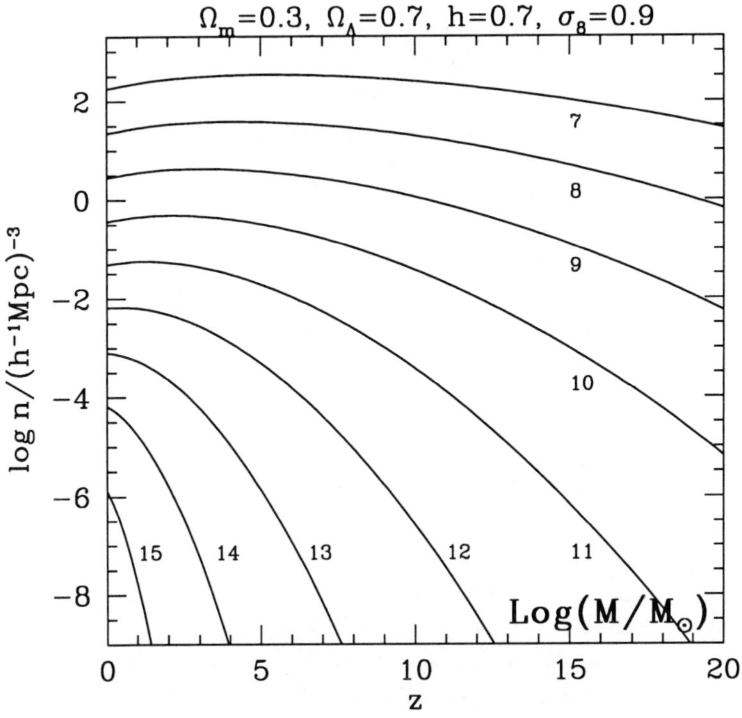

Figure 1. Each curve indicates the comoving number density of dark matter haloes with masses exceeding a specific value M in the standard ΛCDM model. The label on each curve indicates the corresponding value of $\log M/M_{\odot}$

ponds to the observed peak of star formation and quasar activity in the Universe. We will come back to this point later.

2.5. *N-body Simulations: techniques*

In order to obtain an accurate description of the formation of structure in the dark matter component of the Universe in the non-linear regime, N-body simulations are required. These have become a standard tool in the field of cosmology. They allow the simplified analytic models described in the previous sections to be tested and extended, and they are often useful for suggesting new analytic approaches to problems.

An N-body simulation solves the equations of motion for a set of particles that

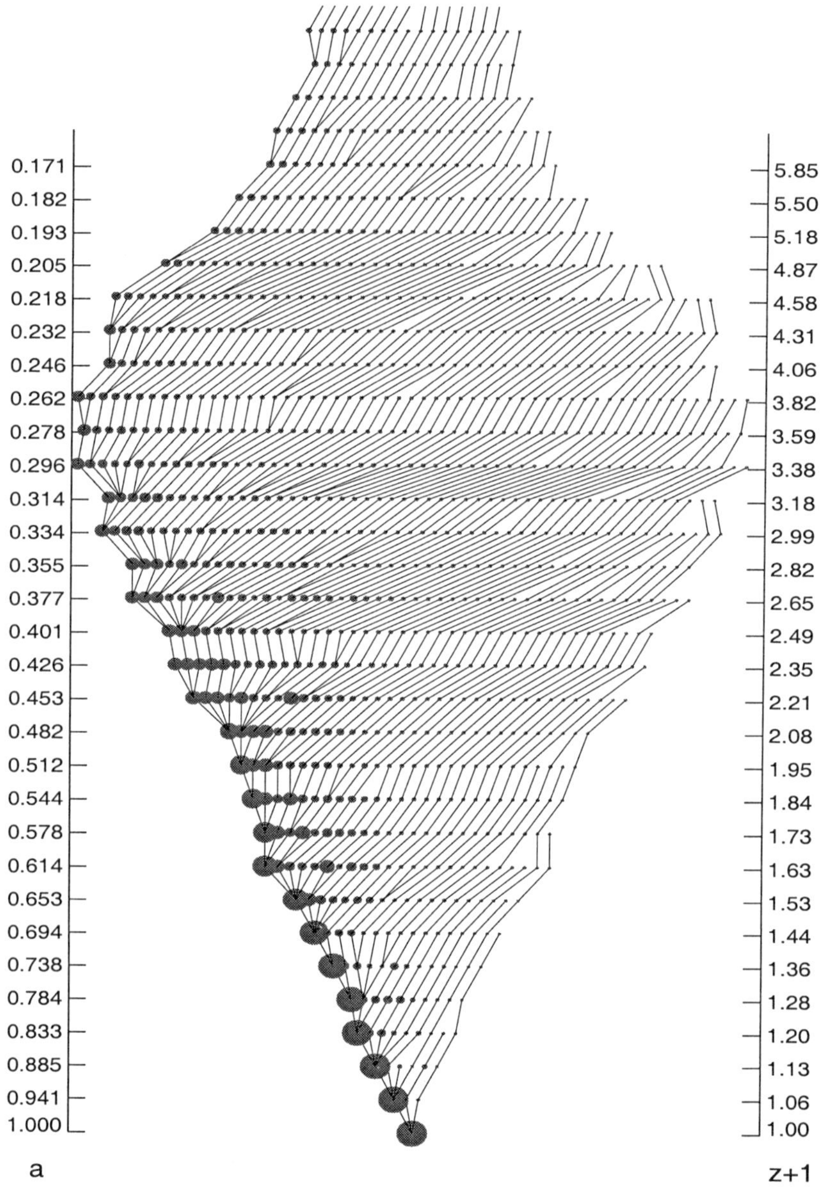

Figure 2. The 'merger tree' for a present-day halo of $10^{15} M_\odot$.

interact only through gravity. These may be written

$$\frac{d^2\vec{x}}{d\tau^2} + \frac{\dot{a}}{a}\frac{d\vec{x}}{d\tau} = \vec{g}, \tag{7}$$

where the accelerations $\vec{g}$ are computed from the positions of all the particles, usually through solution of Poisson's equations,

$$\vec{g} = -\vec{\Delta}\Phi; \quad \Delta^2\phi = 4\pi G a^2[\rho(\vec{x},\tau) - \bar{\rho}(\tau)]. \tag{8}$$

It is important to maintain the accuracy of the integrations of the equations of motion as the Universe expands. A number of different possibilities have been proposed in the literature; one popular scheme is to choose the time variable $p = a^\alpha$ [12]. The proper choice of α enables constant time steps to be used in the integration, but the equations of motion then take a more complicated form. The most critical aspect of integrating the equations of motion is the determination of the gravitational acceleration. All schemes require compromises which attempt to reconcile conflicting demands for speed of execution, for mass resolution (*i.e.*particle number), for linear resolution (determined by the effective "softening" or small-scale modification of the $1/r^2$ law introduced by the scheme used to solve Poisson's equation), for accurate representation of the true pairwise forces between particles, and for efficiency when treating nearly uniform or highly clustered conditions.

In cosmology, one usually wishes to simulate a "representative" region of the Universe or a particular system which is embedded in a dynamically active environment. When studying a typical region of the Universe, the usual choice is to apply periodic boundary conditions on opposite faces of a cubic box. This avoids any artificial boundaries and forces the mean density of the simulation to remain at the same value. For studies of individual galaxy halos or clusters, tree algorithms for solving Poisson's equation allow a straightforward solution to the problem of representing the tidal field of material that always remains outside the object of interest.

The initial conditions for N-body simulations are generated using two steps. The first step is to set up a "uniform" distribution of particles which can represent the unperturbed Universe. The second is to impose growing density fluctuations with the desired characteristics. Most often, the unperturbed Universe is represented by a regular cubic grid of particles. This simple procedure does introduce a strong characteristic length scale on small scales (the grid spacing) and it may also affect the statistical properties of the non-linear point distribution, particularly those that emphasize low-density regions. Modern simulations use tricks to generate very uniform particle distributions with no preferred directions or scale. Given the unperturbed particle distribution, any desired linear fluctuation distribution can be realized using Fourier techniques [12].

2.6. *N-body Simulations: results*

One of the most striking results of N-body simulations of structure formation in cold dark matter-dominated Universe is that structures are not spherical on large scales. The virialised halos tend to be distributed along filaments, which enclose large underdense regions, or "voids". As the Universe evolves, matter tends to stream along the filaments and collect into massive halos, which are often located at the intersections of multiple filaments. This filamentary network has been dubbed the "cosmic web" by simulators.

Figure 3 shows an example of an N-body simulation of a region of 100 Mpc in diameter. This simulation utilises a trick that has often been used in studies of the formation of individual objects, such as rich galaxy clusters. The *constrained realization* technique [14] sets up a Gaussian random field that satisfies certain constraints. In the simulation in Fig. 3, the smoothed linear density field matches that derived from the IRAS 1.2 Jansky survey, a survey of nearby galaxies covering the entire sky. As a result, the simulation reproduces a number of well-known local structures, for example the nearby Coma and Virgo clusters. As can be seen, the nearby clusters are linked by a network of lower-density filamentary structures.

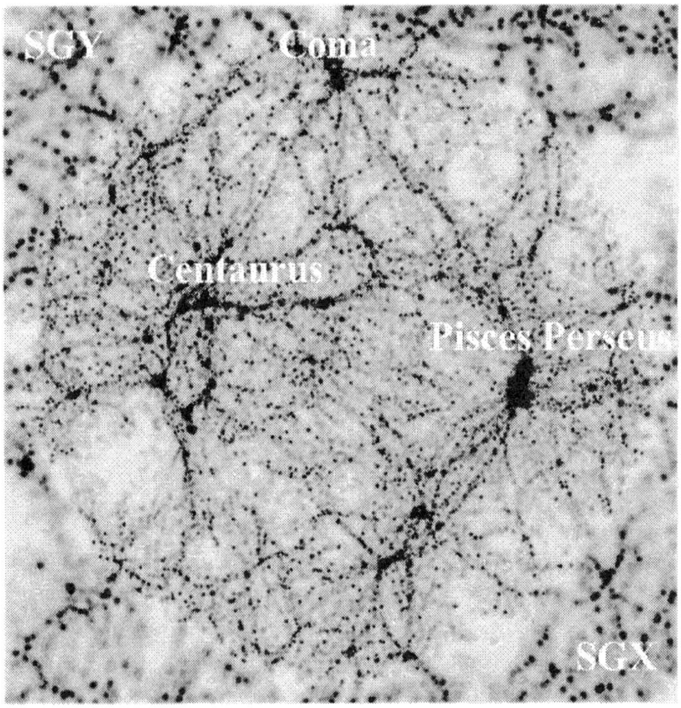

Figure 3. An example of an N-body simulation of the local Universe out to a distance of 8000 km/s from Mathis *et al.*[13]. The smoothed linear density field matches that derived from the IRAS 1.2 Jy galaxy survey and well-known local structures can be seen.

As well as studying structure on large scales, N-body simulations can be used to study the distribution of dark matter within the virialized regions of individual dark matter halos. In the 1980s and early 1990s, simulators found that dark matter halos retained very little 'memory' of their initial conditions. As small dark matter haloes merged to form larger ones, the individual haloes were disrupted very quickly by tidal forces and the majority of present-day halos were found to have very little residual substructure. Because we observe groups and clusters that contain tens to hundreds of galaxies and that are clearly close to virial equilibrium, this lack of substructure was seen as a serious problem for the simulations (although, as discussed by White & Rees [15], dissipational processes such as cooling could cause baryonic material to condense and reach much higher densities at the centres of dark matter halos, where it would be much harder to disrupt).

The idea that dark halos contain no substructure has, however, changed quite dramatically in recent years. The mass resolution of N-body simulations has undergone dramatic improvement. Figure 4 shows an example of an ultra-high resolution simulation of a galaxy cluster of mass $10^{15}M_\odot$ from Springel *et al.*[16]. Within the virial radius, the cluster is resolved with about 20 million dark matter particles and it is found to contain around 5000 dynamically distinct "subhalos" that in total contain about 10% of the mass of the entire halo. The mass function of these subhalos is well approximated by a power-law $dN/dm \propto M^\gamma$, with $\gamma \sim -1.8$. The shape of the subhalo mass function is independent of the mass of the parent halo. As we discuss in Section 4, this may constitute a problem for the model, as the observed mass functions of galaxy systems like our own Local Group appear to be significantly shallower.

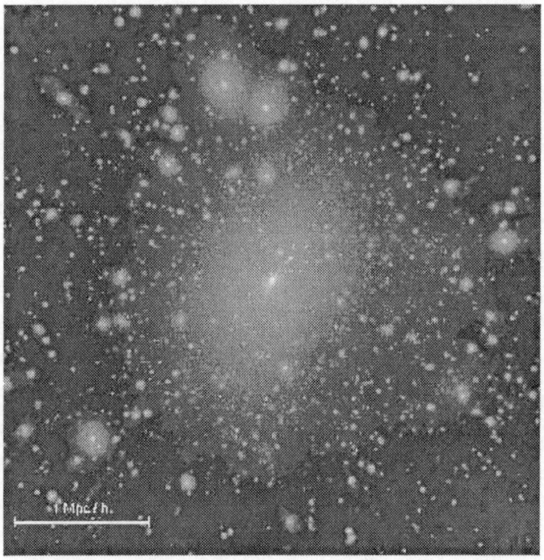

Figure 4. An ultra-high-resolution N-body simulation [16] of a dark matter halo of $10^{15}M_\odot$ (similar to the mass of a rich galaxy cluster).

Another important result from high-resolution N-body simulations is that the density profiles of dark matter halos appear to have a "universal" form [17]

$$\frac{\rho(r)}{\rho_{crit}} = \frac{\delta_c}{(r/r_s)(1 + r/r_s)^2}, \tag{9}$$

where δ_c is a (dimensionless) characteristic density, and r_s is a scale radius. The characteristic density can be shown to scale with the formation time of the halo, as predicted by the extended Press-Schechter theory [8]. At large radii, $\rho(r) \propto r^{-3}$ and at small radii $\rho(r) \propto r^{-1}$. This means that the density of dark matter continues to rise with decreasing radius all the way to the very central regions of dark matter halos. This implies that a substantial fraction of the matter inside many ordinary galaxies ought to be in the form of dark matter. This is something that can, in principle, be tested by direct observation.

3. Baryonic Processes Important in Understanding Galaxy Formation

In this section, I review the physical processes that are important in understanding how galaxies form within a merging hierarchy of dark matter halos in a CDM-dominated Universe.

3.1. *Radiative Cooling of Gas*

The primary cooling processes relevant to galaxy formation are collisional. At temperatures above 10^6 K primordial gas is almost entirely ionized and above a few $\times 10^7$ K, chemically enriched gas is also fully ionized. The only significant radiative cooling mechanism is bremsstrahlung due to the acceleration of electrons as they encounter atomic nuclei. The cooling rate per unit volume is

$$\frac{dE}{dt} \propto n_e n_H T^{1/2}, \tag{10}$$

where n_e and n_H denote the densities of electrons and of hydrogen atoms, respectively.

At lower temperatures, other processes are important. Electrons can recombine with ions, emitting a photon, or partially ionized atoms can be excited by a collision with an electron, thereafter decaying radiatively to the ground state. In both cases, the gas loses kinetic energy to the radiated photon. Both processes depend strongly on the temperature of the gas, in the first case because of the temperature sensitivity of the recombination coefficient and in the second because the ion abundance depends sensitively on temperature. For gas in ionization equilibrium, the cooling rate for both processes can be parameterised as

$$\frac{dE}{dt} = n_e n_H f(T). \tag{11}$$

Collisional excitation is the dominant process and for primordial gas it causes peaks in the cooling rate at 15000 K (for H) and at 10^5 K (for He^+). For gas with solar metallicity there is an even stronger peak at 10^5 K due to oxygen, and a variety of other elements enhance cooling at around 10^6 K (see Figure 5). At temperatures below 10^4 K gas is predicted to be almost completely neutral and its cooling rate drops sharply. Some cooling due to collisional excitation of molecular vibrations is possible if molecules are indeed present.

It should be noted that cooling by collisional excitation and radiative decay can be substantially suppressed in the presence of strong UV backgrounds because the abundance of partially ionized elements is then reduced by photo-ionization and some of the peaks in Fig. 5 may be eliminated. The effectiveness of this mechanism depends strongly on the spectrum of the UV radiation, as well as the ratio of gas density to UV photon density. Suppression is likely to be important at the early stages of galaxy formation, when the background radiation field from quasars was relatively high, and in relatively low mass (and hence low temperature) galaxies.

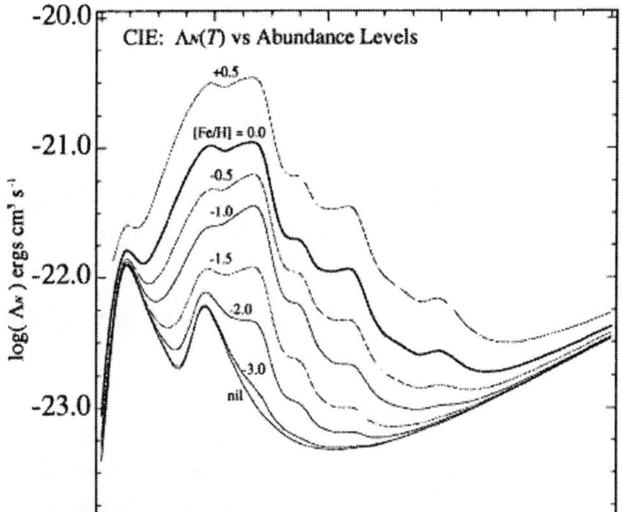

Figure 5. The cooling function from Sutherland & Dopita [18] showing the effect of increasing metallicity on the cooling rate. $n^2\Lambda(t)$ is the cooling rate per unit volume.

3.2. *A Simple Model for Cooling in Dark Matter Halos*

Let us assume that gas is shock heated during the collapse of a dark matter halo and is then in hydrostatic equilibrium with a density profile $\rho_g(r)$ that follows that of the dark matter. The temperature of the gas may be written

$$T = 35.9 \left(\frac{V_{vir}}{100 \text{km s}^{-1}} \right)^2 K \tag{12}$$

The local cooling time $t_{cool}(r)$ can be defined as the ratio of the specific thermal content of the gas, and the local cooling rate per unit volume

$$t_{cool}(r) = \frac{3}{2} \frac{kT\rho_g(r)}{\bar{\mu} m_p n_e^2(r)\Lambda(T,Z)}, \tag{13}$$

where $\bar{\mu} m_p$ is the mean particle mass, $n_e(r)$ is the electron density and $\Lambda(T,Z)$ is the cooling function described in the previous section (as explained, it depends on gas temperature T and metallicity Z).

We define the cooling radius r_{cool} as the radius for which t_{cool} is equal to the age of the Universe at the epoch of interest. If the cooling radius lies within the virial radius of the halo (defined as the radius within which the overdensity is 178), then:

$$\frac{dM_{cool}}{dt} = 4\pi\rho_g(r_{cool}) r_{cool}^2 \frac{dr_{cool}}{dt}. \tag{14}$$

At early times and in low-mass haloes, $r_{cool} > r_{vir}$. The hot gas is then never in hydrostatic equilibrium and the cooling rate is limited by the accretion rate, which can be approximated as

$$\frac{dM_{accr}}{dt} = \frac{M_{hot} V_{vir}}{R_{vir}}. \tag{15}$$

This simple model has been found to be in surprisingly good agreement with full hydrodynamical simulations of cooling within a hierarchy of dark matter halos[19].

3.3. *Angular Momentum and the Dissipative Collapse of Gas within Dark Halos*

Consider a system with mass M, radius R, angular momentum L and energy $E \sim -(GM^2/R)$. The angular velocity of such a system will be about $\omega \sim (L/MR^2)$. The angular velocity for the system to be rotationally supported against gravity is determined by $\omega_{sup}^2 R \sim GM/R^2$. The ratio ω/ω_{sup} represents the degree of rotational support available in the system and can be expressed as

$$\lambda = \frac{\omega}{\omega_{sup}} = \frac{LE^{1/2}}{GM^{5/2}}. \tag{16}$$

The angular momentum in dark matter halos comes from tidal torquing by neighbouring objects as structures collapse. N-body simulations show that the median value of λ for dark matter halos is ~ 0.05 independent of parameters such as redshift, density and the shape of the power spectrum of initial fluctuations. On the other hand, the typical spiral galaxy has $\lambda \sim 0.4$. How does the gas get from an initial spin parameter of 0.05 to a value nearly 10 times larger? The solution is that the gas will 'spin up' as it cools and collapses. However, an argument due to Fall & Efstathiou [20] showed that this does not work unless the gas collapses within a gravitationally dominant dark matter halo.

Let us first consider a gas cloud without any dark matter. The binding energy of the cloud is $E \sim GM^2/R$. Since M is constant during the collapse, we have that $E \propto R^{-1}$ and so $\lambda \propto R^{-1/2}$. In order for the spin parameter to increase by a factor of 10, the gas cloud must collapse by a factor of 100. This process would take a time $t_{coll} = (\pi/2)(R^3/2GM)^{1/2} \sim 5.3 \times 10^{10}$ yr, much longer than the age of the Universe!

Let us now consider a system consisting of both gas and dark matter. Let us write the initial spin parameter of the dark matter plus gas system as

$$\lambda_i = LE^{1/2}/GM^{5/2} \tag{17}$$

and the spin parameter of the resulting disk after collapse as

$$\lambda_d = L_d E_d^{1/2}/GM_d^{5/2}. \tag{18}$$

The energy of the initial system is $E \sim GM^2/R$ and of the disk is $E_d \sim GM_d^2/R_d$. The ratio of the binding energy of the disk to the halo is:

$$\frac{E_d}{E} = \left(\frac{M_d}{M}\right)^2 \left(\frac{R_d}{R}\right)^{-1}. \tag{19}$$

Further, we will assume that angular momentum is conserved during the collapse so that $L_d/M_d = L/M$. We the derive the collapse factor of the gas as

$$\frac{R}{R_d} = \left(\frac{M_d}{M}\right) \left(\frac{\lambda_d}{\lambda}\right)^2. \tag{20}$$

The required collapse factor has been reduced by the factor M_d/M. Even if most of the baryons were to cool, one only now requires collapse by a factor ~ 10 to attain rotational support.

3.4. *Star Formation and Feedback*

The physical processes that regulate the initial cooling and collapse of the gas are relatively well understood. The same cannot be said for the processes that control how rapidly and efficiently the gas is transformed into stars and the effect of energy input from massive stars that explode as supernovae on the interstellar medium of the galaxy.

Modellers typically employ simple prescriptions or "recipes" in order to describe these processes. In the case of star formation, Kennicutt [21] has derived an empirical law for the star formation rate in disk galaxies. Based on Hα, HI and CO measurements of 61 nearby spiral galaxies, Kennicutt has proposed a law of the form

$$\Sigma_{SFR} \propto \Sigma_{gas}/t_{dyn}, \tag{21}$$

where Σ_{SFR} is the star formation rate per unit area averaged within the optical radius of the disk, Σ_{gas} is the surface density of HI and molecular gas within the same radius, and t_{dyn} is the dynamical time scale of the galaxy ($t_{dyn} = r_{opt}/V_c(r_{opt})$).

Kennicutt finds that this star formation law holds over 5 orders of magnitude in gas surface density, from the disks of normal spirals to the circumnuclear star-forming regions of infrared-selected starburst galaxies.

It is also important to consider the effects of supernovae on the conversion of gas into stars in galaxies. So-called "galactic superwinds" have been studied extensively by Heckman and his collaborators [22]. Superwinds are ubiquitous in galaxies where the global star formation rate per unit area exceeds 0.1 $M_{\odot}$ yr^{-1} kpc^{-2}. This is satisfied in the majority of present-day starburst galaxies and in the Lyman break galaxy population at redshifts of ~ 3 , but not in the disks of ordinary spirals such as our own Milky Way. The observations suggest that in starburst galaxies mass is being ejected at a rate that is comparable to the star formation rate and that in these systems, the velocities with which the material is being ejected range from 100–1000 km s^{-1}. This suggests that the ejecta would be able to escape from low mass dark matter haloes where $V_{esc} < V_{wind}$. Modern hydrodynamical simulations of galaxy formation [23] are beginning to incorporate parameterised galactic wind models that are motivated by the empirical data. These numerical experiments show that galactic winds greatly suppress the efficiency of star formation in galaxies that reside in low mass halos. Moreover, outflows from galaxies drive the chemical enrichment of the intergalactic medium.

3.5. *Merging of Galaxies*

As explained in Section 2, dark matter halos are built up through merging of small progenitor halos to form more and more massive systems. When dark halos merge, the accreted galaxies within them remain distinct for some time and these are referred to as "satellite" galaxies. Satellites moving through the background of dark matter will lose energy through the process of dynamical friction. The timescale for the satellite galaxy to sink to the centre of the halo and merge with the central object will depend on the mass of the satellite as well its orbital parameters. A detailed discussion of the dynamical friction process can be found in Chapter 7 of Binney & Tremaine [24].

What happens when two disk galaxies of roughly equal mass merge? This has been studied in detail using numerical simulations (see for example Mihos & Hernquist [25]). The effect of the merger differs substantially according to whether one considers the stars or the gas in the two interacting systems. As the two galaxies encounter one another, the tidal forces from the passing companion cause distortions. The stars form "tidal tails" and bridges that connect the two objects. The inner regions of the disks can form linear barlike structures (see Fig. 6). Eventually, when the two systems merge, relaxation processes transform the stellar component into a $R^{1/4}$ profile that is very reminiscent of the observed profiles of elliptical galaxies.

On the other hand, the gas in the galaxy is subject to strong shocking, dissipation and loss of angular momentum during the merging process. The gas initially shocks at the interface between the two galaxies. At first the gas reacts like the stars and forms a bar, but it then flows inwards. By the end of the merger, a large fraction of the gas has ended up in a compact core in the remnant galaxies. These gas flows are one mechanism for triggering the powerful star formation events or "starbursts" that are often observed in merging or interacting galaxies in the nearby Universe.

Figure 6. A snapshot of two interacting galaxies from a numerical simulation. Strong tidal features and bars are clearly visible.

3.6. *Evolutionary Population Synthesis*

In order to make predictions for the observed properties of galaxies, for example their absolute magnitudes or their colours, galaxy formation models must be coupled to models of evolutionary population synthesis (see for example Bruzual & Charlot[26]).

The main adjustable parameters of these models are

(1) The initial mass function (IMF) $\phi(m)dm$, which specifies the number of stars formed with masses between m and $m + dm$ (with lower and upper cutoffs at typical masses of of 0.1 $M_\odot$ and 100 $M_\odot$).
(2) The star formation rate (SFR) $\psi(t) = dM_*/dt$
(3) The chemical enrichment rate $\chi(t) = dZ/dt$ (where Z is the mass fraction of elements heavier than He).

The models use libraries of stellar evolutionary tracks to follow how stars evolve across the Hertzprung-Russell (HR) diagram, which relates the luminosity of a star to its temperature. Over billions of years, hot high mass stars, which are initially luminous and blue (meaning that most of their energy comes out in the ultraviolet shortwards of ~ 2000 Å), evolve to become cool red giant stars where most of the energy is radiated at infrared wavelengths. The integrated colour of the once blue young stellar population thus becomes red as the giants dominate the light. In order to compute the spectrum of the integrated stellar population, the models make use of 'libraries' of stellar spectra, which are matched to the stars according to their position on the HR Diagram. These spectra are either obtained observationally or are computed using theoretical model atmospheres. Obtaining a stellar library over a wide range in wavelength that samples the full range of temperatures, luminosities and metallicities spanned by stars in different galaxies is a challenging observational and computational task. This remains an important limiting factor for modern population synthesis models.

Fig. 7 shows the evolution of the spectrum of a galaxy following an instantaneous burst of star formation [26]. As can be seen, the integrated luminosity at ultra-violet wavelengths fades considerably during the first Gigayear following the burst. After about ~ 4 Gyr, there is rather little evolution in the overall shape of the spectral energy distribution of the stellar population.

Note that the flux of a galaxy measured at short wavelengths is extremely sensitive to the number of young stars in the galaxy. Even a tiny amount of star formation will boost the UV flux by several orders of magnitude. The flux measured at short wavelengths is thus a poor indicator of the total stellar mass of the galaxy. In order to obtain an estimate of the mass of the galaxy that is largely insensitive to its past star formation history, it is necessary to obtain observations at wavelengths ~ 1 micron.

3.7. *Putting it all Together*

Figure 8 is a schematic representation for how galaxies may be expected to form in the standard ΛCDM Universe. Consider a set of dark matter halos at some early time t. Gas will cool to form a rotationally-supported disk system at the centre of each halo. The size of the disk will be roughly a tenth of the virial radius of its halo.

Later on, some fraction of these halos will merge as structure in the Universe grows by hierarchical clustering. When two halos merge, the lighter "satellite" galaxies will merge with the heaviest "central" galaxy on a dynamical friction timescale. If the satellite and the central galaxies have roughly similar masses, the merging event will destroy the disks and a spheroidal merger remnant will be produced. Gas may be driven to high densities during the merging event and turn into stars in a violent "burst". It has been speculated that the central supermassive black holes found in most galactic bulges may have also been formed in these events.

What happens after the merger? Current models assume that the hot gas component present in the halo is not affected and that it can continue to cool. A composite galaxy consisting of both a spheroidal bulge and a disk accreted at late

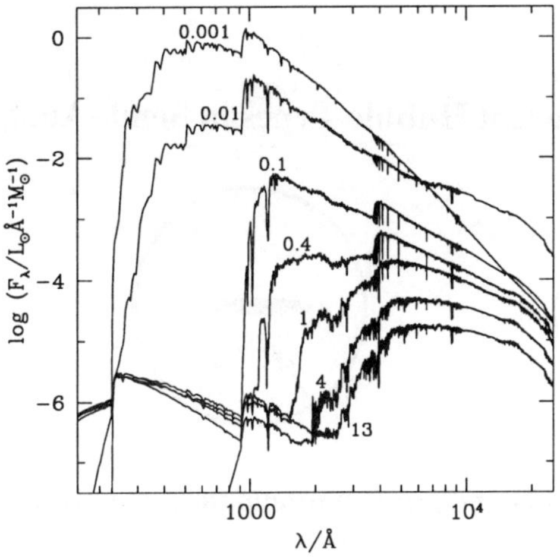

Figure 7. The evolution of the spectral energy distribution of a stellar population following an instantaneous burst. The labels indicate the time after the burst in units of Gigayears.

times is the end product of the galaxy formation process in the majority of cases. Some galaxies are accreted by larger halos before they have time to grow a new disk. These systems will be the classic ellipticals, which have very little disk component.

So how well does this work? In the next section, we will confront these simple models (often called "semi-analytic" models of galaxy formation [27 28 29]) with the observational data.

4. Comparison with the Observations

4.1. *The Galaxy Luminosity Function*

We define the luminosity function $\Phi(L)$ as the number of galaxies per unit volume with luminosity L . In 1976, Schechter [30] proposed a global fitting function to describe the luminosity function

$$\Phi\left(\frac{L}{L_*}\right) = \Phi_* \left(\frac{L}{L_*}\right)^{\alpha} \exp\left(\frac{-L}{L_*}\right), \tag{22}$$

with typical values (averaged over large volumes) $L_{B,*} \sim 10^{10} L_{B,\odot} h^{-2}$, $\alpha \sim -1.2$ and $\Phi_* = \Phi(L_*) \sim 0.01\,\mathrm{Mpc}^{-3}\,h^3$ (B refers to the photometric band centred around 4400Å).

The galaxy luminosity function thus looks like a power law at low luminosity and cuts off exponentially for galaxies with high luminosities. The galaxy luminosity

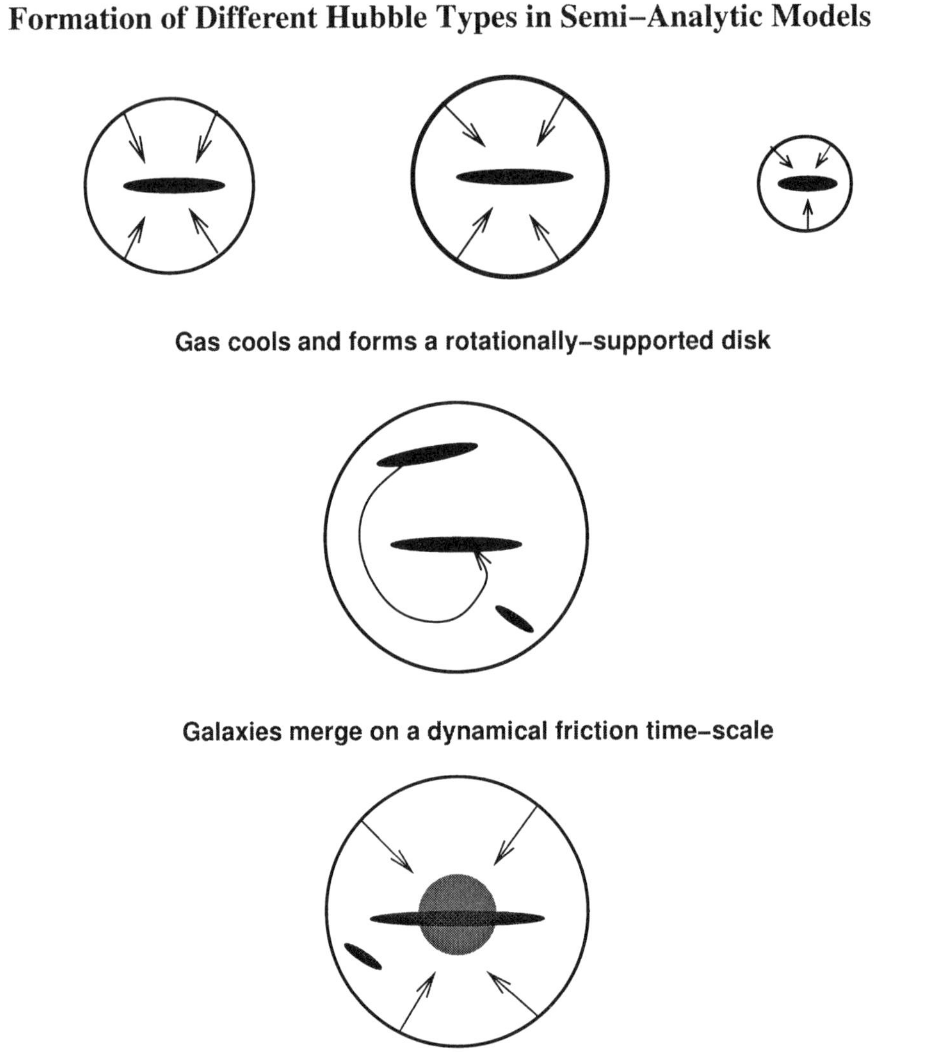

Figure 8. A schematic representation of how galaxies form in current semi-analytic models of galaxy formation.

function is now extremely accurately determined in the nearby Universe [31] and the original fitting function proposed by Schechter has stood the test of time very well.

The shape of the galaxy luminosity function turns out to be non-trivial to understand in the context of the formation picture outlined above. This is illustrated in Figure 9 , which compares the shape of the observed galaxy luminosity function to that of the mass function of dark matter halos for a range of cold dark matter (CDM) cosmologies [29]. The halo mass function has been scaled by multiplying the mass of each halo by the ratio of baryons to dark matter. This brings the abundance of galaxies and halos into reasonable agreement at luminosities around L_* (*i.e.*at the knee of the luminosity function). However, the shapes of the two functions are extremely different. The halo mass function is well approximated by a power law over a large range in mass, but the slope of the power law ($\alpha \sim -2$) is considerably steeper than that observed for galaxies. In addition, the exponential cutoff occurs at much higher mass scales.

Figure 9 illustrates that *baryonic processes* are critical in understanding the shape of the luminosity function. In low mass halos, both photo-ionization by external sources of radiation and supernovae feedback act to prevent gas from cooling and forming stars as efficiently as in high mass haloes. The inclusion of feedback processes tends to *flatten* the faint-end slope of the luminosity function. In high mass haloes, the cooling times become longer and a smaller fraction of the baryons are predicted to cool and form stars. Nevertheless, most attempts to model the luminosity function produce too many very bright galaxies unless cooling is heavily suppressed in massive haloes by some other physical mechanism. There has been recent speculation that the jets produced in radio galaxies may impart enough energy to the surrounding medium to substantially reduce the amount of gas cooling at the centres of some rich clusters.

4.2. *The Two-Point Correlation Function*

The two-point correlation function $\xi(r)$ is a quantitative measure of galaxy clustering and is defined via the probability to find pairs of galaxies at a distance r:

$$dN_{pair} = N_0^2(1 + \xi(r))dV_1 dV_2 \tag{23}$$

where N_0 is the mean background density and dV_1 and dV_2 are volume elements around the two points under consideration.

Observationally, the two-point correlation function averaged over all galaxy types is a power-law:

$$\xi(r) = \left(\frac{r}{r_0}\right)^{-\gamma} \tag{24}$$

with $\gamma = 1.8$ and $r_0 = 5h^{-1}$ Mpc on scales between 100 kpc and 10 Mpc. Beyond 10 Mpc, the correlation function falls more rapidly.

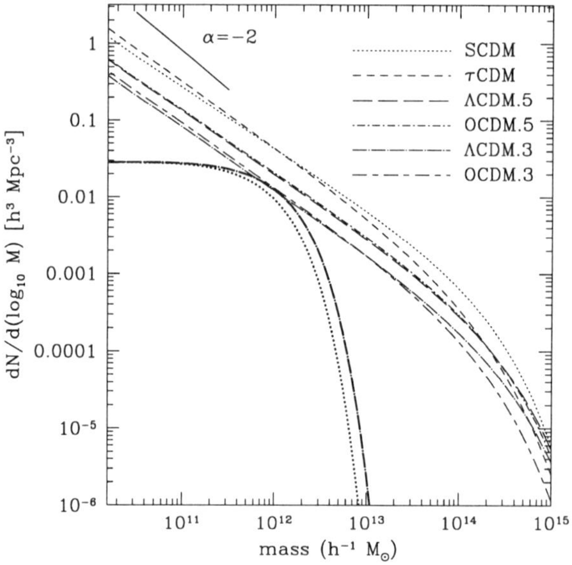

Figure 9. The shape of the observed galaxy luminosity function is compared with that of the halo mass function for a variety of CDM cosmologies.

Attempts to model the luminosity function have been quite successful [32] (see Figure 10). The two point correlation function of the dark matter is not well-described by a power-law and is considerably *steeper* than the galaxy correlation function on scales between 500 kpc and a few Mpc. Nevertheless, the galaxy correlation function predicted by the model agrees very well with the observational data. It is possible to show why this is the case using rather simple analytic arguments. First, one assumes that galaxies are always located within dark matter halos. Galaxies of given luminosity are found in halos with a certain "occupation number", which scales approximately linearly with the mass of the halo. Second, one galaxy is always found at the halo centre and the other galaxies are distributed with a density profile that is the same as that of the dark matter (the "universal" profile of Navarro *et al.*[17]). These assumptions are motivated by the physical model of galaxy formation outlined in the previous section. If one combines these assumptions with analytic models of the halo-halo correlation function, one can explain the power-law form seen in Figure 10. This is often referred to as the *halo model* for galaxies and was first proposed by Benson *et al.*[32].

An important corollary of the halo model is that it should be possible to see deviations away from a $\gamma = 1.8$ power-law correlation function by selecting galaxies according to colour or morphological type, so that one obtains a different form for the halo occupation function [33].

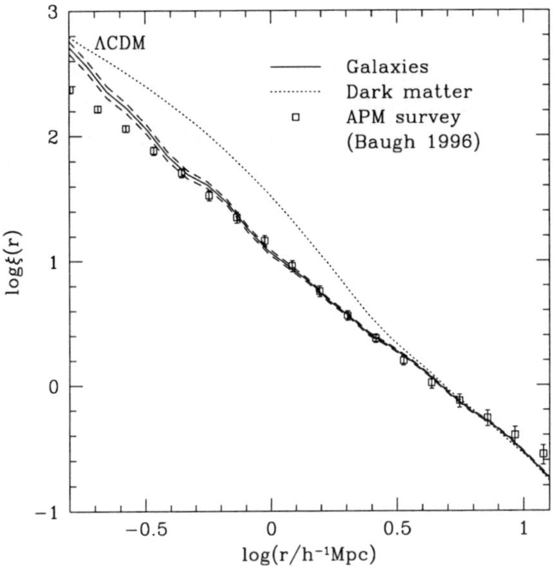

Figure 10. The two point correlation function of dark matter (dotted) in a ΛCDM Universe is compared to that predicted for galaxies (solid). The squares indicate the observational measurements.

4.3. *Different Types of Galaxies*

More than 75 years ago, Edwin Hubble introduced a galaxy classification system that is still widely in use today. Hubble arranged galaxies into a sequence according to bulge-to-disk ratio and the presence and the opening angle of their spiral arms. Spirals were also sub-divided into those with bars and those without. Elliptical galaxies have a large bulge and no obvious disk component. S0 or lenticular galaxies have a dominant bulge component and a disk with no significant spiral structure. Spiral galaxies are arranged in a sequence from Sa to Sd according to decreasing importance of the bulge component. Irregular galaxies do not exhibit regular spiral structure and usually do not have a significant bulge. In addition, there exists a zoo of different types of low mass dwarf galaxies.

One of the reasons why the Hubble classification system has proven so durable is that the physical parameters of galaxies correlate very strongly with Hubble type:
1)The stellar mass of the galaxy increases from irregulars to ellipticals.
2)The specific angular momentum J/M increases from ellipticals to spirals.
3)The mean stellar age (as deduced from galaxy colours and stellar mass-to-light ratios) increases from irregulars through spirals to ellipticals.
4)The mean surface brightness increases from irregulars to spirals to ellipticals.
5)The cold gas content of galaxies decreases from irregulars through to ellipticals.
It is an important challenge for theoretical models of galaxy formation to explain the origin of these different galaxy types and their properties.

4.4. *The Formation of Galactic Disks*

In the standard picture, disk galaxies form when hot gas in a dark matter halo cools and contracts until it forms a rotationally supported structure.

Let us consider the case of a dark matter halo with an isothermal density profile

$$\rho(r) = \frac{V_c^2}{4\pi G r^2}. \tag{25}$$

Based on the spherical collapse model, we define the limiting radius of the dark halo to be the radius r_{200} within which the mean mass density is $200\rho_{crit}$. The radius and mass of a halo of circular velocity V_c seen at redshift z are

$$R_{200} = \frac{V_c}{10H(z)}; \quad M = \frac{V_c^2 r_{200}}{G} = \frac{V_c^3}{10GH(z)}, \tag{26}$$

where

$$H(z) = H_0[\Omega_\Lambda + (1 - \Omega_\Lambda - \Omega_0(1+z)^2 + \Omega_0(1+z)^3]^{1/2} \tag{27}$$

is the Hubble constant at redshift z. We assume that the mass which settles into the disk is a fixed fraction m_d of the halo mass and that the angular momentum of the disk is a fixed fraction j_d of that of the halo. We further assume the disks to have exponential surface density profiles,

$$\Sigma(R) = \Sigma_0 \exp(-R/R_d), \tag{28}$$

where R_d and Σ_0 are the disk scalelength and central surface density, and are related to the disk mass through

$$M_d = 2\pi\Sigma_0 R_d^2. \tag{29}$$

If the gravitational effect of the disk is neglected, its rotation curve is flat and its angular momentum is just

$$J_d = 2\pi \int V_c \Sigma(R) R^2 dR = 4\pi\Sigma_0 V_c R_d^3 = 2M_d R_d V_c. \tag{30}$$

Using $J_d = j_d J$ and $\lambda = JE^{1/2}G^{-1}M^{-5/2}$, we have that

$$R_d = \frac{\lambda G M^{3/2}}{2V_c E^{1/2}} \left(\frac{j_d}{m_d}\right). \tag{31}$$

From the virial theorem, the total energy of the isothermal sphere is

$$E = -\frac{GM^2}{2r_{200}} = -\frac{MV_c^2}{2}, \tag{32}$$

Inserting this into equation (33) and using equations (28) and (31) we obtain an expression for the predicted exponential scale length of the disk as a function of the circular velocity of the halo, the spin parameter λ of the halo, and the Hubble constant H:

$$R_d = \frac{1}{2^{1/2}}\left(\frac{j_d}{m_d}\right)\lambda r_{200} \simeq 8.8h^{-1}\mathrm{kpc}\left(\frac{\lambda}{0.05}\right)\left(\frac{\mathrm{V_c}}{250\mathrm{kms}^{-1}}\right)\left[\frac{\mathrm{H}}{\mathrm{H_0}}\right]^{-1}\left(\frac{\mathrm{j_d}}{\mathrm{m_d}}\right). \tag{33}$$

We have illustrated the case of the isothermal density profile because the equations are particularly easy to deal with, but the same analysis can be carried out for more realistic dark matter halo density profiles [34]. Figure 11 shows the resulting predictions for the scale length of the disk as a function of V_c for two different CDM cosmologies at $z = 0$ and at $z = 1$. Motivated by the results of N-body experiments, the spin parameter λ is assumed to have a lognormal distribution centred on $\lambda = 0.05$ with r.m.s. dispersion $\sigma_\lambda = 0.5$. The solid line shows the median value of R_d, while the short and long-dashed lines indicate the 10th and 90th percentiles of the distribution. The crosses in the diagram are data points drawn from a sample of low-redshift spiral galaxies with measured rotation curves.

Figure 11 shows that the observed sizes of disk galaxies in the local Universe are in remarkably good agreement with the predictions of the model outlined above. Not that the model assumes that angular momentum is *conserved* during the collapse of the disk. In practice, gas-dynamical simulations have shown that this assumption is very easily violated. Angular momentum is transferred from the baryons to the dark matter during mergers and this tends to lead to the formation of disk galaxies that are too small in comparison with the observations [35]. The model also predicts that at given circular velocity, disks are *smaller* at higher redshifts. The shift in size is smaller in the now-standard ΛCDM cosmology, but should still be detectable with a large enough sample of high redshift galaxies.

4.5. *The Formation of Spheroids and Bulges*

As we have discussed, in models where structure assembles through hierarchical clustering, mergers between galaxies occur frequently, particularly at high redshifts, and galactic spheroids are believed to form when two disk galaxies of near-equal mass merge with each other. It can be shown that a model of this type can explain the observed abundance of bulge-dominated galaxies at the present day [27] and the fact that early-type galaxies occur preferentially in dense environments where mergers occur more frequently [36].

It has proved more difficult to explain trends in the stellar populations of these objects. Elliptical galaxies have very uniform, old and metal-rich stellar populations. In hierarchical structure formation models, massive galaxies assembled at lower redshifts than less massive systems. However, several lines of observational evidence appear to suggest that star formation occurred over a shorter timescale in the most massive ellipticals [37]. If this is the case, it follows that the star formation cannot be closely linked with the assembly of mass in these systems. It is not yet understood why this should be the case.

4.6. *Dwarf Galaxy Crisis?*

Dwarf galaxies are by definition located in the lowest mass dark matter haloes. In these systems, the gravitational binding energy is low and feedback processes are

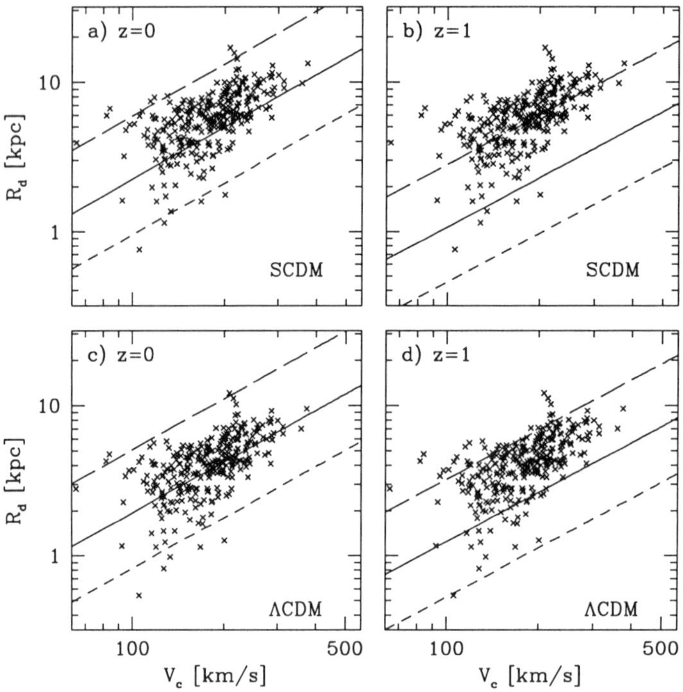

Figure 11. The relation between disk scale length and halo circular velocity predicted by a simple model in which disks form from gas which cools and contracts until rotational support is achieved, while conserving angular momentum[34].

likely to play a major role in regulating how these systems evolve. Dwarf galaxies are also observed to have rather irregular star formation histories. Instead of proceeding in a continuous fashion, star formation is apparently episodic and occurs in bursts separated by periods of relative quiescence.

Most theoretical work on dwarf galaxies has focused on explaining their *abundances.* As discussed above, the ΛCDM model produces a halo mass function with a rather steep faint end slope ($\alpha \sim -1.8$). If this is to be reconciled with the observational data, an important prediction of the model is that a large fraction of low mass halos do not contain detectable galaxies.

4.7. *Evolution of Galaxies to High Redshift*

Over the past decade, a new generation of ground-based and space-based telescopes has made it possible for astronomers to study galaxies out to redshifts when the Universe was less than a tenth its current age. Some of the most notable results are the following:
1) The "star formation rate density" (mass of stars forming per unit time per unit comoving volume) increases by a factor ~ 10 from the present day out to $z \sim 1$. At higher redshifts, it remains approximately constant out to $z \sim 5$.
2) The stellar mass density (total stellar mass in galaxies per unit volume) decreases by a factor of ~ 10 from the present day to $z \sim 2$.
3) At redshifts $z \sim 3$, the brightest star-forming galaxies detected at rest-frame UV wavelengths are as strongly clustered as L_* galaxies today.
4) At high redshifts ($z \sim 2-3$) a substantial fraction of star formation is occurring in dusty galaxies, where most of the UV radiation is absorbed and re-radiated at infrared wavelengths.
5) The Hubble sequence as we know it appears to have been mostly in place at $z < 1$. At higher redshifts, galaxies are significantly smaller, many appear highly disturbed and it is no longer possible to identify classical elliptical or spiral systems.

In short, high redshift observations are beginning to provide a census of how star formation has occurred in galaxies as a function of cosmic time, and this now serves as an important constraint on theoretical models [38]. In coming years, the observational data will attain a level of detail where it should become possible to begin disentangling the complex physical processes that have determined how galaxies have formed their stars. The combination of new data with sophisticated simulations that include the important gas-physical processes will no doubt shape progress in the field of galaxy formation over the next decade.

References

1. P.J.E. Peebles, *The Large Scale Structure of the Universe* (Princeton University Press, Princeton , 1980).
2. W.H. Press and P. Schechter, *Astrophys.J.* **187**, 425 (1974).
3. R.K. Sheth, H.J. Mo and G. Tormen, *Mon.Not.Roy.Astr.Soc.* **323**, 1 (2001).
4. A.R Jenkins *et al.*, *Mon.Not.Roy.Astr.Soc.* **321**, 372 (2001).
5. H.J. Mo and S.D.M. White, *Mon.Not.Roy.Astr.Soc.* **336**, 112 (2002).
6. J.R. Bond, S. Cole, G. Efstathiou and N. Kaiser, *Astrophys.J.* **379**, 440 (1991).
7. R.G. Bower, *Mon.Not.Roy.Astr.Soc.* **248**, 332 (1991).
8. C. Lacey and S. Cole, *Mon.Not.Roy.Astr.Soc.* **262**, 627 (1993).
9. G. Kauffmann and S.D.M. White, *Mon.Not.Roy.Astr.Soc.* **261**, 921 (1993).
10. R.K. Sheth and G. Lemson, *Mon.Not.Roy.Astr.Soc.* **305**, 946 (1999).
11. R. Somerville and T.S. Kolatt, *Mon.Not.Roy.Astr.Soc.* **305**, 1 (1999).
12. G. Efstathiou, M. Davis, S.D.M. White and C.S. Frenk, *Astrophys.J.Supp.* **57**, 241 (1985).
13. H. Mathis *et al.*, *Mon.Not.Roy.Astr.Soc.* **333**, 739 (2002).

14. Y. Hoffman and R. Ribak, *Astrophys.J.* **380**, 5 (1991).
15. S.D.M. White and M.J. Rees, *Mon.Not.Roy.Astr.Soc.* **183**, 341 (1978).
16. V. Springel, S.D.M. White, G. Tormen and G. Kauffmann, *Mon.Not.Roy.Astr.Soc.* **328**, 726 (2001).
17. J.F. Navarro, C.S. Frenk and S.D.M. White, *Astrophys.J.* **490**, 493 (1997).
18. R.S. Sutherland and M.A. Dopita, *Astrophys.J.Supp.* **88**, 253 (1993).
19. N.Yoshida, F. Stoehr, V. Springel and S.D.M. White, *Mon.Not.Roy.Astr.Soc.* **335**, 762 (2002).
20. M.J. Fall and G.P. Efstathiou, *Mon.Not.Roy.Astr.Soc.* **193**, 189 (1980).
21. R.C. Kennicutt, *Astrophys.J.* **498**, 541 (1998).
22. T.M. Heckman, in *Extragalactic Gas at Low Redshift*, eds J.S. Mulchaey and J.T. Stocke, (ASP Conf. Series, San Francisco, 2002).
23. V. Springel and L. Hernquist, *Mon.Not.Roy.Astr.Soc.* **339**, 289 (2003).
24. J. Binney and S. Tremaine, *Galactic Dynamics* (Princeton University Press, Princeton , 1987).
25. C. Mihos and L. Hernquist, *Astrophys.J.* **464**, 641 (1996).
26. G. Bruzual and S. Charlot, *Astrophys.J.* **405**, 538 (1993).
27. G. Kauffmann, S.D.M. White and B. Guiderdoni , *Mon.Not.Roy.Astr.Soc.* **264**, 201 (1993).
28. S. Cole, A. Aragon-Salamanca, C.S. Frenk, J.F. Navarro and S. Zepf , *Mon.Not.Roy.Astr.Soc.* **271**, 781 (1994).
29. R. Somerville and J. Primack, *Mon.Not.Roy.Astr.Soc.* **310**, 1087 (1999).
30. P. Schechter, *Astrophys.J.* **203**, 297 (1976).
31. M. Blanton, *Astrophys.J.* **592**, 819 (2003).
32. A. Benson, S. Cole, C.S. Frenk, C.M. Baugh and C.G. Lacey, *Mon.Not.Roy.Astr.Soc.* **311**, 793 (2000).
33. G. Kauffmann, J.M. Colberg, A. Diaferio and S.D.M. White, *Mon.Not.Roy.Astr.Soc.* **303**, 188 (1999).
34. H.J. Mo, S. Mao and S.D.M. White, *Mon.Not.Roy.Astr.Soc.* **295**, 319 (1998).
35. J.F. Navarro and M. Steinmetz, *Astrophys.J.* **528**, 607 (2000).
36. A. Diaferio, G. Kauffmann, M. Balogh, S.D.M. White, D. Schade and E. Ellingson, *Mon.Not.Roy.Astr.Soc.* **323**, 999 (2001).
37. D. Thomas, L. Greggio and R. Bender, *Mon.Not.Roy.Astr.Soc.* **302**, 537 (1999).
38. L. Hernquist and V. Springel, *Mon.Not.Roy.Astr.Soc.* **341**, 1253 (2003).

THE PHYSICS OF GALAXY FORMATION

MICHAEL A. DOPITA

Research School of Astronomy & Astrophysics, The Australian National University, Cotter Road, Weston Creek, ACT 2611, Australia
E-mail: Michael.Dopita@anu.edu.au

The epoch of galaxy formation, occurring between one and six billion years after the Big Bang, was initiated by the collapse of over-dense regions of matter, resulting in extraordinary bursts of star formation, rapid growth of massive nuclear black holes, and the rapid structural evolution of the early universe. To understand these phenomena and thus gain insight into the evolution of galaxies in general is a central objective of modern astrophysics. Here, we summarize some important theoretical problems in galaxy formation and review recent data obtained on the ultra-steep spectrum radio sources. These are found in the densest regions of the early universe, and are associated with AGN in the most massive galaxies embedded in what will become the most massive clusters of the present-day universe.

1. Introduction

The mystery of the formation of galaxies is one of the central problems of modern astrophysics. We know that the seeds of galaxies are found in the tiny fluctuations of the cosmic microwave background (CMB) radiation formed $\sim 1 - 3 \times 10^5$ years after the Big Bang. Roughly 0.3 billion years (0.3 Gyr) later, over-dense regions started to collapse under their own gravity, heralding the start of the epoch of galaxy formation which runs from about redshift $z \sim 15$ down to as little as $z \sim 1$ for the smallest dwarf galaxies. The exact age of the universe at these redshifts is determined by the cosmological parameters, but roughly, it lies in the range $0.3 - 6$ Gyr. Exact figures for any cosmology can be computed at: http://www.astro.ucla.edu/~wright/CosmoCalc.html.

The epoch of galaxy formation was the most active period in the gravity-dominated evolution of the universe. The Universe itself was still relatively dense and both the pressure and density of the protogalactic gas was high. Only 3 billion years after the Big Bang, about 80% of all the stars now seen in our local universe had already been formed along with most of the heavy elements now present in the interstellar gas. In addition, the super-massive black holes which now lurk in the centres of most, if not all, elliptical galaxies had grown rapidly to attain much of their current mass. Relativistic jets ejected by these black holes interacted with the interstellar gas in their host galaxies, generating strong radiative shocks, and inducing powerful bursts of star formation. If continued, such bursts would convert all the gas contained in the protogalaxy into stars in roughly a dynamical (orbital or collapse) timescale ($\sim 10^8$ years). To understand the basic observational parameters of galaxy formation we need to be able to understand how much of the line or continuum emission we see at any wavelength is coming from star formation, how much from jet-driven shocks, and how much from the photons or photoionization produced by the central engine.

In these lectures, I will briefly address some of the outstanding problems of the interstellar physics of galaxy formation, emphasizing the physics of star formation, black hole growth and jet production and the importance of dust in determining what we can see of these processes operating in the high redshift universe.

2. How Did Galaxies Get the Way They Are?

It is commonly supposed that the elliptical galaxies in the nearby universe have largely resulted from mergers of disk or satellite galaxies at earlier times. Thus disk galaxies are the more "fundamental" building blocks. However, there are some properties of modern-day disk galaxies that demand attention, and which may prove capable of providing an intimate insight into the way galaxies were formed.

2.1. *The Bulge : Black Hole Connection*

There is increasing evidence that the formation of black holes, and the formation of galactic stellar bulges are intimately related: both forming at the epoch of galaxy collapse. For example, Boyle & Terlevich (1998) showed that the quasi-stellar object (QSO) luminosity density evolution is essentially the same as the star formation rate evolution. This suggests that galactic bulges and their associated massive black holes grew together (coevally).

Further evidence that the black hole "knows" about its galaxian environment comes from the amazingly good correlation between stellar velocity dispersion and black hole mass discovered by Ferrarese & Merritt (2000) and Gebhardt *et al.* (2000). This relationship applies to both elliptical and disk types. A good, but weaker correlation exists between the black hole mass and the total bulge luminosity relationship. Provided that the optical luminosity is a good measure of the rate of accretion of matter onto the central black hole, such relationships demonstrate that the central black hole and the stellar bulge of its host galaxy are intimately connected.

This connection is most likely to have been established at the epoch of galaxy collapse. For example, Silk & Rees (1998) have suggested that outflows driven by radiation pressure limit the black hole masses by ejecting the residual gas. The point at which this occurs depends on the ratio of the radiation pressure force and the attractive force due to the combined galaxian and black hole potential. This mechanism would provide a black hole mass that is proportional to the line of sight stellar velocity dispersion raised to the fifth power. This is close to the observed relationship. Although other interpretations are possible, the model finds support through observations of high-redshift radio galaxies (see below).

2.2. *The Tully-Fisher Relationship*

The Tully-Fisher (1977) relationship is a connection between the absolute luminosity, L, and the maximum disk rotational velocity, v_m. One of the most comprehensive studies of this relationship is that by Giovanelli *et al.* (1997).

The Tully-Fisher relationship can easily be understood as a natural consequence of the structure of spiral galaxies. In these galaxies, the rotational velocity rises quickly to its maximum value, v_m, inside (roughly) a scale length, R_0, of the exponential disk, before becoming flat into the outer regions of the galaxy as dark matter comes to dominate the contributions to the galaxian potential. This implies that the mass of the galaxy as a function of r is given by $M(r) \propto r^2$. If spiral galaxies are characterized by a similar scale length and central surface density, Σ_0, then the Keplerian velocity of the disk can be written as:

$$v_m^2 = GM/r = 4\pi G\Sigma_0 r. \tag{1}$$

Thus, to the extent that the mass to light ratio of the stars remains constant within the luminous disk of the galaxy, usually defined in terms of the Holmberg radius, which defines a limiting surface brightness, then we would expect $v_m \propto L^{1/4}$. We see that there are a number of assumptions that go into this relationship, and so we should not be too concerned that the observational slope determined by Giovanelli *et al.* (1997) is somewhat different: $v_m \propto L^{1/3.1}$. This reflects the fact that dwarf disk galaxies are more dominated by dark matter, even in their central regions.

3. Simulations of Galaxy Formation

Galaxy formation, understood as the formation of dense, rotationally supported agglomerations of gas, dust and stars occurs as an inevitable consequence of the ΛCDM simulations. However, the extent to which these agglomerations resemble real galaxies is strongly dependent upon the physics they contain, and on the resolution of the simulation. Perhaps the finest "state-of-art" simulations currently available are those of Abadi *et al.* (2002), following the work of Navarro & Steinmetz (1997) and Steinmetz & Navarro (2002).

So, what physics is included in such simulations? This physics must (at least) cover the dynamics of both the baryonic and dark matter components, provide a model of star formation and of heavy element production by these stars, and to properly account for the dynamical feedback between the winds and explosions of the stars and the surrounding interstellar gas. This is a fairly tall order, since many of these processes are still imperfectly understood.

The dark matter is usually assumed to be both cold and non-self interacting, and to be constrained to move in potential defined by all matter, baryonic or otherwise.

As far as the baryonic matter is concerned, the advanced models include not only the gravitational potential of the non-baryonic component, but also contain self-gravity, gas pressure and shock fronts and radiative processes. The thermal balance of the gas is computed using both Compton and radiative cooling and photoelectric heating resulting from the photoionizing UV background (Navarro & Steinmetz, 1997, 2000).

Star formation is more difficult to deal with. However, what is usually done is to develop a "prescription" for the star formation rate. In the local universe, there is an empirically-derived connection between the local star-formation rate in the disk, and the local disk properties. This is usually expressed in terms of a Schmidt (1959) relationship connecting the star-formation rate per unit area of disk, $\Sigma_{\rm SFR}$, with the surface density of gas, $\Sigma_{\rm g}$:

$$\Sigma_{\rm SFR} = A\Sigma_{\rm g}^{\beta}, \tag{2}$$

where the power law index is determined observationally as $0.9 < \beta < 1.8$.

What is the physical meaning of the Schmidt Law above? The simplest theoretical scenario is one in which the star-formation rate is presumed to scale with the growth rate of gravitational perturbations within the disk. In this case, the local star-formation rate (per unit volume) will scale as the local gas density divided by the growth timescale of the gravitational instabilities,

$$\rho_{\rm SFR} \propto \rho_{\rm g}/\left(G\rho_{\rm g}\right)^{-1/2} \propto \rho_{\rm g}^{3/2}. \tag{3}$$

The scaling to surface quantities depends upon the local scale height of the gas layer, but it is plausible that this may produce a β in the right range.

A simpler approach is to suppose that star formation scales as the gas density divided by a local dynamical (orbital or infall) timescale (Larson, 1988; Wyse, 1986; Silk, 1997; Elmegreen, 1997; Kennicutt 1998). For example the Abadi *et al.* (2002) ΛCDM modelling adopts a star formation density, $\rho_{\rm SFR}$,

$$\rho_{\rm SFR} = \frac{\kappa\rho_{\rm g}}{\max(\tau_{\rm cool}, \tau_{\rm dyn})} \tag{4}$$

where $\rho_{\rm g}$ is the local gas density, $\tau_{\rm cool}$ is the local cooling timescale for the gas and $\tau_{\rm dyn}$ is the local dynamical timescale. This ensures that gas which is heated to high temperatures must first cool before it can become effective in forming stars. The cooling timescale can be obtained from the cooling functions given by Sutherland & Dopita (1993). The efficiency factor κ is small, ~ 0.05, this number chosen so that the gas is transformed into stars only over a timescale much longer than the dynamical timescale.

If star formation is difficult to deal with, then properly accounting for feedback is well-nigh impossible. When young massive stars are formed, they produce highly energetic stellar winds until they finally explode as supernovae. This process liberates about 10^{49} ergs ${M_{\odot}}^{-1}$ of energy. The effect this energy injection has depends critically on the local environment. If the local interstellar medium (ISM) is dense, much of this energy is radiated away locally. However, collective effects can be very important. The velocity of a shock, v_s, in a medium with density ρ is given approximately by $P = \rho {v_s}^2$, where P is the driving pressure. Shocks become radiative when the cooling timescale behind the shock, $\tau_{\rm cool}$, is comparable to the dynamical timescale, $\tau_{\rm dyn}$. The cooling timescale given by radiative shock models (Dopita & Sutherland: 1995, 1996) in the velocity range 200 – 900 km s^{-1} can be approximated as:

$$\tau_{\rm cool} \approx 23000 \left[\frac{n_{\rm ISM}}{\rm cm^{-3}}\right]^{-1} \left[\frac{v_{\rm s}}{\rm 300km\ s^{-1}}\right]^{3.9} \ {\rm yrs} \tag{5}$$

where $n_{\rm ISM}$ is the density of the local ISM. Thus, the cooling efficiency drops precipitously as the density decreases. In regions where supernova winds and remnants are able to collide with each other, the local region is quickly swept clear of the original ISM and therefore bubble and remnants merge into a region of very low density and very high pressure which is only cooled by adiabatic expansion. This tends to produce a two-phase medium in which the star-forming ISM is pressure confined by a much more tenuous hot gas with relatively large volume filling factor. If the pressure in the star formation region can be maintained at a high enough value, then the bubble of hot gas may eventually "burst" releasing the chemically-enriched gas it contains into the inter-galactic medium (IGM). These physical processes have not been well-described by theoretical models up to the present. The effect of feedback is crudely accounted for by Abadi *et al.* (2002) by assuming that a certain fraction, $\epsilon_e \sim 0.05$, of the kinetic energy released by the young stars is available to heat the large-scale ISM and IGM. The fraction is estimated by seeing what best simulates the relationship between star formation rate and density determined by Kennicutt (1998). However, it is not at all certain that parameters determined in this way can be applied to the density, pressure and abundance regime found in collapsing galaxies in the early universe. Much more theoretical work is required.

Finally, let us note that chemical evolution by nuclear synthesis is included in some codes (Sommer-Larsen, Götz & Portinari 2002; Marri & White 2002). The chemical yields as a function of mass are derived from stellar evolution codes. In using these, we should be constantly aware that these codes are rather suspect in determining both the energy input and the nucleosynthetic products of the very massive (Population III) stars which are thought to present in the first generations of stars in collapsing galaxies and their satellites (Marigo *et al.* 2002; Schaerer 2002). The production of heavy elements is particularly important in determining the cooling timescale of the ISM, and so this bears upon the feedback parameters and the star-formation rates that we have discussed above. The diffusion and mixing of the nucleosynthetic products is also an important parameter which has been very little studied up to the present.

In conclusion, even the most sophisticated particle models are not a definitive representation of the physics of galaxy formation. Certainly, they are now much more sophisticated than the "toy" models that were current a few years ago, but we need considerably more complexity in our treatment of star formation rates, feedback, the phase structure of the ISM and in chemical enrichment and stellar evolution of supermassive stars at low metallicity.

4. Problems with the Simulations

4.1. *Cusps and Cores*

Because of their negligible primordial velocity dispersion, the phase density in the cores can become very high (Navarro, Frenk & White 1996; Moore *et al.* 1999). This gives a "cusp" in the simulations which can be represented by the Navarro-Frenk-White(NFW) profile:

$$\rho(r) = \rho_0 \frac{r_s^3}{r(r + r_s)^2} \tag{6}$$

where r_s is a scale radius at which the radial power-law density function changes from a slope of –1 to –3. When compared with the accurate mass distributions determined for the cores of spiral galaxies in the local universe (*see* Freeman, this volume), this distribution implies too great a central concentration of matter in the model galaxies.

A related problem is that of the angular momentum of the dark matter. High resolution simulations of cold dark matter halos (Bullock *et al.* 2001) suggests that there exists a universal distribution of specific angular momentum (j) fitting:

$$M(\leq j) = \frac{\mu M_{vir}}{[1 + (\mu - 1) j_{max}/j]} \tag{7}$$

where M_{vir} is the virial mass, j_{max} is the maximum specific angular momentum of the halo material and μ is a factor which determines the overall shape of the profile. This distribution implies too much dark matter at low angular momentum.

Likewise, the baryonic matter in the simulations also suffers an angular momentum problem. The ΛCDM models show that the gas loses too much angular momentum by dynamical friction with the halo (Navarro & Benz 1991, Navarro & White 1999, Navarro & Steinmetz 2000). This gives disks that are systematically too small *cf* observations. This problem was dubbed the "angular momentum catastrophe" by Navarro & Benz (1991).

All of these problems can be ameliorated by slowing the collapse of the cores, by allowing more interaction and torques, and by having smaller baryonic clumps of matter in the simulations which suffer a much smaller dynamical friction. This could be a natural consequence of higher resolution simulations, or it could be a consequence of the feedback from the first generation of star formation resulting in shredding of the primordial clumps of baryonic matter.

An entertaining way out of these problems is to postulate that the dark matter is self-interacting with a cross section which depends as a power of the velocity of the interaction, $\sigma_{DM}(v) = \sigma_0 (v/v_0)^p$. And why not, since we do not know what it is? In an elegant paper, Hennawi & Ostriker (2002) suggest that the accretion of self-interacting dark matter onto seed black holes can build up the supermassive black holes seen in the cores of current-day galaxies. If the dark matter cusp is represented by $\rho(r) \propto r^{-\alpha}$ (with $\alpha \approx 1.3 \pm 0.2$), then they find that the mass of the central black hole built up is strongly dependent on the value of α. This is because, in a cusp with steeper α, more matter comes within the interaction radius and is trapped in the black hole. This model may kill two or three birds with one stone, by explaining the bulge-mass : black hole mass relationship, removing the low-angular momentum material into the black hole, and softening the central cusp by the self-interaction of the dark matter (see also Colín *et al.* (2002). Self-interacting dark matter may also help to explain the Tully-Fisher relationship (Mo & Mao, 2000), provided that $\langle \sigma_{DM}(v) \rangle / m_{\rm DM} \sim 10^{-16}$ cm^{3}s^{-1}GeV^{-1}, where $m_{\rm DM}$ is the mass of the dark matter particle.

4.2. *The Satellite Problem*

Simulations of gas collapse in ΛCDM models show that the halo should contain many dark matter sub-condensations (amounting to several hundred in the case of the Milky Way, which should show as current-day "satellites". These are not observed. What ancient objects are observed — and these in nearly these quantities — are the globular clusters. From their inferred age, and their low metallicities, these are certainly representative of a population of objects formed during the initial collapse of the galaxy. Currently the ΛCDM models have little to say about their mode of formation.

The current-day stellar masses of the globular clusters (roughly $\sim 10^5 - 10^6 \mathrm{M}_\odot$) are likely to be only a few percent, at most, of their primordial baryonic mass. This comes about for two reasons. First, a large fraction of this matter would have been ejected in gaseous form at the time of formation, as hot gas from supernova explosions and stellar winds swept out from their feeble gravitational potential wells. Second, tidal interactions with the galaxy since their formation would have reduced the mass, leaving only the tightly bound cores.

There is no evidence for dark matter in any globular cluster today. This outcome could have come to pass in the following way. First, the baryonic matter falls into the cores of the primordial "mini-halos" of dark matter, where it forms the first generation of stars in the galaxy. The baryonic matter is dissipative, and as a consequence of radiative shocks driven by photoionization, stellar winds and supernova explosions, a portion of the gas becomes more tightly bound to the central cusp, and deepens the gravitational potential there. This gas may form the low-mass stars that persist in the globular cluster up to this day while the remainder of the gas is ejected into the galactic medium. At a later stage, the more weakly bound halo of dark matter is stripped by tidal interactions, the residual stellar cores dynamically relax, and a globular cluster is born. If more massive globular cluster precursors were formed in the early galaxy, then these would have settled into the bulges by dynamical friction, where again they could be dispersed by tidal stripping. The implication of this is that the initial mass of the proto-globular cluster (the dark plus baryonic mass) may have been as high as $10^8 - 10^{10} M_\odot$ since, at best, only a few percent of the baryonic mass is transformed into globular cluster stars, and the initial baryonic mass fraction was itself only $\sim 20\%$ of the total. The current-day mass function of the globular clusters is therefore limited by tidal disruption of these "mini-halos" at the low mass end, and by their dynamical friction with the halo matter at the high-mass end.

Whilst this proposed sequence of events is merely speculative at present, it potentially provides a relationship between globular clusters, and the nucleated dwarf elliptical galaxies, which would then represent the un-stripped form of satellite galaxy, as has been suggested many years ago (Zinnecker *et al.* 1988). There clearly exists the potential to use globular clusters more effectively to unravel the physics of the formation epoch of our galaxy.

5. Looking Under The Lamppost: Observing Galaxy Formation

Radio galaxies result from jets of relativistic particles being shot out of an Active Galactic Nucleus (AGN) in the centre of a (usually) massive host galaxy. They are characterized by roughly power-law radio spectra which are the result of synchrotron emission by relativistic electrons spiralling in the local magnetic field. Typically, the power-law index, α, lies in the range $-0.4 \geq \alpha \geq -0.9$. However, sources have been identified with spectral indices much greater than this, up to $\alpha \sim -2.2$. In 1979 it was discovered that, as the spectral index becomes steeper, the rate of identification of the galaxy hosts of these radio sources on Palomar Sky Survey Plates (with a limiting red magnitude of $R \sim 20$) becomes progressively lower (Tielens *et al.* 1979; Blumenthal & Miley, 1979). This suggests that the host galaxies are very distant, a conclusion which has been abundantly confirmed by further observations (Röttgering *et al.* 1994, De Breuck *et al.* 2002). A rough idea of the distance of the host galaxy can be obtained simply by measuring the K magnitude, since the host galaxies of radio sources are not only the most massive, but also the most luminous at any epoch, and consequently they fall on an almost linear relation in the K : z plane (De Breuck *et al.* 2002). Currently, host galaxies have been identified and confirmed spectroscopically out to $z = 5.19$ (van Breugel *et al.* 1999).

Why should high z radio galaxies (Hi-zRGs) be ultra-steep spectrum radio sources? The answer lies in both cosmology and the physics of these sources. If observed out to sufficiently high frequencies, all radio galaxies exhibit a spectral break to steeper spectral indices. This is the result of a break in the energy distribution of the relativistic electrons, caused either by the maximum energy of the injected electrons, or by synchrotron ageing of the electron population which preferentially removes the most energetic electrons from the population. These synchrotron losses scale as the magnetic field pressure in the jet and its cocoon, which is controlled by the density of the surrounding medium. Thus, in a proto-galactic environment where the gas density and gas fraction is high, synchrotron losses are high, pushing the steep-spectrum break to lower frequencies.

A second reason for a lower spectral break is the inverse Compton losses experienced by the relativistic electrons. In this case, the IR photons pervading the jet are up-scattered to X-ray or even γ-ray energies by the relativistic electrons, which consequently "age" much more rapidly. This process depends on the energy density of the IR photons, which arise from two sources: the emission by warm dust in the galaxy, and the cosmic microwave background. The first of these is enhanced in these strongly star-forming galaxies, and the second is enhanced by a factor of $(1+z)^4$ over the local value — a factor of almost 1000 for a $z = 4.5$ galaxy. Ultimately, we would expect these factors to age the electrons so fast as to limit the observability of radio sources at high redshift, even assuming that the massive black holes necessary for their production have had the time to form.

Finally, whatever the frequency of the spectral break, it is shifted towards lower frequencies by a factor $(1 + z)^{-1}$ by the Cosmic expansion.

Because the ultra-steep spectrum radio sources are contained in the most massive young galaxies being built up in the early universe, they also lie in the most over-dense regions of space built up from the initial density fluctuations. Therefore, a search of their environment is likely to reveal evidence for the formation of the earliest clusters of galaxies. For this reason, and for the intrinsic interest of studying the host galaxies themselves, "looking under the lamppost" provided by the radio source is proving to be a rich and interesting field of research providing a great deal of observational insight into the physics of galaxy formation.

5.1. *Shocked Lobes & Lyman-α Halos*

The distant radio galaxies often are associated with bright extended (100 − 200 kpc) emission-line nebulae, most often detected in Ly-α. This extended gas has three possible origins, it is primordial gas cooling infalling galaxy (Steidel *et al.* 2000), it is photoionized gas left over from the earlier merging events which formed the host radio galaxy, or it is gas which is being shocked and possibly expelled by strong interactions with the radio jets of the host galaxy.

The study by Best *et al.* (2000a,b) studied powerful 3C radio galaxies with $z \sim 1$. They showed that, when the radio lobes are still able to interact with the gas in the vicinity of the galaxy, they are predominantly shock-excited, but when the lobe has burst out into intergalactic space, the ionized gas left behind is predominantly photoionized. The ratio of fluxes in the different classes of source suggests that the energy flux in the UV radiation field is about 1/3 of the energy flux in the jets. Thus, both shocks and photoionization are important in the overall evolution of radio galaxies. This result, confirmed by Inskip *et al.* (2002), proves that that the properties of the radio jet are intimately connected with the central engine.

The Hi-zRGs have been recently studied by De Breuck (2000). He finds that diagnostic diagrams involving C IV, He II and C III] fit to the pure photoionization models, but that the observed C II]/C III] requires there to be a high-velocity shock present. He argues that composite models would be required to give a self-consistent description of all the line ratios, and that these may require a mix of different physical conditions as well.

Such sources are uniquely associated with massive gas-rich multi-L_* galaxies in the early universe ($< 2 - 3$Gyr). They display a strong "alignment effect", with regions of very high star formation rate (> 1000 $M_\odot$ yr^{-1}), and emission line gas having the spectral characteristics of the NLR extended along the direction of the steep-spectrum radio lobes. In these objects, the radio jet is driving strong shocks into the galaxian ISM (evidenced by extensive Ly-α haloes; (Reuland *et al.* 2003), triggering enormous rates of star formation in the surrounding cocoon.

A fine example is provided by the $z \sim 3.8$ radio galaxy 4C 41.17 which has recently been studied in detail by Bicknell *et al.* (2000). This object consists of a powerful "double-double" radio source embedded in a 190×130 kpc Ly-α halo (Reuland *et al.* 2003) and shows strong evidence for jet-induced star formation at 3000 $M_\odot$ yr^{-1} associated with the inner radio jet. This is apparently induced by the strong dynamical interaction of the inner jet with the shocked and compressed gas in the wall of the cocoon created by the passage of the outer jet. Shock-induced star formation in jet walls was proposed in the context of Seyfert galaxies by Steffen, 1997). In 4C 41.17, the outer jet also appears to have induced a large-scale outflow with velocities in excess of 500 km s^{-1} in the line-emitting gaseous halo. Thus we may be seeing the "end of the beginning" in which the central super-massive black hole has finally become large enough to drive the whole accreting envelope of gas into outflow, triggering a last and spectacular burst of star formation in the process.

The Ly-α halo of 4C 41.17 is not alone. Reuland *et al.* (2003) describe two other examples in which gas is found aligned with the radio jets, and in which star formation rates of ~ 1000 $M_\odot$ yr^{-1} are inferred.

The observation of such violent ejection events in Hi-zRGs lends credence to the self-regulating scenario advanced by Silk & Rees (1998) and Haiman & Rees (2001) to explain the tight Black Hole/Bulge correlations discussed above. In this model, the black hole may grow in the nuclear regions until its energy input becomes sufficient to heat and expel both the circum-nuclear gas and any material still being accreted towards the galaxy, thus effectively terminating both the galaxy and the black hole growth, as appears to be happening in 4C 41.17.

5.2. *Cluster Environments*

The search for star-forming young galaxies in the vicinity of Hi-zRGs using narrow-band filters is a young field of research which is starting to reveal the richness of these environments. In this way, Keel *et al.* (1999) found 14 candidate Ly-α emitters within 3.2 Mpc of the radio galaxy 53W002 at $z = 2.39$, and Le Fèvre *et al.* (1996) confirmed two Ly-α emitters near a source at $z \sim 3.14$. However, the best study is that of Kurk *et al.* (2000) who found 50 objects with EW > 20 angstroms in the field of PKS 1138-262 at $z = 2.156$, which is itself associated with a vast Ly-α halo. Many of these objects have been subsequently confirmed spectroscopically as being associated with the environment of the Hi-zRG. This was done both in the optical, and at Hα redshifted into the IR (Pentericci *et al.* 2000; Kurk et al. 2003a,b). Star formation rates of $6 - 44$ $M_{\odot}$ yr^{-1} are inferred, which implies a star formation rate density an order of magnitude larger than in the Hubble Deep Field North. This proves unequivocally that rapid star formation is occurring in the over-dense (cluster) environment of this radio galaxy.

Finally, Reuland *et al.* (2003) have identified galaxies whose K-band images are actually seen in absorption against the extensive Ly-α halo of 4C 41.17, suggesting that the space density of collapsing and star-forming galaxies associated with radio sources remains high to very high redshifts, consistent with our belief that, in the Hi-zRGs, we are sampling the most over-dense portions of the early universe.

Acknowledgments

Mike Dopita acknowledges the support of the Australian National University and the Australian Research Council through his ARC Australian Federation Fellowship, and under the ARC Discovery project DP0208445.

References

1. Abadi M. G., Navarro, J. F., Steinmetz, M. & Eke, V. R. 2003, ApJ, 591, 499.
2. Best, P. N., Röttgering, H. J. A. & Longair, M. S. 2000a, MNRAS, 311, 1.
3. Best, P. N., Röttgering, H. J. A. & Longair, M. S. 2000a, MNRAS, 311, 23.
4. Bicknell G. V. *et al.* 2000, ApJ, 540, 678.
5. Blumenthal, G. & Miley, G. 1979, A&A, 80, 13.
6. Boyle, B. J. & Terlevich, R. J. 1998, MNRAS, 293, L49.
7. Bullock, J. S. *et al.* 2001, ApJ, 555, 240.
8. Colín, P., Alvila-Reese, V. & Valenzuela, O. 2000, ApJ, 542, 622.
9. De Breuck, C. W. *et al.* 2000, A&AS, 143, 303.
10. De Breuck, C. W. *et al.* 2002, AJ, 123, 637.
11. Dopita, M. A. & Sutherland, R. S. 1995, ApJ, 455, 468.
12. Dopita, M. A. & Sutherland, R. S. 1996, ApJS, 102, 161.
13. Elmegreen, B. G. 1997, Rev. Mex. Ast y Ap. Conf. Ser. 6, 165.
14. Ferrarese, L. & Merritt, D. 2000, ApJ, 539, L9.
15. Gebhardt, K. *et al.* 2000, ApJ, 539, 13.
16. Giovanelli, R. *et al.* 1997, AJ, 113, 22.
17. Haiman, Z. & Rees, M. J. 2001, ApJ, 556, 87.
18. Hennawi, J. E. & Ostriker, J. 2002, ApJ, 572, 41.
19. Inskip, K. J. *et al.* 2002, MNRAS, 337, 1381.
20. Inskip, K. J. *et al.* 2002, MNRAS, 337, 1407.
21. Keel, W. C. *et al.* 1999, AJ, 118, 2547.
22. Kennicutt, R. C. Jr. 1998, ApJ, 498, 541.
23. Kurk, J. D., *et al.* 2000, A&A, 358, L1.
24. Kurk, J. D. *et al.* 2003a, A&A (in press).
25. Kurk, J. D. *et al.* 2003b, A&A (in press).

26. Larson,R. B. 1988, in *Galactic & Extragalactic Star Formation*, eds R.E Pudrich & M. Fitch, Kluwer: Dordrecht, NATO ASI v232, p5.
27. Le Fèvre, O. *et al.* 1996, ApJ, 471, L11.
28. Marigo, P., Chiosi, C., Giradi, L & Kudritski, R.-P. 2003, in Proc. IAU Symp 212 *A Massive Star Odyssey*, eds K.A. Van Der Hucht, A. Herrero & C. Esteban (ASP), p334.
29. Marri, S. & White, S. D. M. 2003, MNRAS, 345, 561.
30. Mo, H. J. & Mao, S. 2000, MNRAS, 318, 163.
31. Moore, B. *et al.* 1999, ApJ, 524, L19.
32. Navarro, J. F. & Benz, W. 1991, ApJ, 380, 320.
33. Navarro, J. F., Frenk, C. S. & White, S. D. M. 1996, ApJ, 462, 563.
34. Navarro, J. F. & Steinmetz, M. 1997, ApJ, 478, 13.
35. Navarro, J. F. & Steinmetz, M. 2000, ApJ, 538, 477.
36. Navarro, J. F. & White, S. D. M. 1999, MNRAS, 267, 401.
37. Pentericci, L. *et al.* 2000, A&A, 361, L25.
38. Reuland, M. *et al.* 2003, ApJ, in press.
39. Röttgering, H. *et al.* 1994, A&AS, 108, 79.
40. Schaerer, D. 2003, A&A, 397, 527.
41. Schmidt, M. 1959, ApJ, 129, 243.
42. Steffen, W. *et al.* 1997, MNRAS, 286, 1032
43. Steinmetz, M. & Navarro 2002, NewA, 7, 155.
44. Silk, J. 1997, ApJ, 481, 703.
45. Silk J. & Rees, M. J. 1998, A&A, 331, L1.
46. Sommer-Larsen, J. Götz, M. & Portinari, L. 2003, ApJ, 596, 47.
47. Sutherland, R. S. & Dopita, M. A. 1993, ApJS, 88, 253.
48. Tielens, A., Miley, G., & Willis, A. 1979, A&AS, 35, 153.
49. Tully, R. B. & Fisher, J. R. 1977, A&A, 54, 661.
50. van Breugel, W. *et al.* 1999, ApJ, 518, 61.
51. Wyse, R. F. G. 1986, ApJ, 311, L41.
52. Zinnecker, H. *et al.* 1988 in *Globular Cluster Systems in Galaxies*, eds J.A. Grindlay & A.G. Davis Philip (Kluwer: Dordrecht), p603.

DARK MATTER IN GALAXIES

K.C. FREEMAN
Research School of Astronomy & Astrophysics, Mount Stromlo Observatory, The Australian National University, Canberra

These lectures present a brief overview of what we know about dark matter in galaxies. I will stress some of the current problems.

1. Introduction

We believe that galaxies formed through a hierarchy of merging. The merging elements were a mixture of baryonic and dark matter. The dark matter settled into a partially virialized spheroidal halo, while the baryons (in disk galaxies) settled into a rotating disk and bulge.

What can we learn about the properties of dark halos ? Do the properties of dark halos predicted by simulations correspond to what is inferred from observational studies ?

These lectures will primarily be about dark matter in disk galaxies. Disk galaxies are flat systems, supported against gravity by their rotation, and they are the simplest galaxies for studying the properties of the dark halos.

2. Rotation of Spirals

Most spirals do not rotate like rigid bodies. They show a wide range of rotation curve morphology, depending on the radial distribution of stars. The extremes range from almost solid body rotation, as seen for some lower luminosity disks, to rotation curves in which the rotational velocity is almost constant with radius throughout the galaxy which is more typical of the brighter disks like the Milky Way. See Figure 1 for some extreme examples.

What keeps the disk in equilibrium (this is an important question to ask for any stellar system)? Most of the kinetic energy is in the rotation. In the radial direction, gravity provides the radial acceleration needed for the approximately circular motion of gas and stars in the disk. In the vertical direction, gravity is balanced by the vertical pressure gradient associated with the random vertical motions of the disk stars.

For the gas in a disk galaxy, the radial potential gradient provides the accelera-

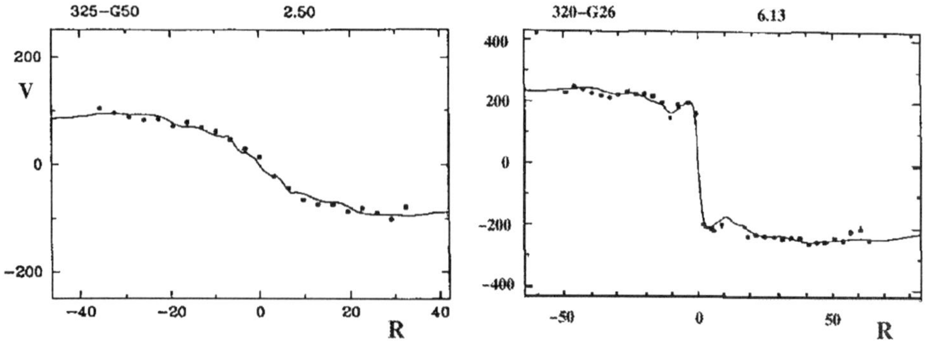

Figure 1. Optical rotation curves for two spiral galaxies from Buchhorn (1991), showing the wide variety of rotation curve morphology seen among spiral galaxies.The units of V and R are km s^{-1} and arcsec respectively. The points show the rotation data. See the text for explanation of the curves.

tion for the circular motion.

$$\frac{V^2}{R} = -\frac{\partial \Phi}{\partial R} \simeq \frac{GM(R)}{R^2}$$

where $V(R)$ and $\Phi(R)$ are the rotational velocity and potential at radius R in the plane of the disk, and $M(R)$ is the enclosed mass within radius R. As shown in Figure 1, the shape of $V(R)$ can be anything from solid body to $V \simeq$ constant (flat). For the larger spirals like our Galaxy, $V(R)$ is usually close to flat, so the enclosed mass increases linearly with R, at least out to the maximum extent of the rotation curve.

$M(R) \propto R$ is not what we would expect for a gravitating system of stars. We would expect $M(R)$ to tend to some asymptotic mass M for large R. Is $M(R) \propto R$ evidence for a dark halo ? Not necessarily. It depends on how far the observed rotation curve extends. Most spirals have a light distribution that is roughly exponential: $I(R) \propto \exp(-R/h)$ where the scale length h is about 4 kpc for a large galaxy like the Milky Way. Rotation curves measured optically from the spectra of ionized gas typically extend to about $r = 3h$.

Now assume that the surface *density* distribution of stars in our disk galaxy is proportional to the optical surface *brightness* distribution. Can this surface density distribution, with its associated gravitational potential $\Phi(R)$, explain the observed rotation curve $V(R)$? The answer to this question is yes and no.

The answer is yes for optical rotation curves extending out to about 3 radial scale lengths. In Figure 1, the points are the observed rotational velocities and the curve is the expected curve derived from the surface density distribution, assuming that mass follows light. Despite the very different shapes of the rotation curves, the light distribution can explain the observed optical rotation curves out to about 3 scale lengths. The only scaling is in the velocity coordinate, through the adopted mass to light ratio M/L.

The answer is no for galaxies with 21 cm neutral hydrogen (HI) rotation curves that extend out to $R \gg 3h$. Figure 2 shows a decomposition of the rotation curve of the spiral NGC 3198, adopting the maximum value for the M/L ratio for the stellar disk that is consistent with the observed rotation curve (*i.e.* the adopted M/L ratio cannot be so high that the calculated rotation curve is higher anywhere than the observed rotation curve. In this galaxy the HI rotation curve extends to about $11h$. With the maximum possible M/L ratio for the stars, the expected $V(R)$ from the stars and gas falls well below the observed rotation curve in the outer region of the galaxy. This kind of shortfall is seen for almost all spirals with rotation curves that extend out to many scale lengths. We conclude that the luminous matter dominates the radial potential gradient $\partial\Phi/\partial R$ for $R \lesssim 3h$ but beyond this radius the dark halo becomes progressively more important.

Typically, out to the radius where the HI data ends, the ratio of dark to luminous mass is 3 to 5. Values of 10 to 20 are found in a few examples.

For the decomposition of NGC 3198 described above, the stellar M/L ratio was taken to be as large as possible without leading to a hollow dark halo. This kind of decomposition is known as a maximum disk (or minimum halo) decomposition. Many galaxies have been analysed in this way. The decomposition usually works out as for NGC 3198, with comparable peak circular velocity contributions from disk and dark halo. This is believed to be at least partly due to the adiabatic compression of the dark halo by the baryons as they dissipate and condense to form the disk.

3. The Maximum Disk Question

The inferred stellar M/L ratios from maximum disk decompositions are usually consistent with those expected from synthetic stellar populations, at least for the brighter spirals like the Milky Way. Nevertheless, some people still do not believe that the maximum disk approach is correct. They argue that the dark halo is probably more significant gravitationally than the maximum disk / minimum halo hypothesis would indicate; this is equivalent to adopting a smaller stellar M/L ratio for the disk. One reason for this belief comes from the Milky Way itself: the apparent surface density of the galactic disk and halo near the sun is only about 50 $M_\odot$ pc^{-2}, which may be too low to be consistent with a maximum disk (Kuijken & Gilmore 1989).

The maximum disk question is important for us here, because inferences about the properties of dark halos from rotation curves depend so much on the correctness of the maximum disk interpretation. For example, if the maximum disk decompositions are correct, the contribution to $V(R)$ from the halo is approximately solid-body in the inner parts of the galaxy, so the dark halos have approximately uniform density cores which are much larger than the scale length of the disk. In

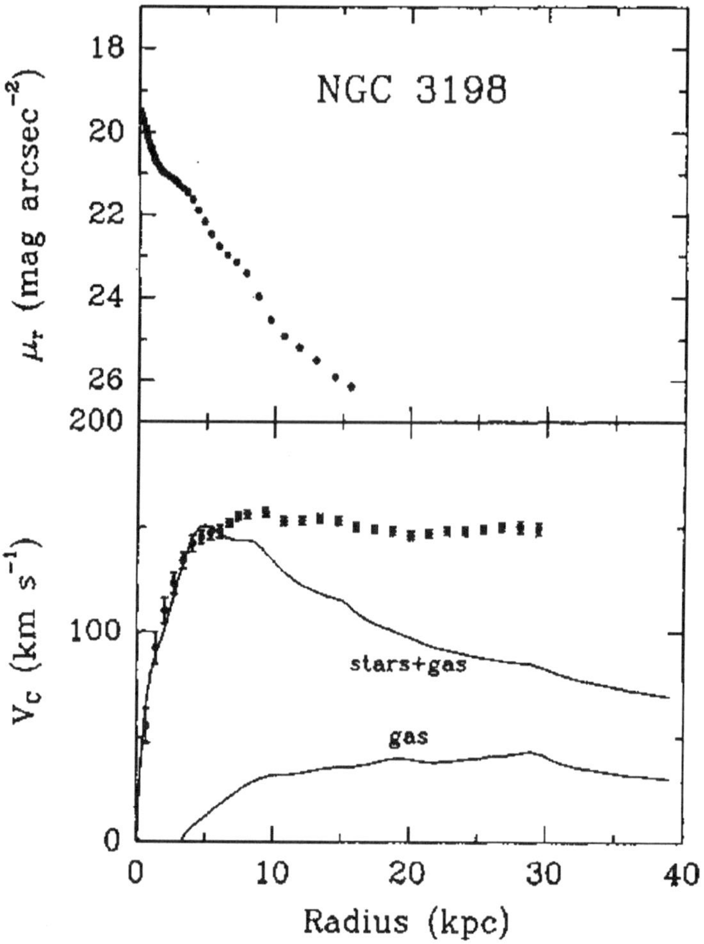

Figure 2. The *upper panel* shows the surface brightness distribution of the spiral galaxies NGC 3198, from Begeman(1989). The *lower panel* shows the large discrepancy between the HI rotation curve (points) and the expected contribution to the rotation curve from the stars plus gas, adopting the maximum disk hypothesis as explained in the text (§3).

contrast, the halos that form in cosmological simulations have *steeply cusped* inner halos with density distributions $\rho \sim r^{-1}$ or even steeper near the center.

Optical rotation curves favor the maximum disk interpretation. In the inner regions of the disks of larger spirals, the rotation curves are well fit by assuming that mass follows light. For example, Buchhorn (1991) analysed about 500 galaxies with I-band surface brightness distributions and a wide range of optical rotation curve morphologies spanning the extremes shown in Figure 1. He was able to match the observed and expected rotation curves well for about 97% of his sample, with realistic M/L ratios. The implication is that either the stellar disk dominates the gravitational field in the inner parts of the disk, or the potential gradient of the halo faithfully mimics the potential gradient of the disk in almost every spiral.

3.1. *Other support for the maximum disk interpretation*

Athanassoula *et al.* (1987) used the dynamical theory of spiral structure to give a dynamical constraint on the stellar M/L ratio for the disk. From the number of spiral arms observed in each of their galaxies, they argue that most of the disks are indeed close to maximum.

Bell & de Jong (2001) and Perez (2003) compared the M/L ratio from synthetic stellar populations with those derived dynamically from maximum disk rotation curves. They find good agreement when they use a stellar mass function like that for the solar neighborhood.

Debattista & Sellwood (1998) showed that a dense halo (as in a submaximal disk decomposition) would rapidly slow down the rotation rate of the bars in barred spiral galaxies. In a low density halo (as in a maximum disk system), the bar rotation stays high. See Athanassoula (2002) for a more detailed study of the interaction of bars and dark halos. Evidence from gas flows in barred galaxies (*e.g.* Weiner *et al.* 2001; Perez & Fux 2004) indicates that bars do rotate rapidly, with corotation just beyond the end of the bar.

I conclude that the maximum disk picture is probably correct, at least for galaxies of normal surface brightness. (We will discuss low surface brightness galaxies later).

4. Modelling the Dark Halo

Our goal is to estimate the typical parameters for dark halos (*e.g.* their density, scale length, velocity dispersion, shape) to compare with the properties of halos from cosmological simulations. Since about 1985, observers have used model dark halos with constant density cores to interpret rotation curves. Commonly used models include the nonsingular isothermal sphere, which has a well defined core radius and central density; its density falls off as $\rho \sim r^{-2}$ at large r so $V(r) \sim$ constant as often observed.

A simple analytical form is the pseudo-isothermal sphere

$$\rho = \frac{\rho_\circ}{1 + (r/r_c)^2}$$

which again has a well defined core radius and central density and $\rho \sim r^{-2}$ at large r. Using this model for the dark halos of large galaxies like the Milky Way, we find that $\rho_\circ \sim 0.01\ M_\odot \mathrm{pc}^{-3}$ and $r_c \sim 10$ kpc. For comparison, the density of the galactic disk near the sun is about $0.1\ M_\odot \mathrm{pc}^{-3}$. We will see later that the values of $\rho_\circ$ and r_c depend strongly on the luminosity of the galaxy.

Why were these models with central cores used ? I think it was because (1) rotation curves of spirals do appear to have an inner solid-body component which indicates a core of roughly constant density, and (2) hot stellar systems like globular clusters had been successfully modelled by King models, which are modified non-singular isothermal spheres (with cores). On the other hand, CDM simulations consistently produce halos that are cusped at the center. This has been known since the 1980s and has been popularized by Navarro *et al.* (1996) with their NFW density distribution which parameterizes the CDM halos:

$$\rho(r) = \rho_\circ \frac{r_s}{r} \frac{1}{[1 + (r/r_s)]^2}$$

These are cusped at the center, with $\rho(r) \sim r^{-1}$.

The last several years have seen a long controversy on whether the observed rotation curves imply cusped or cored dark halos. This continues to be illuminating. Galaxies of low surface brightness (LSB) are important in this debate. The disks of normal (or high surface brightness) spirals have a fairly well defined characteristic central surface brightness of about 21.5 B mag arcsec^{-2} (*e.g.* Freeman 1970). In the LSB galaxies, the disk surface brightness can be more than 10 times lower than in the normal spirals. These LSB disks are fairly clearly sub-maximal, and the rotation curve is believed to be dominated everywhere by the dark halo. So the rotation curves of these LSB galaxies potentially give a fairly direct estimate of the structure of the inner parts of the dark halo. The observational problem is to determine the shape of the rotation curve near the center of the galaxies. Near the center, a cored halo gives a solid body rotation curve, while the rotation curve for a cusped halo rises very steeply.

Observationally, it is not easy to tell. HI rotation curves have limited spatial resolution, so the beam smearing can mask the effects of a possible cusp. Optical rotation curves, including the 2D optical rotation data with Fabry-Perot interferometers, have much better spatial resolution and favor a cored halo with a power law slope near zero (de Blok *et al.* 2001). The recent HI study of the very nearby LSB galaxy NGC 6822, with 20 pc linear resolution (Weldrake *et al.* 2003), also clearly favor a cored halo.

What is wrong: observations or theory ? Does it matter ? Yes: the density distribution of the dark halos provides a critical test of the nature of dark matter and of galaxy formation theory. For example, the proven presence of cusps can exclude some dark matter particles (*e.g.* Gondolo 2000). The halo density profiles can also provide some constraints on the fluctuation spectrum (*e.g.* Ma & Fry 2000).

Maybe CDM is wrong. For example, self-interacting dark matter can give a flat central $\rho(r)$ via heat transfer into the colder central regions. But further evolution can then lead to core collapse (as in globular clusters) and even steeper r^{-2} cusps (*e.g.* Burkert 2000; Dalcanton & Hogan 2001).

Alternatively, there are ways to convert CDM cusps into flat central cores, so that we do not see the cusps now. For example, bars are very common in disk

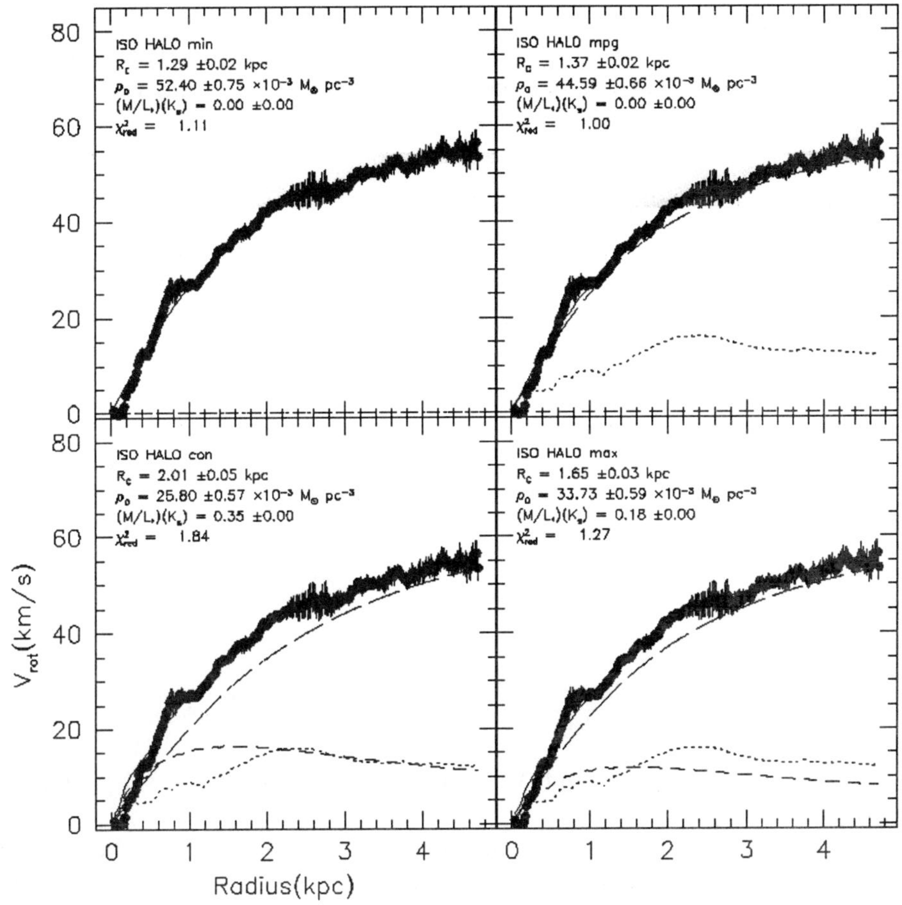

Figure 3. The rotation curve of the nearby LSB galaxy NGC 6822. The panels show fits of models with isothermal halos and different adopted stellar M/L ratios. Excellent fits are achieved with low M/L ratios, favoring the presence of a cored halo (Weldrake *et al.* 2003).

galaxies: about 70% of disk galaxies show some kind of central bar structure. Many galaxies that do not appear to be barred from their optical images show clear central bars in near-infrared images which are dominated by older stars and are less affected by dust absorption. The bars are believed to come from gravitational instability of the disk. Weinberg & Katz (2002) showed that the angular momentum transfer and dynamical heating of the inner halo by the bar can remove a central cusp in about 1.5 Gyr.

This issue is far from settled. I think that the current belief is that the cusp structure may be flattened by the effect of blowout of baryons in early bursts of star formation as the halo is built up (*e.g.* Dekel *et al.* 2003). This idea has a couple of additional major attractions. Before discussing these, a short digression is needed on two important dynamical processes involved in hierarchical galaxy formation: dynamical friction and tidal disruption. The discussion follows Binney & Tremaine (1987).

4.1. *Dynamical friction*

Dynamical friction is the frictional effect on a mass M moving through a sea of stars of mass m. Assume that the smaller masses m are uniformly distributed, and adopt the "Jeans Swindle" (*i.e.*ignore the potential of the uniform distribution of the m objects.) Then the motion is determined only by the force of M and the *disturbances* that M produces on the distribution of the m objects.

M raises a response in the sea of smaller objects, and this response acts back on M itself. Summing the effects of the individual encounters of M and m, we see that M suffers a steady *deceleration* parallel to its velocity $\mathbf{v}$. If the velocity distribution of m is Maxwellian

$$f = \frac{n_\circ}{(2\pi\sigma^2)^{3/2}} \exp(-v^2/2\sigma^2)$$

then the drag is

$$\frac{d\mathbf{V}_M}{dt} = -\frac{4\pi \ln\Lambda G^2 \rho_m M}{V_M{}^3} \left(\text{erf}\chi - \frac{2\chi}{\sqrt{\pi}} \exp(-\chi^2) \right) \mathbf{V}_M$$

for $M \gg m$. $\chi = V_M/\sqrt{2}\sigma$ and Λ = (maximum impact parameter) $\times$ (typical speed)$^2/GM$: $\Lambda \gg 1$.

So (i) the drag acceleration is $\propto \rho_m$ and $\propto M$ and (ii) the drag force $\propto M^2$. This comes about because stars deflected by M generate a downstream density enhancement: the enhancement $\propto M$, and the force back on $M \propto M^2$.

This estimate neglects the self-gravity of the density enhancement; i.e. it includes the attraction of m on M, but not m on m. The estimate seems to be fairly consistent with the results of N-body simulations, as long as the ratio of M to the total mass of the m objects $\lesssim$0.2 and the orbit of M is not confined to the core or to the exterior of the larger system. The estimate also neglects resonances between the orbit of M and the orbits of m objects within their system: such resonances enhance dynamical friction.

For example, consider the likely fate of the LMC, now located at about 60 kpc from the Galaxy. For circular orbits, the torque from dynamical friction due to the dark halo of our Galaxy gives a decay time

$$t_{fric} = \frac{10^{10}}{3} \left(\frac{r}{60 \text{ kpc}}\right) \left(\frac{V_c}{220 \text{ km s}^{-1}}\right) \left(\frac{2 \times 10^{10}}{M_\odot}\right) \text{ yr}$$

so if the galactic halo extends out beyond a radius of 60 kpc and the LMC orbit is approximately circular (both of which are true), then the LMC (and SMC) will sink into the Galaxy in a time less than the Hubble time.

4.2. *Tidal disruption*

Consider a satellite of mass m in a circular orbit around a host of mass M at a distance D. The angular speed around the common center of mass is $\Omega^2 = G(m+M)/D^3$. In this rotating frame, we have the Jacobi integral $E_J = E - \mathbf{\Omega}.\mathbf{L} = \frac{1}{2}v^2 + \Phi_{eff}(\mathbf{r})$, where Φ_{eff} is the effective potential of the gravity plus the centrifugal force. The contours of Φ_{eff} have a saddle point between the two masses, where $\partial\Phi_{eff}/\partial x = 0$ (see Binney & Tremaine, Figure 7.8). Beyond this saddle point the contours open out. For $m \ll M$, the distance of the saddle point from the smaller mass is

$$r_J = \left(\frac{M}{3M}\right)^{1/3} D$$

This is a measure of the *tidal radius* of m. It is a rough estimate because

(1) the zero velocity (ZV) surface is not spherical
(2) orbits do not necessary escape because the ZV surface is open
(3) the orbit of m is not usually circular
(4) m often lies within M, so the point mass approximation is poor

but the main point here is that tidal removal of matter can occur at a radius from m such that $\rho_m(r_J) \sim \rho_M(D)$. For example, we expect an infalling satellite to remain intact in to a distance D from the larger galaxy, such that $\rho_M(D) \sim$ the mean density of the satellite.

To summarise the merger preliminaries:

- *Dynamical friction* (M in a sea of m). The drag force $\propto \rho_m M^2$, neglecting resonances and the selfgravity of the wake.
- Tidal disruption. Occurs when the mean density of the host within the satellite orbit $\sim$ the mean density of the satellite. Very dense satellites can survive accretion while low density satellites are broken up.

5. Galaxy Formation Problems

In simulations of galaxy formation (*e.g.* Moore *et al.* 1999), the virialized halos are quite lumpy, with much substructure, corresponding to many more satellites and dwarf galaxies than observed in the environment of the Milky Way. The simulations suggest that a galaxy like the Milky Way should have about 500 satellites with bound masses $> 10^6\ M_\odot$. These are not seen optically and probably not in HI. What is wrong ? Maybe there are a large number of baryon-depleted dark satellites, or there is some problem with the details of CDM (*e.g.* that the short wavelength end of the fluctuation spectrum needs modification).

The baryons also clump and, as they settle to the disk, the clumps suffer dynamical friction against the halo and so lose angular momentum. The resulting disks then have smaller angular momentum than those observed: they are therefore smaller in radius and spinning more rapidly than real galaxies. This remains one of the more serious problems in the current theory of galaxy formation (*e.g.* Abadi *et al.* 2003). We need to find ways to suppress the loss of angular momentum of the baryons to the dark halo.

One way to avoid this loss of angular momentum is by blowout of baryons early in the galaxy formation process. For example, Sommer-Larsen *et al.* (2003) made N-body + SPH simulations with a star formation prescription. Star formation begins early in the galaxy formation process. Small elements of the hierarchy (dwarf galaxies) form stars long before the whole system has virialized. The stellar winds and SN from the forming stars temporarily eject most of the baryons from the forming galaxy. The halo virializes and then the baryons settle smoothly to the disk. Because they settle smoothly, the loss of angular momentum via dynamical friction is much reduced.

The blowout process (§4) can also contribute to reducing the problem of too much substructure and to the cusp problem in another way (*e.g.* Dekel *et al.* 2003). Because the smaller elements of the hierarchy grow first, they are denser (we will see observational evidence for this later). This means that they are less likely to be tidally disrupted as they settle to the inner parts of the halo via dynamical friction, so they can contribute to the high density cusp in the center of the virialized halo. Blowout of the baryon component of these dense small elements can contribute to unbinding them. Their chances of survival against the tidal field of the virializing halo are then reduced, so (1) the substructure problem (*i.e.* too many small elements) is reduced, and (2) the cusp problem is reduced.

6. How Large are Dark Halos

Flat rotation curves imply that $M(r) \propto r$, like the isothermal sphere with $\rho(r) \sim r^{-2}$ at large r. This cannot go on forever: the halo mass would be infinite. Halos must have a finite extent, and their density distribution is probably steeper than $\rho(r) \sim r^{-2}$ at very large r. For example, the NFW halo with

$$\rho(r) = \rho_\circ \frac{r_s}{r} \frac{1}{[1 + r/r_s]^2}$$

has $\rho(r) \sim r^{-3}$ at large r.

Tracers of dark matter in the Milky Way (the rotation curve observed out to a radius of about 20 kpc, kinematics of stars and globular clusters in the stellar halo, and kinematics of satellites out to $R > 50$ kpc) all indicate that the enclosed mass rises linearly as in other galaxies, and is well approximated by $M(r) = r(\text{kpc}) \times 10^{10}$ $M_\odot$. This is what we would expect if the galactic rotation curve stays flat out to $r > 50$ kpc. This still does not tell us how far the dark halo extends. Other arguments are needed.

6.1. *Timing arguments*

M31 is now approaching the Galaxy at about 118 km s^{-1}. Its distance is about 750 kpc. Assuming that their initial separation was small, we can estimate a lower limit on the total mass of the Andromeda + Galaxy system such that they are now approaching at the observed velocity. The Galaxy's share of this mass is $(13 \pm 2) \times 10^{11}\ M_{\odot}$. A similar argument for the Leo I dwarf at a distance of about 230 kpc gives $(12 \pm 2) \times 10^{11}\ M_{\odot}$. Our relation for $M(r)$ for the galactic halo, derived for $r \sim 50$ kpc, then indicates that the dark halo extends out beyond a radius of 120 kpc, if the rotation curve remains flat, and possibly much more if the density distribution declines more rapidly at large radius. This radius is much larger than the extent of any directly measured rotation curves, so this "timing argument" gives a realistic lower limit to the total mass and radial extent of the galactic dark halo (Zaritsky 1999). This argument was originally due to Kahn & Woltjer (1959).

For our Galaxy, the luminous mass (disk + bulge) is about $6 \times 10^{10}\ M_{\odot}$. The luminosity is about $2 \times 10^{10}\ L_{\odot}$. The ratio of total dark mass to stellar mass is then at least $120/6 = 20$ and the total mass to light ratio is at least 60 in solar units.

Satellites of disk galaxies can also be used to estimate the total mass and extent of the dark halos. Individual galaxies have only a few observable satellites each, but we can make a super-galaxy by combining observations of many satellite systems and so get a measure of the mass of a typical dark halo. For example, Prada *et al.* (2003) studied the kinematics of about 3000 satellites around about 1000 galaxies. With a careful treatment of interlopers, they find that the velocity dispersion of the super satellite system decreases slowly with radius. The halos typically extend out to about 300 kpc but their derived density distribution at large radius is steeper than the isothermal: $\rho(r) \sim r^{-3}$, like most cosmological models including the NFW halos. The total mass to light ratios are typically $100 - 150$, compared with the lower limit from the timing argument of 60 for our Galaxy. (Note that the Prada galaxies are bright systems, comparable to the Galaxy).

7. The Shapes of Dark Halos

What do we expect from simulations ? Dark halos from simulations are typically triaxial, with mean axial ratios 1 : 0.85 : 0.65 (*e.g.* Steinmetz & Muller 1995). What do we see ? The shapes of halos are difficult to measure, because the shape of the equipotentials (which affect the observed kinematics) is more spherical than the shape of the density distribution itself. Many different attempts have been made to measure the shapes of the dark halos. I will briefly review some of them.

7.1. *Flaring of the HI layer in the Galaxy*

The HI layer has an approximately isothermal velocity dispersion of about 8 km s^{-1}. In a spherical dark halo the outer HI layer will then flare vertically more than if the dark halo is flattened. For our Galaxy, Olling & Merrifield (2000) use this flaring to estimate that the axial ratio of the dark halo is about 0.8.

7.2. *Polar ring galaxies*

Polar ring galaxies like NGC 4650A have matter rotating in two approximately orthogonal planes, so we can measure the potential gradient in these two planes. For example, in NGC 4650A, optical kinematics indicate that the dark halo has an axial ratio of about 0.3 to 0.4 (Sackett *et al.* 1994). However an HI study of this system shows that the halo could be flattened to either of the two orbital planes (Arnaboldi & Combes 1996). We should also be aware that polar ring galaxies are unusual systems; it is possible that the survival of a well-developed polar ring may *require* a flattened and triaxial halo.

7.3. *IC 2006*

The elliptical galaxy IC 2006 is surrounded by a ring of HI at a radius of about 6.5 effective (*i.e.* half light) radii. The mass to blue light ratio at this radius is about 16, compared with the M/L ratio of about 5 in the inner regions. This is a good indication that IC 2006 has a dark halo like most galaxies. The kinematics of the HI ring show that the ring is almost perfectly circular (within 2%; Franx *et al.* 1994), which suggests that the halo of this elliptical galaxy is very close to axisymmetric (*i.e.*two equal axes in the plane of the ring).

7.4. *Carbon stars in the galactic halo*

Ibata *et al.* (2001) studied the kinematics of carbon stars in the galactic halo. At least half of them appear to be associated with the debris of the disrupting Sgr dwarf which extends in an almost polar great circle from a galactocentric radius of about 16 kpc to 60 kpc. The fact that the debris lies on a great circle suggests that the galactic halo does not exert a significant torque on the stream of debris. The distribution of carbon stars favors a nearly spherical galactic halo in the region $16 < R < 60$ kpc. Simulations of the precessing Sgr debris in potentials of different flattening show that an axial ratio as flat as 0.75 is very unlikely.

In summary, the evidence so far indicates that dark halos are fairly close to spherical.

8. Rotation of Dark Halos

Halos are believed to acquire angular momentum through tidal interactions with other halos as they form. The dimensionless parameter $\lambda = J|E|^{\frac{1}{2}}M^{-\frac{5}{2}}G^{-1}$ where J is the angular momentum of a system and E and M are its binding energy and mass, is a measure of the ratio of (rotational velocity)/(virial velocity). For example, for a disk in centrifugal equilibrium, $\lambda \simeq 0.45$. Cosmological simulations give well-defined and similar distributions of λ, with a mean $\lambda \simeq 0.05$. So the simulated halos are relatively slowly rotating (*e.g.* Bullock *et al.* 2001).

If baryons and dark matter are initially well mixed and have similar specific angular momentum J/M, and if the baryons conserve their angular momentum as they collapse to a disk in centrifugal equilibrium, then the radial collapse factor for the disk is $R_{\text{halo}}/h_{\text{disk}} = \sqrt{2}/\lambda \simeq 30$ (Fall 2002) where R_{halo} is the radius of the halo and h_{disk} is the exponential scalelength of the equilibrium disk. For example, for our Galaxy, the optical scale length of the disk is about 4 kpc, and the halo extends out to at least 120 kpc, consistent with the factor 30.

Galaxies with higher λ-values are initially closer to centrifugal equilibrium, so would typically form disks of lower surface brightness. This is supported by the observation that the distribution of surface brightness has a similar shape to the distribution of λ from the simulations (*e.g.* Bullock *et al.* 2001).

So far we have discussed the angular momentum of dark halos in general terms. The shape or figure of a rotating body may be axisymmetric or triaxial. If it is triaxial *and* the triaxial figure itself is rotating, then the torque of the rotating figure may be important for galactic dynamics. For example, Bekki & Freeman (2002) argued that the figure rotation of a triaxial dark halo could be important for stirring up spiral structure in the outer regions of galaxies where self-gravity appears to be too low to sustain spiral structure. NGC 2915 is an example of a galaxy with HI spiral structure extending far beyond the optical galaxy (Meurer *et al.* 1996). For some other spectacular examples, see www.nfra.nl/~oosterlo.

9. Dwarf Spheroidal Galaxies

These are faint satellites of our Galaxy (seen also around M31). Their absolute magnitudes are as low as $M_V = -8$. They have very low surface brightnesses and masses that are typically about $10^7\ M_\odot$. Radial velocities of individual stars in several of these dSph galaxies show that their M/L ratios can be very high. Some of the faintest dSph galaxies have $M/L \sim 100$. Figure 4 shows M/L values for the Local Group dSph galaxies. Figure 5 shows the radial variation of the velocity dispersion in the Fornax galaxy, which is the largest of the Galactic dSph galaxies; the velocity dispersion is approximately constant with radius, and the inferred M/L ratio is about 10, significantly higher than the value of about 2 expected for an old metal-poor population.

10. The Tully-Fisher Law

Simple centrifugal equilibrium arguments for a self-gravitating disk give a relation between the luminosity and rotational velocity known as the Tully-Fisher law:

$$L \propto V^4/[I_\circ(M/L)^2]$$

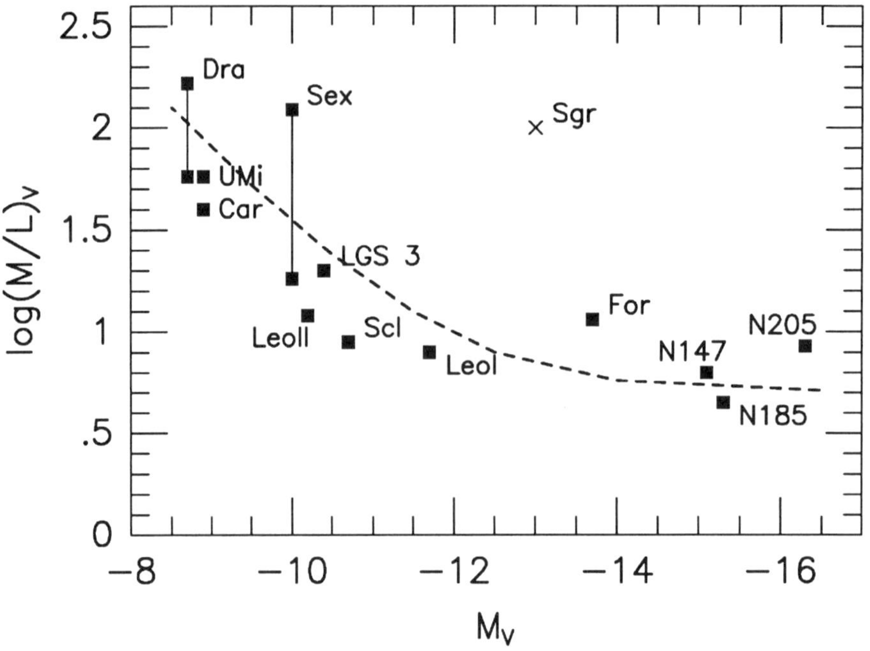

Figure 4. The correlation between M/L_V and M_V for Local Group dSph galaxies with good kinematic data. The dashed line shows a model in which each galaxy has a dark halo mass of $2.5 \times 10^7 M_\odot$ plus a luminous component with $M/L_V = 5$. (From Mateo 1997).

where L is the luminosity of the galaxy, V is its rotational velocity and the central surface brightness $I_\circ$ and M/L are roughly constant from galaxy to galaxy for spirals of normal surface brightness.

Observationally, the exponent of V in the Tully-Fisher law depends on the measured wavelength of the luminosity: it varies from about 3.2 at B to about 4.5 at H. This probably reflects a weak dependence of $I_\circ$ and M/L on L, analogous to the tilt of the fundamental plane for elliptical galaxies. Figure 6 shows how the observed slope varies, and also how the scatter in the Tully-Fisher law becomes smaller as the wavelength increases, due to the reduced effect of dust and star forming regions on the luminosity.

The zero point of the Tully-Fisher law needs explaining. For example, in the I-band, the Tully-Fisher law is

$$M_I = -10.00(\log W_{50} - 2.5) - 21.32$$

(Sakai *et al.* 2000). Here W_{50} is the HI profile width at half peak height corrected for inclination, which is a measure of the rotational velocity. This equation states

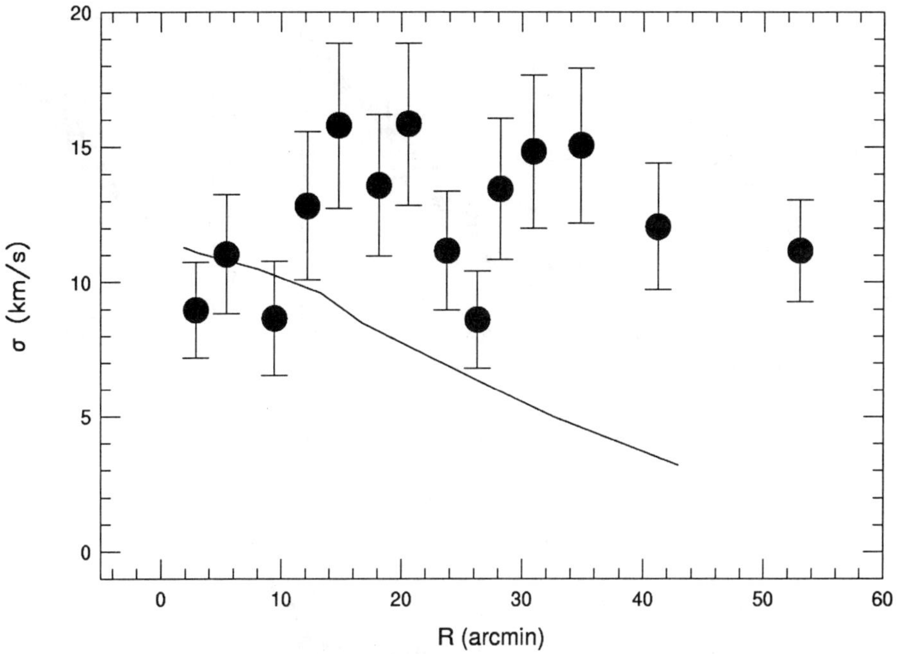

Figure 5. The radial variation of velocity dispersion in the Fornax dSph galaxy, from Mateo (1997). The curve shows the velocity dispersion expected if the mass were distributed like the light.

that a galaxy with $M_I = 21.32$ has a velocity width of 316 km s^{-1}, not 500 km s^{-1}. For a self-gravitating disk alone, *e.g.* an exponential disk, the zero point depends on the product $I_\circ(M/L)^2$. M/L is determined by the stellar population. The central surface density $\Sigma_\circ = I_\circ(M/L)$ depends on the mass M and angular momentum J for the disk: simple arguments show that $\Sigma_\circ = M^7/J^4$. The $J(M)$ relation is defined by the dynamics of galaxy formation and evolution. It determines the zero point of the Tully-Fisher law. This is a current problem in understanding galaxy formation (see §5): simulations show that too much angular momentum is lost from the baryons to the dark halo during the galaxy formation process. Because of the conspiracy for disks of normal surface brightness (*i.e.* the approximate equality of the rotation curve contributions from disk and halo, as seen in Figure 2), this argument is not much changed by the presence of the dark halo.

Now consider low surface brightness (LSB) disks. Here the gravitational field is believed to be dominated by the dark halo everywhere. Yet the Tully-Fisher law for LSB galaxies is almost identical in slope and zero point to the Tully-Fisher law for the high surface brightness galaxies (Zwaan *et al.* 1995). In the LSB galaxies, we believe that the dark halo determines W_{50}, while the baryons determine the absolute magnitude. We then infer that the baryon mass is related to the halo

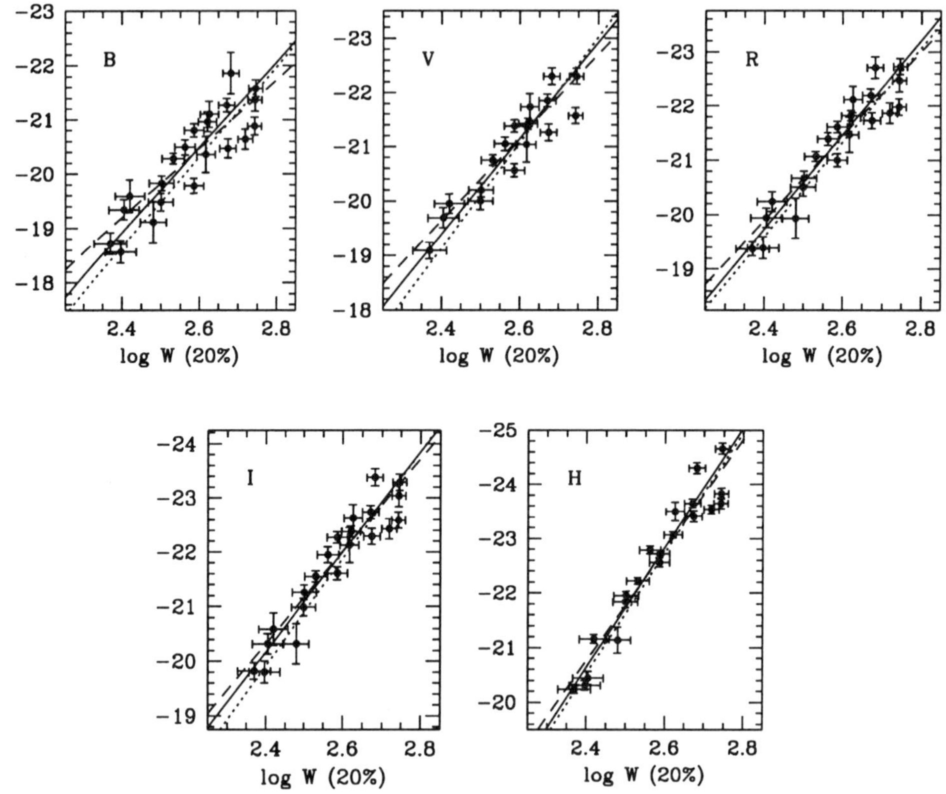

Figure 6. The observed Tully-Fisher law: note how the slope and the scatter change with wavelength (from Sakai *et al.* 2000).

dynamics. Why should this be ?

The reason may be found in the scaling laws for dark halos, *i.e.* the relationship between parameters for the dark halos, like the central density $\rho_\circ$ and the core radius r_c, and the absolute magnitude of the galaxy. Kormendy & Freeman (2003, to be published) derived values for $\rho_\circ$ and r_c for a sample of galaxies with absolute magnitudes M_B ranging from -8 to -23. They found that the central density *decreases* with increasing luminosity, by about 3 orders of magnitude, while the core radius increases by about the same amount. In the mean, the product $\rho_\circ r_c$ is approximately constant for the dark halos. This means that the *surface density* of the halos is approximately constant, which is equivalent to a Faber-Jackson law for halos

$$M_{\rm halo} \propto \sigma^4 \simeq V_{\rm halo}^4/4$$

where $V_{\rm halo}$ is the rotational velocity in the gravitational field of the halo. Then, if the ratio of baryon mass to dark mass is constant from galaxy to galaxy, a Tully-Fisher law between the baryon mass and the halo rotational velocity $V_{\rm halo}$ would follow.

Why should the dark halos follow a Faber-Jackson law ? Fall (2002) describes how the index k of the mass-velocity relation $M_{\rm halo} \propto V_{\rm halo}^k$ for the dark halos depends on the initial spectrum of density perturbations, the comological parameters and the range of masses considered. A slope of 4 corresponds to an effective index $n \simeq -2$ of the CDM spectrum on galactic scales.

Some very gas-rich galaxies are under-luminous for the HI line widths. For example, for NGC 2915 and DDO 154 the order-of-magnitude of the ratios of dark matter mass to gas mass to stellar mass are 100 : 10 : 1. These two galaxies lie 2 to 3 magnitudes below the Tully-Fisher relation. However, if we notionally convert the gas into stars with a M/L ratio of about unity, these galaxies rise to the standard Tully-Fisher relation. This shows again how the Tully-Fisher law is about the relationship of *total baryon content* to the circular velocity of the dark halos (see Freeman 1999, McGaugh *et al.* 2000).

11. How Much Galactic Dark Matter is There ?

Current estimates of the density of (stars + cold gas) and of the total baryon density from big bang nucleosynthesis arguments are $\Omega_{\rm stars+cold\ gas} = 0.0042$ and $\Omega_{\rm BBNS} = 0.04$, so the luminous mass in galaxies is only about 10% of the baryon mass. The rest of the baryons are believed to be hot gas, probably in groups of galaxies. See Fukugita *et al.* (1998), Table 3.

The current estimate of the total matter density of the universe $\Omega_{\rm matter}$ is about 0.27. Recent weak lensing studies indicate that the dark matter within the virial radii of halos is about 37 % of the total matter density of the universe; *i.e.* $\Omega_{\rm dark\ halos} \simeq 0.11$ (Hoekstra *et al.* 2003). If this is correct, then the typical ratio of dark matter to baryonic matter *within galaxies* is $0.11/0.0042 \simeq 25$. This is consistent with the independently derived lower limit of about 20 for our own Galaxy: see §6.1.

References

1. Abadi, M. *et al.* 2003. ApJ, 591, 499.
2. Arnaboldi, M., Combes, F. 1996. A&A, 305, 763.
3. Athanassoula, E., Bosma, A., Papaioannou, S. 1987. A&A, 179, 23.
4. Athanassoula, E., 2002. In "The Dynamics, Structure and History of Galaxies", ASP Conference Series, Vol. 273, ed G. Da Costa & H. Jerjen.
5. Bell, E., de Jong, R., 2001. ApJ, 550, 212.

6. Begeman, K.G. 1989. A&A 223, 47.
7. Bekki, K., Freeman, K. 2002. ApJ, 574, L21.
8. Binney, J, Tremaine, S. 1987. "Galactic Dynamics", Princeton University Press.
9. Buchhorn, M, 1991. ANU thesis.
10. Bullock, J. *et al.* 2001. ApJ, 555, 240.
11. Burkert, A. 2000. ApJ, 534, L143.
12. Dalcanton, J., Hogan, C. 2001. ApJ, 561, 35.
13. Debattista, V, Sellwood, J. 1998. ApJ, 493, L5.
14. de Blok, E., Bosma, A., Rubin, V. 2001. ApJ, 552, L23.
15. Dekel, A. *et al.* 2003. ApJ, 588, 680.
16. Fall, S.M. 2002. In "The Dynamics, Structure and History of Galaxies" ASP Conference Series Vol. 273, ed G. Da Costa & H. Jerjen, p 289.
17. Franx, M., van Gorkom, J., de Zeeuw, P. 1994. ApJ, 436, 642.
18. Freeman, K. 1970. ApJ 160, 811
19. Freeman, K. 1999. In "The Low Surface Brightness Universe", ed. J. Davies, C. Impey, S. Phillipps, p 3.
20. Fukugita, M. *et al.* 1998. ApJ, 503, 518.
21. Gondolo, P., P. 2000. Physics Letters B, 494, 181.
22. Hoekstra, H. *et al.* 2004. ApJ, 606, 67.
23. Ibata, R. *et al.* 2001. ApJ, 551, 2941.
24. Kahn, F., Woltjer, L. 1959. ApJ, 130, 705
25. Kuijken, K., Gilmore, G., 1991. MNRAS 239,605
26. Ma, C-P., Fry 2000. ApJ, 543, 503.
27. McGaugh, S. *et al.* 2000. ApJ, 533, L99.
28. Mateo, M. (1997). In "The Nature of Elliptical Galaxies" ASP Conference Series Vol. 116, ed M. Arnaboldi, G. Da Costa, P. Saha, p 259.
29. Meurer, G. *et al.* 1996. AJ, 111, 1551.
30. Moore, B. *et al.*1999. ApJ, 524, L19
31. Navarro, J., Frenk, C., White, S. 1996. ApJ, 462, 563.
32. Olling, R., Merrifield, M. 2000. MNRAS, 311, 361.
33. Prada, F. *et al.* 2003. ApJ, 598, 260.
34. Perez Martin, I., 2003, ANU thesis.
35. Perez Martin, I., Fux, R. 2002. In preparation.
36. Sackett, P.D. *et al.* 1994. ApJ, 436, 629.
37. Sakai, S. *et al.* 2000. ApJ, 529, 698.
38. Sommer-Larsen, J. *et al.* 2003. ApJ, 596, 47.
39. Steinmetz , M., Muller, E. 1995. MNRAS, 276, 549.
40. Weiner, B., Sellwood, J., Williams, T. 2001. ApJ, 546, 931.
41. Weinberg, M., Katz, N. 2002. ApJ, 580, 627.
42. Weldrake, D., de Blok, E., Walter, F. 2003. MNRAS, 340, 12.
43. Zaritsky, D. 1999. In "The Third Stromlo Symposium" ASP Conference Series Vol. 165, ed B.K. Gibson, T.S. Axelrod, M.E. Putman, p 34.
44. Zwaan, M. *et al.* 1995. MNRAS, 273, L35.

NEUTRAL HYDROGEN IN THE UNIVERSE

F. H. BRIGGS

Australian National University, Mount Stromlo Observatory,
Cotter Road, Weston Creek, ACT 2611, Australia
and
Australia National Telescope Facility
P.O. Box 76, Epping, NSW 1710, Australia
E-mail: fbriggs@mso.anu.edu.au

Neutral atomic hydrogen is an endangered species at the present age of the Universe. When hydrogen is dispersed at low density in the intergalactic medium, the gas is vulnerable to photoionization, and once ionized, the time for recombination exceeds the Hubble time. If hydrogen clouds are confined to sufficient density that they are self-shielding to the ionizing background, they are vulnerable to instability, collapse and star formation, which over time, locks the hydrogen into long lived stars. When neutral clouds do exist after the Epoch of Reionization, they associate closely with galaxies; in these locations, they provide valuable kinematical tracers of the gravitational potentials that bind galaxies and groups.

1. Introduction

Although hydrogen is always portrayed as "the most abundant" of the elements in the Universe, atoms of hydrogen are actually rare. Most of the hydrogen spends most of its time in an ionized state — namely, in a plasma of protons and electrons, accompanied by the ionized nuclei of helium and traces of heavier elements. Here and there, clouds of neutral, atomic hydrogen do exist, but these clouds find themselves confined to large gravitational potential wells, which they share with stars; the clouds rely on the gravity that holds galaxies together to also confine the hydrogen to relatively high density, which makes the clouds less vulnerable to photoionization. But in this environment, they become more vulnerable to instability, collapse and star formation, and for that reason there is a close association of neutral-gas-richness with star formation.

Astronomers study the kinematics of the hydrogen clouds in galaxies, since their motion is a tracer of the depth and shape of the gravitational potential. Observations that inventory the neutral gas content of galaxies provide a measure of the reservoir of fuel that is readily available for forming new stars.

Figure 1 gives an overview of the history of neutral gas clouds over the age of the Universe. It begins at the phase transition corresponding to the release of the Cosmic Microwave Background photons (at $z \sim 1100$), when the ionized baryons and electrons combine to become a neutral gas, commonly labelled HI by

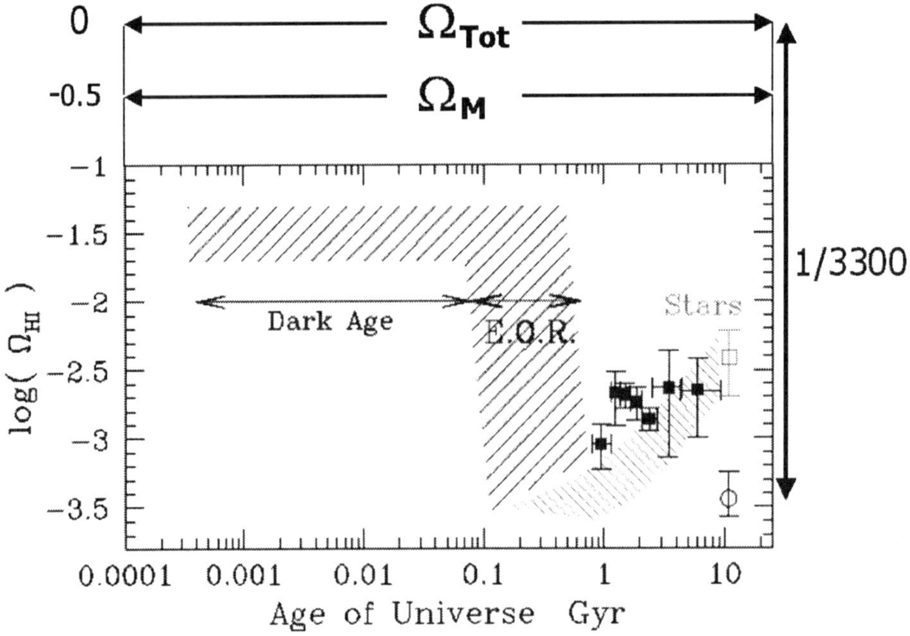

Figure 1. History of the neutral hydrogen content of the Universe. The logarithm of the neutral gas density normalized to the 'closure density' necessary to close the Universe is plotted as a function of the age of the Universe. Square filled points are measurements from Damped Lyman-α QSO absorption-line statistics. The open circle at far right represents the neutral gas content of the present day ($z = 0$) Universe. For comparison, the rising trend of stellar mass content appears as a hatched envelope, which increases to the value measured at $z = 0$ from the optical luminosity density of stars.

astronomers, and which is composed of H^o atoms (in chemical notation). Along with the hydrogen, the primordial mix includes some helium and a trace of lithium. There follows the only period, lasting about 100 million years, when the majority of the Universe's atoms are neutral. This period, known as the 'Dark Age,' ends when the first objects collapse as a result of gravitational instability, providing sources of ionizing energy. We refer to the end of the Dark Age as the 'Epoch of Reionization' (EoR), when the H^o atoms become H^+ (and the HI becomes HII). We associate the EoR with the onset of the first generation of stars (which form in the most over-dense regions) and the appearance of protogalactic objects, which become the building blocks of structure — leading to galaxies and clusters of galaxies, as the forces of gravity run their course.

In the diagram of Fig. 1, the EoR is also marked by the appearance of a second shaded region that indicates schematically the beginnings of the build up of mass in stars, as subsequent generations of star formation gradually lock increasing numbers of baryons into low mass, long lived stars. The stellar mass content of the Universe rises steadily from the EoR to the present, where we have precise measurements through meticulous inventories of the numbers of galaxies and their luminosities (*i.e.*, the galaxy luminosity function and the integral luminosity density)(see for instance, Madgwick *et al.* 2002).

Astronomers can also make accurate measures of the neutral gas content at the present epoch (for instance, Zwaan *et al.* 2003). These result from the direct detection of the radio spectral line emission from atomic hydrogen at 21cm wavelength, and the observations lead to an HI Mass Function for neutral gas clouds (which is analogous to the optical luminosity function for galaxies) to quantify the relative numbers of small and large clouds.

Through the period following the EoR, astronomers have statistical measures of the HI content as a function of time through the observation of QSO absorption lines. Any gas rich object that populates the Universe has a random chance of intervening along the line of sight to distant objects. Quasi-stellar objects are especially useful as background sources since they have strong optical and UV continuum emission against which intervening gas clouds can imprint a distinctive absorption line spectrum. In the case of thick clouds of neutral gas, the Lyman-α line of HI is so strong that it presents an easily recognized 'damping wing' profile, which has led to the Damped Lyman-α (DLA) class of QSO absorption line (Wolfe *et al.* 1986); in the minds of most astronomers, the DLAs are associated with gas-rich protogalaxies, which are the precursors of the larger galaxies that we observe around us at present (Prochaska & Wolfe 1997, Haenelt *et al.* 1998).

The measure of Ω_{HI} during the Dark Age is substantiated by the remarkable agreement of two very different techniques: (1) the measurement of the abundances of the light elements (deuterium, helium, and lithium) and the constraints they impose on primordial nucleosynthesis (Olive *et al.* 2000), and (2) the measurement of the fluctuation spectrum of the CMB, which specifies a number of cosmological parameters, including the baryon number density (Spergel *et al.* 2003). For purposes of constructing Fig. 1, all of the Universal baryons are assumed to be locked into their neutral atomic form throughout the Dark Age.

A further consequence of the precise cosmological measurements that have resulted from studies of the CMB is that we can compare the relative importance of atomic hydrogen throughout history with the dominant constituents: the dark matter and the dark energy (Spergel *et al.* 2003). As indicated in Fig.1, the current best cosmological model has a flat Universe ($\Omega_{tot} = 1$), with the mass density contributing $\Omega_M \approx 0.3$ and a dark energy providing $\Omega_\Lambda \approx 0.7$. The mass density is dominated by the dark matter component, which accounts for 84 percent of Ω_M. Especially at present, the Ω_{HI} amounts to a tiny fraction of the mass-energy budget of the Universe.

The following sections elaborate the conditions that hydrogen gas experiences, focusing on why there are so few HI clouds remaining once the EoR has occurred, the use of HI as a kinematic tracer, and the expectation that radio observations of the 21cm line will help to elucidate the processes that ended the Dark Age.

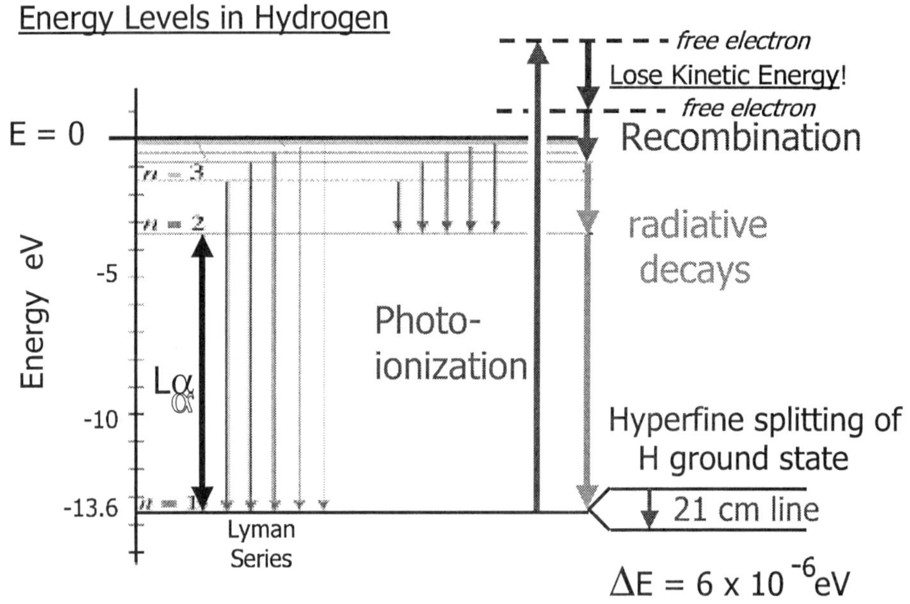

Figure 2. The Energy Level diagram for the Hydrogen atom, with annotations for (1) The Lyman series, with Lyman-α (Lα) marked, (2) the photoionization-recombination cycle indicated, with photoionization from the ground state followed by the free electron heating the surrounding plasma by losing kinetic energy to collisions, and the radiative recombination leading to emission of photons through radiative decay, and (3) the small hyperfine splitting of the ground state to give rise to the 21cm line.

2. Observing Hydrogen

Astronomers can observe hydrogen because it emits and absorbs light. The internal structure of the atom allows only discrete energy levels, and this limits the photon energies that can be exchanged with the atom, and it also makes clear under what conditions various spectral lines would be expected to occur. Figure 2 sketches the energy levels for atomic hydrogen.

Hydrogen clouds have long been observed in our galaxy in HII regions and planetary nebulae, where the Balmer series lines are seen in emission. The energy levels that produce the Balmer lines must be populated, in order for them to radiatively decay (by emitting a photon) to reach the $n = 2$ first excited state. In Galactic nebulae, this is accomplished by photoionizing the nebulae with ionizing UV photons from hot stars, followed by recombination and radiative decay. Also important in this process is the energy lost by the photoelectron, as it is scattered in the nebula, since this is the source of heating for the gas. Clearly, they are the ionized hydrogen clouds — not the neutral ones — that radiated effectively.

Neutral hydrogen in galaxies is cool with temperatures ranging from ~50 to a few hundred degrees for the clouds to a few thousand degrees for the warm phase, intercloud medium (Wolfire *et al.* 2003, Liszt 2001). These temperatures are too low to excite the atoms to the $n = 2$ level or above, so there are seldom excited atoms capable of emitting or absorbing Balmer wavelength photons. (This situation is clearly very different from the hydrogen in the atmospheres of stars where temperatures and densities are high enough to excite the $n = 2$ level, allowing the Balmer lines to have a long history in helping to classify stars through absorption line spectroscopy at optical wavelengths.) Cool hydrogen cannot absorb optical wavelengths, but it is very effective at absorbing in the ultra-violet Lyman lines and in the "Lyman continuum," which is the wavelength range corresponding to ionizing photons with energies greater than 13.6 eV.

Fortunately, atomic hydrogen has another low lying energy level that arises from a tiny, "hyperfine" splitting of the $n = 1$ ground state. This allows hydrogen to emit and absorb photons with the radio wavelength 21.1 cm. A qualitative interpretation of this splitting is that it arises from the relative alignment of the magnetic moments of spinning charges of the electron and proton; the quantum mechanics of the hydrogen atom allow for only two possible alignments, and there are therefore only two energy levels in the split ground state. The energy required to change the alignment is so small that weak collisions can excite and de-excite the hyperfine levels. This means that the kinetic temperature of the gas cloud, T_K, is effective at setting the hydrogen spin temperature, T_S, which governs the hyperfine level populations according to

$$\frac{N_+}{N_-} = \frac{g_+}{g_-} \exp\left(-\frac{\Delta E}{kT_S}\right) \approx \frac{g_+}{g_-} \exp\left(-\frac{h\nu}{kT_K}\right) \tag{1}$$

where g_+ and g_- are the degeneracies of the upper and lower levels ($g_+/g_- = 3$), $\Delta E \sim 6 \times 10^{-6}$eV (the energy of a $\lambda = 21$cm photon), and k is the Boltzmann constant. Under dilute conditions where atomic collisions become infrequent, then collisions with photons may dominate in setting the N_+/N_- ratio. For example, at the end of the Dark Age, the intergalactic medium has become sufficiently diffuse that the CMB photons will pin $T_S \approx T_{CMB} = 2.73(1 + z)$K; once substantial overdensities evolve, the gas again becomes coupled to the gas kinetic temperature.

In summary, neutral hydrogen clouds are always capable of emitting 21cm line photons. If they chance to fall between the observer and a bright radio continuum source, then 21cm line absorption lines may be seen. Neutral hydrogen clouds do not absorb optical or infrared wavelength hydrogen lines (the Balmer or Paschen series for example), but they are strong absorbers of the ultraviolet Lyman lines, and they are effective at absorbing photons with energies greater than 13.6 eV (wavelengths $\lambda < 911$Å). All neutral clouds observed so far have traces of "metals" — elemental species heavier than helium, such as NaI and CaII — that may allow the clouds to be detected in optical wavelength absorption lines when they are observed against sufficiently bright background stars or QSOs; neutral clouds also show very strong absorption in UV absorption lines by species such as MgI, MgII, FeII, SiII, CII, OI, AlII, among others. Neutral clouds do not emit optical or UV photons, unless they are bathed in a radiation field of energetic, ionizing photons, in which case they may become detectable in the recombination lines.

3. Hydrogen in the nearby Universe

Even now, some 13.7 billion years after the Big Bang according to the WMAP "concordance cosmology" (Spergel *et al.* 2003), hydrogen remains the most abundant element. Estimates for where the baryons are located in the present day Universe are plotted in Fig. 1, where the stellar mass and neutral gas components account for $\Omega_*(z = 0) \sim 0.004$ and $\Omega_{HI}(z = 0) \sim 0.0004$. The WMAP cosmology has a baryonic mass density of $\Omega_{baryon} = 0.0044$ with total matter $\Omega_m = 0.27$.

Roughly 90% of the Universal baryons in Fig. 1 remain unaccounted for at the present epoch, since the sum of the mass in stars and the neutral gas clouds at $z = 0$ is far less than that produced in the Big Bang. A more complete census (Shull 2003, Penton *et al.* 2000) finds that many of the missing baryons fill the vast ionized sea that comprises the intergalactic medium (IGM). At present, the IGM is of such low density that, once ionized, the recombination time greatly exceeds a Hubble time (see discussion in Section 5.1). Their presence is observable through the small fraction $n^0/(n^+ + n^0) \sim 10^{-3}$ of neutrals in the Lyman-α forest clouds and through the traces of highly ionized species (CIV and OVI) indicating low level metal pollution of the IGM due to stellar mass loss over the age of the Universe (Shull 2003).

Here and there within the sea of ions and electrons, there are "condensations" where higher densities of baryons have undergone gravitational collapse that led to star formation. Each of the condensations formed within the confining potential well of a dark matter halo. A consistent picture of structure formation has the baryonic material being carried along into the evolving halos in constant proportion to the dark matter. Once confined in the halo, the baryons cool and gravo-thermal instabilities cause the gas clouds to collapse and form stars, leading to the objects we call galaxies.

The neutral baryons are but a small fraction of the total mass. Their distribution among galaxies of different types and sizes has been carefully measured (for example, Roberts & Haynes 1994). The general rule is that the late-type spiral and irregular galaxies are the richest in HI, consistent with their blue colours and populations of young stars. The elliptical and S0 galaxies are generally devoid of HI, in accord with their older stellar populations, although they occasionally have outlying HI clouds of substantial mass (Oosterloo *et al.* 2003).

The HI mass function (HIMF) quantifies the relative number of galaxies with different HI masses in the same way that the optical luminosity function gives the numbers of galaxies of different luminosities. The main features of both the HIMF and the luminosity function are described by an analytic form called a Schechter Function (Schechter 1976). The HIMF has a functional dependence $\Theta(M_{HI})$ on HI

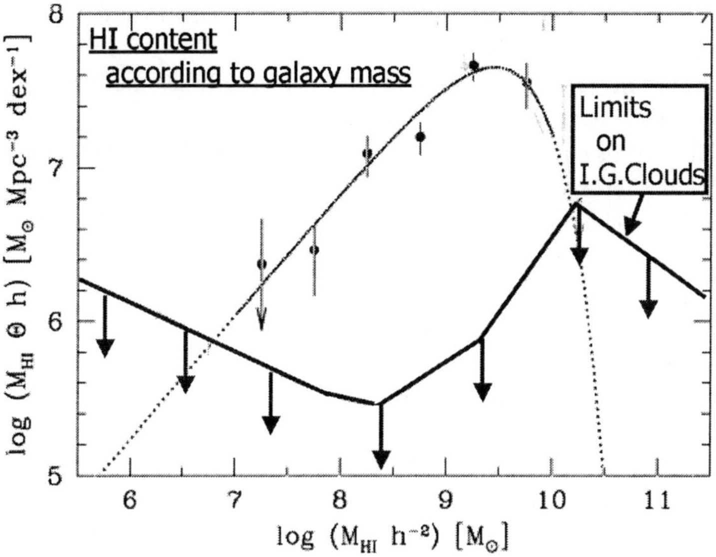

Figure 3. The integral neutral gas content of galaxies as a function of HI mass, showing that the more massive systems around $M^*_{HI} \sim 10^{9.55}h^{-2}M_\odot$ (for $h = H_o/100$ km s^{-1}) are the dominant repositories of neutral gas at $z \approx 0$. Current limits on the abundances of intergalactic HI clouds permit no competitive amounts of neutral gas anywhere in the mass range characteristic of galactic systems (Zwaan *et al.* 1997).

mass M_{HI}:

$$\Theta(M_{HI})dM_{HI} = \Theta^* \left(\frac{M_{HI}}{M^*_{HI}}\right)^{-\alpha} \exp\left(-\frac{M_{HI}}{M^*_{HI}}\right) dM_{HI} \qquad (2)$$

with three parameters Θ^*, α, and M^*_{HI} that fix the shape and normalization. Plots of these functions on log-log axes make clear that the M^*_{HI} is the break point or "knee" that sets the high mass cutoff to the distribution; an exponential becomes a fairly hard cutoff on a log-log plot. The distribution below the cutoff is set by the power law slope α, and Θ^* specifies the normalization of the curve. The HIPASS survey with the Parkes Telescope has provided recent determinations of the parameters: $\Theta^* = (8.6 \pm 2.1) \times 10^{-3} h_{75}^{-3}$Mpc^{-3}, $\alpha = 1.30 \pm 0.08$, and $M^*_{HI} = (6.1 \pm 0.9) \times 10^9 h_{75}^{-2}$

While the HIMF specifies the number of galaxies per Mpc^3 as a function of mass, a more useful plot for assessing the relative importance of the different mass ranges in the HI census is a plot of $\psi(M_{HI}) = \Theta(M_{HI})dM_{HI}/d\log_{10}M_{HI} = M_{HI} \ln 10\ \Theta(M_{HI})$, which compares the total amount of HI mass per Mpc^3, showing the galaxy population in each logarithmic interval of M_{HI}. Fig. 3 has an example, where the HI density M$_\odot$Mpc^{-3} is calculated per decade of HI mass. The peak near $10^{9.4}h_{100}^{-2}M_\odot$ indicates that these galaxies with HI masses near the knee are the most important contributors of HI mass. Although the HIMF has a greater number of small masses per Mpc^3, the rarer large galaxies add up to a larger integral mass density. The sharp exponential cutoff to the HIMF indicates a very low contribution from galaxies with $M_{HI} > 10^{10.5}\mathrm{M}_\odot$.

A number of radio surveys in the 21cm line have blindly scanned the sky in search of intergalactic hydrogen clouds. To qualify as an "intergalactic cloud," a cloud must be isolated from any galactic system that emits starlight. The goal has been to find HI clouds that are confined to their own dark matter potential well without an accompanying stellar population. The surveys are considered "blind" when the region for the study has been chosen without regard for any prior knowledge of the numbers or types of optically identified galaxies in the region.

More than 20 years ago, Fisher and Tully (1981) deduced that the amount of mass in HI clouds was not cosmologically significant. That is to say, the integral mass content of a possible intergalactic cloud population did not come close to being enough to close the Universe by bringing its mass density up to the critical density. They arrived at this deduction by noting that every 21cm line observation made to catalogue the HI mass in a nearby galaxy also includes a comparable amount of integration on blank sky nearby the galaxy. These blank sky observations are taken to calibrate the instrumental spectral passband shape on a galaxy by galaxy basis. Fisher and Tully found no HI signals in the off-source scans that were not associated with galaxies in the off-galaxy calibration spectra. Ten years later, Briggs (1990) made a similar analysis of the large number of new observations that had been obtained using the same observing technique, and he concluded that in the HI mass range of $\sim 10^8$ to $10^{10} M_\odot$ intergalactic HI clouds must be rare; they had to be outnumbered by galaxies with HI masses in this range by at least 100:1.

Since 1990, radio spectrographs have become better suited for making truly blind surveys of large areas of sky, resulting in a number of studies: Zwaan *et al.* (1997), Spitzak & Schneider (1998), Kraan-Korteweg *et al.* (1999), Rosenberg & Schneider (2000), Koribalski, B.S. *et al.* (2003). Despite detecting thousands of galaxies in the hydrogen line, these surveys have turned up no "free-floating" HI clouds (i.e., clouds that are not associated with the gravitational potential containing a population of stars).

Blitz *et al.* (1999) and Braun and Burton (1999) have explored the possibility that the infalling population of small HI clouds associated with the halo of the Milky Way Galaxy — the "High Velocity Clouds" — are remnants of a primordial extragalactic population. In this scenario, the HI masses of the clouds would typically be larger than $\sim 10^7 M_\odot$, and every large galaxy should be surrounded by a similar halo of a few hundred of these objects if the phenomenon is a genuine and common feature of galaxy formation and evolution. The fact that nearby galaxies and groups do not possess such a halo of small clouds (Zwaan & Briggs 2000, Zwaan 2001) has ruled out this idea, requiring that the clouds must be an order of magnitude less massive and fall at distances within $\sim$200 kpc of the Milky Way, well within our Galaxy's halo.

The clear association of neutral gas clouds with star-bearing galaxies implies that the HI relies on the confinement of the galaxies' gravitational potentials for their survival (see Sect. 5.1).

4. Redshifted HI in Evolving Galaxies

Radio astronomers would like to extend these kinds of 21cm emission line studies to higher redshifts, in order to monitor the amount of HI as a function of time and its relation to star forming regions. Unfortunately, the inverse-square law very quickly takes its toll, and the current generation of radio telescopes cannot detect individual galaxies at redshifts much beyond 0.2. For this reason, much of what we know about the neutral gas content as a function of age of the Universe comes from the statistical analysis of the QSO absorption-lines. The next generation of radio telescopes has the design goal of being able to detect individual galaxies in the 21cm line to redshifts around three.

4.1. *QSO absorption lines*

Much of what we know about the gas content — both neutral and ionized — in evolving galactic systems over the redshift range from 6 to close to the present comes from the study of QSO absorption-lines. The strong ultra-violet continua of active galactic nuclei make fine sources of fairly clean background spectrum against which the intervening gas clouds imprint their distinctive absorption signatures. The QSOs themselves are marked by characteristic, broad emission lines that indicate the emission redshift; occasional "associated" narrow-line absorption occurs in the QSO host galaxy, and outflowing material from the nucleus causes broad absorption lines (BALs) in some 5–10% of QSOs.

The class of QSO absorption-line that occurs when intervening protogalaxies chance to fall along the sightline to a higher redshift QSO has much to tell us about the amounts of neutral and ionized gas as a function of time, the metal abundances, and kinematics in the intervenor.

The statistics for QSO absorption lines are typically analysed by keeping track of the rate of intervention per unit redshift for each of the species (like triply ionized carbon, CIV, or singly ionized magnesium MgII) separately. This interception rate as a function of redshift is named $n(z) = dN/dz$ and called D-N-D-Z. Clearly it is inversely proportional to the mean-free-path between absorptions. The mean-free-path is related to the number density and cross-section of the absorbers $l_o = 1/n_o\sigma_o$. For a distribution of galaxy sizes, the expression generalizes to an integral, where n_o becomes the luminosity function $\Phi(L)$, and σ_o adopts a dependence on galaxy properties, including luminosity $\sigma(L)$. The fiducial luminosity L^* is the common reference for comparison, so QSO absorption-line discussions often quote cross-sections as though they were computed for L^* galaxies with non-evolving co-moving density. Fig. 5 illustrates this idea by presenting the cross-sections that non-evolving L^* galaxies would need to have to explain the intervention statistics for the species

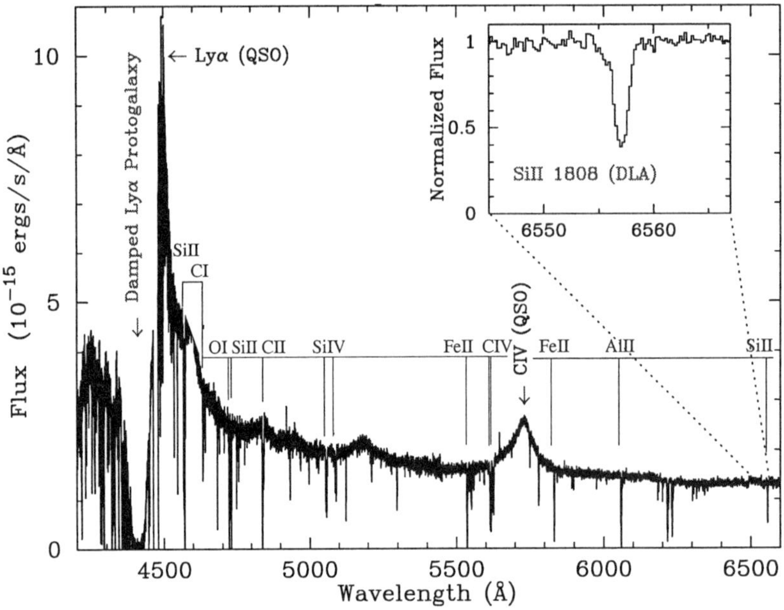

Figure 4. Spectrum of the $z_{em} = 2.701$ QSO FJ081240.6+320808, showing broad emission lines of the QSO (Ly-α and CIV are labelled) and absorption lines in a DLA system at $z = 2.626$, including the damped Lyman-α line and narrow metal lines. The inset box shows a zoom-in on one of the weaker lines (SiII 1808) in this system. (figure courtesy of Prochaska *et al.* 2003).

HI, CIV and MgII in the redshift range approximately 1 to 2.5. In fact, the rest wavelengths of these ions are substantially different so that the extensive ground-based observations monitor the $dN/dz(z)$ dependence over different redshift ranges for different ions. Indeed, the the statistics show that the different species have different redshift dependencies over these ranges, so that the figure serves only as a rough illustration that the cross sections in CIV and MgII are substantially larger than the sizes of galaxy disks at $z = 0$, a conclusion that has led to the hypothesis of "metal-rich gaseous halos around galaxies." A variety of processes could fill halos with gas after metal enrichment by galactic stars; these include winds from star forming regions and tidal effects during merging and interactions with companion galaxies. The MgII gas arises in predominantly neutral gas clouds, although the column densities can be as low as $N_{HI}{\sim}10^{17}\text{cm}^{-2}$; this same column density is the critical level where gas clouds become optically thick to photons capable of ionizing hydrogen, so there is a direct association of MgII with the QSO absorption systems that are "optically thick at the Lyman limit," *i.e.*, the systems known as either Lyman Limit or Lyman Continuum absorbers.

The statistics that give rise to the cross sections in Fig. 5 are based on strong absorption line complexes of the sort expected along lines of sight through galaxies with metal rich halo gas. More recent studies using the high resolution spectrographs at Keck and VLT are sensitive to weaker equivalent width thresholds. These new studies have been effective at tracing the rise in metallicity of the intervening

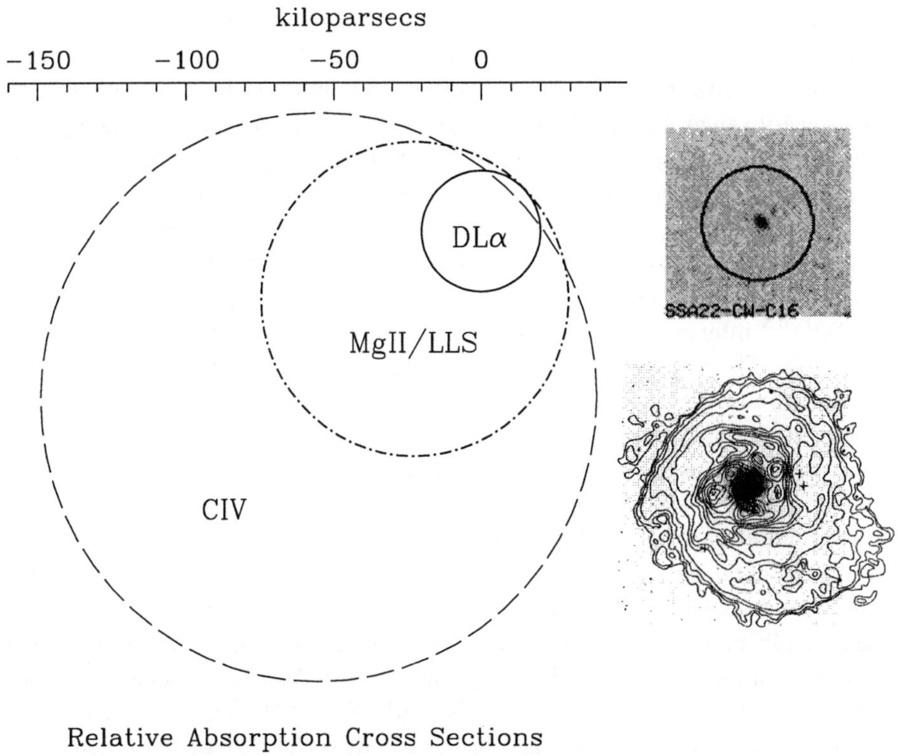

Figure 5. Comparison of quasar absorption-line cross sections for CIV, MgII-Lyman Limit, and damped Lyman−α lines with the physical size of the optical emission from a colour-selected galaxy at $z \approx 3$ *top right* (Giavalisco *et al.* 1996a) and the HI extent of a nearby, large $L \sim L_*$ galaxy M74=NGC628 *lower right* (Kamphuis & Briggs 1993). The absorption cross-sections are taken from Steidel 1993 and adapted to $H_o = 75$ km s^{-1}Mpc^{-1}. The $z \approx 3$ galaxy is centred in a 5″ diameter circle that subtends 37.5 kpc ($\Omega_M = 0.2, \Omega_\Lambda = 0$). The Holmberg diameter of NGC628 is ~36 kpc at a distance of 10 Mpc; the outermost contour is $1.3{\times}10^{19}$cm^{-2} and over half of the absorbing cross section is above 10^{20}cm^{-2}.

gas clouds in evolving galaxies with increasing age of the Universe (Pettini *et al.* 2002, Prochaska *et al.* 2003). In addition, they have discovered weak metal lines even in the Lα forest clouds (Lu 1991, Pettini *et al.* 2003).

Figure 5 also compares the absorption cross sections with the observed sizes of the colour-selected "Lyman break" galaxies at redshifts $z \sim 3$ and a large L^* spiral, M74, observed in the 21cm line nearby at $z \sim 0$. Although the large HI extent shown by M74 is not rare among nearby galaxies, such large cross sections are certainly in the minority, implying that cross sections of neutral gas were larger in the past. The Lyman break galaxies are somewhat less common than the comoving number density of L^* galaxies, implying that for every tiny, but highly luminous star forming system of the sort seen in the HST imaging, there must also be roughly double the gas-cloud cross-section drawn in the figure, which must exist as low surface brightness or non-luminous material at these redshifts.

Absorption line observers also quantify the relative numbers of low and high column density absorbers. The distribution function $f(N_{HI})dN_{HI}$ (the "F-of-N" distribution) specifies the number of absorbers per unit redshift with N_{HI} in the column density range N_{HI} to $N_{HI} + dN_{HI}$. Over nearly ten orders of magnitude of column density, $f(N_{HI})$ can be approximated as $f(N_{HI}) = N_o N_{HI}^{-1.5}$, a single power-law which applies surprisingly well throughout the Lyman-α forest through to the DLA lines. When speaking of the relative frequency of occurrence of different column densities, it is convenient to use the number per logarithmic interval (say, per decade) and define an $F(N_{HI})d(\log_{10}N_{HI}) = f(N_{HI})dN_{HI}$. Then $F(N_{HI}) = N_o \ln(10) N^{-0.5}$ interceptions per decade, a shallower decline with column density than $f(N_{HI})$. The implication is that absorption lines with HI column density in the decade around 10^{17} are one-tenth as frequent as column densities in the decade centred on 10^{19}, for example.

A natural question to ask under these circumstances when the "$f(N)$" statistics indicate that low N_{HI} clouds are more common than high N_{HI} is: Which column densities contain more mass? The total neutral mass contained in the distribution comes from integrating $\int N f(N) dN = \int N F(N) d(\ln N) = \int N^{0.5} d(\ln N)$, implying that the high mass end of the distribution dominates in the amount of *neutral* gas per logarithmic interval. The lower N_{HI} clouds are numerous, but when integrated up, they contain less neutral HI. However, since the low N_{HI} forest clouds are *highly ionized*, the HI that is seen in these clouds (with $N_{HI} < 10^{17}\mathrm{cm}^{-2}$) is just the tip of the iceberg of total mass contained in the clouds – the ionized hydrogen in the Lyman-α forest clouds accounts for many of the missing baryons of Fig.1. It is also clear that the expression for integral mass diverges at the large N limit, so that there must be some physical cutoff to the high column $f(N)$ distribution (Boissier *et al.* 2003).

4.2. *21cm line studies*

The next generation of radio telescopes will be able to measure HI in galaxies to high redshifts (Taylor & Braun 1998, van Haarlem 1999), but our present telescopes are limited to doing absorption line studies that are similar to those done optically. The 21cm HI line is much weaker than the HI Lyman-α line: the optical depths at line centre are in the ratio $\tau_{L\alpha}/\tau_{21cm} \approx 3{\times}10^8 (T_s/100)$, explaining why only the highest column densities of cool hydrogen are detected in absorption at in the 21cm line. The dependence on spin temperature is a consequence of the correction for stimulated emission that arises because the upper levels of the hyperfine splitting are always populated under normal astrophysical conditions (Spitzer 1978). A further implication is that the gas temperature can be measured through a combination of the measurements of the two optical depths (one at UV/optical wavelengths and one at radio frequencies). When this remote sensing "thermometer" is applied to QSO absorption line systems as a function of redshift, a strong trend is observed that higher redshift systems become significantly warmer with increasing redshift

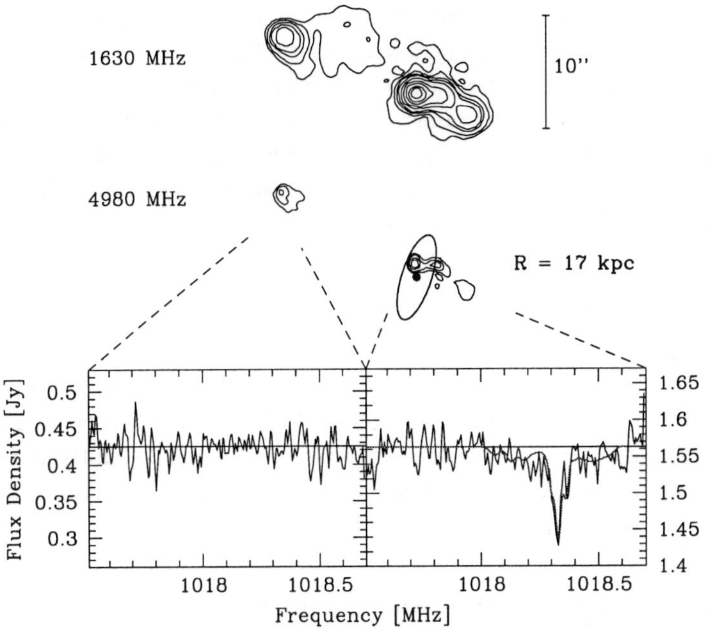

Figure 6. Radio 21cm HI absorption against the extended radio source PKS1229-021. The 21cm absorption occurs at $z = 0.395$, corresponding to 1018 MHz. As an interferometer, the WSRT has just enough resolution to decompose the absorption spectrum into the separate spectra for the two principal components of the radio emission (Briggs, Lane & de Bruyn, in prep.). The VLA contour maps shown here for the higher frequencies (Kronberg *et al.* 1992) have better angular resolution but poor sensitivity to extended emission at 4980 MHz. The absorption, which only occurs against the righthand component, may have broad wings corresponding to absorption by a rotating system (the disk in the schematic representation), giving rise to opacity that is distributed across the face of the western component of the radio source. The oval is centred on the known location of an optically luminous galaxy in HST imaging (Le Brun *et al.* 1997).

and that this effect is strongly correlated with lower metallicity and the associated lower gas cooling rates (Kanekar & Chengalur 2003).

A virtue of 21cm absorption line studies against high redshift radio sources is that some background radio sources have very large physical extent, allowing them to backlight large areas of the foreground absorbing galaxy. Several such cases have been studied (Briggs *et al.* 1989, Briggs *et al.* 2001), and hundreds more will be accessible with future radio telescopes.

The principal question to be addressed is whether the gas-rich galaxies (such as the systems selected through DLA surveys) are large systems in orderly rotation like spiral galaxies or are aggregates of numerous smaller dwarfs systems with more random velocities that are in the process of merging, or are somewhere in between. Gas tracers like the 21cm line, which senses cold gas even in the absence of stars, have an important role to play in analysing the physical sizes and dynamical masses of primitive systems, prior to their having established themselves as optically luminous galaxies.

Fig.6 illustrates how a disk galaxy leaves its imprint on the background radio source. When observed with a radio telescope of sufficient sensitivity and resolution, we expect to see the signs of rotation in the velocity field in the disk galaxy at $z_{abs} = 0.395$ that is absorbing against the background radio source at $z_{em} = 1.045$. The present resolution is only adequate to confirm that the HI optical depth is only significant against the western lobe of the source, which is consistent with the presence of an optically luminous galaxy close to this sight line.

5. Ionization, Reionization, and Re-reionization

Figure 1 summarizes the principal historical phases in the evolution of the neutral gas content of the Universe. Recombination at the time of the release of the CMB photons led to a period when the vast majority of the Universal baryons found themselves in neutral atoms. Once sources of ionization formed in the earliest astrophysical structures, the survival of neutral clouds has been a competition between ionization and recombination rates.

5.1. *The ionization/recombination competition*

Since ionization is such a common hazard to the existence of neutral atoms, it is natural to ask, "how rapidly can an ion recover through recombination, if it does chance to become ionized?" For hydrogen, the recombination rate R is easily computed (for instance Spitzer 1978), and the time t_{recomb} it takes for recombination to eliminate the electrons in a cloud of electron density n_e is

$$t_{recomb} = \frac{n_e}{R} = \frac{n_e}{n_e n_p \alpha_{recomb}} = \frac{1}{n_p \alpha_{recomb}} \approx \frac{T^{1/2}}{2\times 10^{-11} n_p} \quad \text{sec} \tag{3}$$

where n_p is the proton density and α_{recomb} is the recombination coefficient. To get a feeling for the vulnerability of the bulk of the baryons that populate the intergalactic medium, the number density of baryons n_{baryon} forms an estimate of n_p; over-dense regions will have relatively shorter recombination times. In an expanding Universe, $n_p \sim n_{baryon} \sim (1+z)^3$, so that

$$t_{recomb} \propto \frac{T^{1/2}}{(1+z)^3} \tag{4}$$

The recombination time of the IGM at mean density has a strong dependence on age of the Universe through the $(1+z)^3$, and a modest dependence on temperature T. Fig. 7 provides a rough illustration of how the IGM temperature varies with time and the net influence of the dependencies in Eqn. 4 on the ionization state of the Universe.

If the expansion of the Universe would allow a completely uniform expansion of the IGM without the growth of gravitationally-driven density instabilities, the gas kinetic temperature would decline in the adiabatic expansion with dependence $T_k \propto (1+z)^2$. At the same time, the CMB radiation temperature declines as

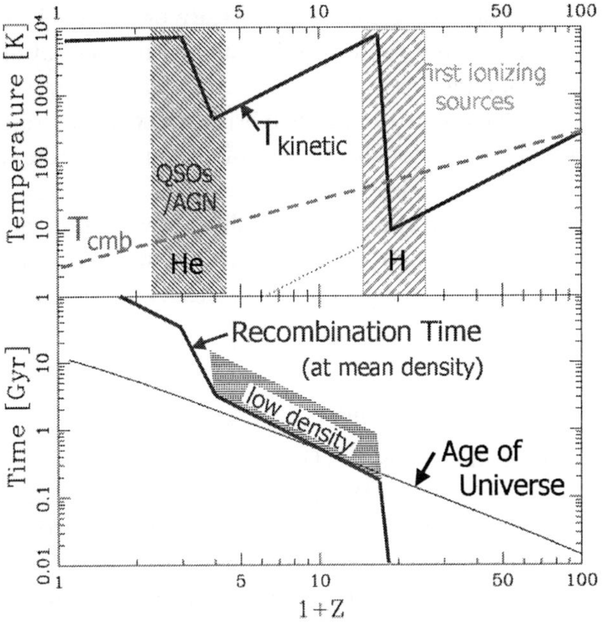

Figure 7. Recombination time in the intergalactic medium as a function of redshift z. *Upper Panel:* Kinetic temperature T_k and CMB temperature T_{CMB} vs. redshift. Episodes of heating through photoionization of hydrogen occur during the Epoch of Reionization and during the reionization of helium at a later time by the harder radiation from active galactic nuclei. *Lower Panel:* Recombination Time for an intergalactic medium of mean baryonic density, compared with the Age of the Universe as a function of redshift.

$T_{CMB} \propto (1+z)^{-1}$, causing the two temperatures to decouple after $z \approx 100$, when electron scattering ceases to be effective. The IGM is reheated when photoionization spreads through the medium generating energetic photoelectrons that deposit their kinetic energy through scattering. Once the IGM is fully ionized, there is no effective means of adding energy to the gas, since the photons generated by the stars can now flow uninhibited through a transparent medium, and the IGM again cools adiabatically due to Universal expansion.

A similar heating event can occur during the age around $z \sim 2$ when QSOs are most common. QSOs, as well as lesser AGN, radiate photons that are capable of ionizing helium, and these harder photons generate photoelectrons throughout the IGM, providing a second round of localized heating.

The two heating events impact on the ability of the Universe to recombine. The lower panel of Fig. 7 compares the recombination time t_{recomb} of an IGM of mean density to the age of the Universe t_{age} as a function of redshift. If t_{recomb} is long compared to t_{age}, the IGM would never recover from its ionized state, even if the source of ionizing photons were turned off completely. The figure shows that there is a period between the two heating events, when recombination can compete with ionization, depending on 1) the intensity of the ionizing flux and

2)the local density. Under-dense regions would already be destined to stay forever ionized. Over-densities, especially those clouds confined in gravitational potential wells, may be able to recombine.

At low redshifts, the density of the mean IGM has become so dilute, that the IGM will remain ionized, even though the photoionizing background from AGN tails off.

The existence of atomic hydrogen clouds at all at low redshift is due to their confinement to high density (greater than ~0.1 cm^{-3}) where the recombination times are $< 10^5$yrs, and recombination can compete effectively to make self-shielding clouds.

5.2. *EoR: The end of the Dark Age*

The Epoch of Reionization that ends the Dark Age is now the subject of intense observational and theoretical interest. When and how do the first stars light up and begin the process of ionization and reheating? Several fine review articles summarize the current views (see for example, Barkana and Loeb 2001, Miralda-Escudé 2003). One of the findings of the Wilkinson Microwave Anisotropy Probe (WMAP) has been a measurement of the optical depth to electron scattering between us and the so-called "surface of last scattering" at redshift around $z = 1089$. This optical depth in turn specifies a minimum redshift ($z_{reion} = 17 \pm 4$ according to Spergel *et al.* 2003) when the bulk of the reionization must have taken place. This value for z_{reion} is at odds with measurements of the Gunn-Peterson effect that are consistent with the bulk of reionization occurring at $z_{gp} = 6.2$ (Gnedin 2001, Pentericci *et al.* 2002). This conflict of the two measurements has led to a variety of models that invoke a "smouldering" or even double reionization (Gnedin & Shandarin 2002, Cen 2003). The idea as outlined in Section. 5.1 is that the IGM density is high enough at redshifts $z > 10$ that recombination is still effective. A continuing source of ionizing photons is required to maintain the ionization of the IGM. To continue to do this with stellar sources carries the implication of ongoing metal production, which runs the danger of generating more metals than are observed in the IGM.

The LOFAR (Low Frequency Array) and SKA (Square Kilometre Array) radio telescopes, whose design and construction are taking place over the next 15 years, promise to allow astronomers to look into the EoR in the redshifted 21cm line at frequencies of 80 to 200 MHz, corresponding to the redshift range $z = 17$ to 6.2 discussed above. Unlike the WMAP result, which is an integral measurement of the electron content on a large angular scale, the 21cm observation will map the structure defined by the neutral clouds in three dimensions, resolving the neutral IGM both in angle on the sky and in depth through spectral resolution (Tozzi *et al.* 2000, Furlanetto & Loeb 2002, Furlanetto *et al.* 2003, Chen & Miralda-Escudé 2003). Thus, these instruments will not only clarify the timing of when the first stars form, but they will also monitor the growth of structure in the neutral component of the IGM through a period that promises to be complex and highly dependent on the astrophysics of material of primordial composition. Therefore, the star formation mechanisms at work will be unlike those we can study easily in the nearby star forming regions at $z \approx 0$.

References

1. Barkana, R., Loeb, A. 2001, ARA&A, 39, 19
2. Blitz, L. *et al.* 1999, ApJ, 514, 818
3. Boissier, S., Peroux, C., Pettini, M. 2003, 338, 131
4. Braun, R. and Burton, W.B. 1999, A&A, 341, 437
5. Briggs, F.H. *et al.* 1989, ApJ, 341, 650
6. Briggs, F.H. 1990, AJ, 100, 999
7. Briggs, F.H., de Bruyn, A.G., Vermeulen, R.C. 2001, A&A, 373, 113
8. Cen, R. 2003, ApJ, 591, 12
9. Chen, X., Miralda-Escudé, J. 2004, ApJ, 602, 1
10. Fisher, J.R., & Tully, B. 1981, ApJ, 243, L32
11. Furlanetto, S., Loeb, A. 2002, ApJ, 579, 1
12. Furlanetto, S. *et al.* 2003, astro-ph/0305065
13. Giavalisco, M., Steidel, C.C, Macchetto, F.D. 1996, ApJ, 470
14. Gnedin, N.Y. 2001, astro-ph/0110290
15. Gnedin, N.Y., Shandarin, S.F. 2002, MNRAS, 337, 1435
16. van Haarlem, M., ed. 1999, "Perspectives on Radio Astronomy: Science with Large Antenna Arrays," Proceedings of a conference held in Amsterdam in April 1999, (ISBN: 90-805434-1-1)
17. Haehnelt, M., Steinmetz, M., Rauch, M. 1998, ApJ, 495, 647
18. Kamphuis & Briggs 1993, A&A, 253, 335
19. Kanekar, N., Chengalur, J.N. 2003, A&A, 399, 857
20. Koribalksi, B. *et al.* 2003, submitted
21. Kraan-Korteweg, R.C., *et al.* 1999, A&AS, 135, 225
22. Kronberg, P.P., Perry, J.J., Zukowski, E.L.H. 1992, ApJ, 387, 528
23. Le Brun, V., *et al.* 1997, A&A, 321, 733
24. Liszt, H. 2001, A&A, 371, 698
25. Lu, L. 1991, ApJ, 379, 99
26. Madgwick, D.S. *et al.* 2002, MNRAS, 333, 133
27. Miralda-Escudé, J. 2003, Sci, 300, 1904
28. Olive, K.A., Steigman, G., Walker, T.P. 2000, PhR, 333, 389
29. Oosterloo, T., *et al.* 2003, IAUS, 217, 108
30. Pentericci, L., *et al.* 2002, AJ, 123, 2151
31. Penton, S.V., Shull, J.M., Stocke, J.T. 2000, ApJ, 544, 150
32. Pettini, M., *et al.* 2002, A&A, 391,21
33. Pettini, M., *et al.* 2003, ApJ, 594, 695
34. Prochaska, J.X., Wolfe, A.M. 1997, ApJ, 487, 73
35. Prochaska, J.X., Howk, J.C., Wolfe, A.M. 2003, Nature, 423, 57
36. Prochaska, J.A., *et al.* 2003, ApJ, 595, L9
37. Roberts, M.S., Haynes, M.P. 1994, ARA&A, 32, 115
38. Rosenberg, J.L. & Schneider, S.E. 2000, ApJS, 130, 177
39. Schechter, P. 1976, ApJ, 203, 297
40. Shull, J.M. 2003, in The IGM/Galaxy Connection: The Distribution of Baryons at z=0, ASSL Conference Proceedings Vol. 281, J.L. Rosenberg & M.E. Putman, eds, Kluwer Academic Publ, p.1

41. Spergel, D.N. *et al.* 2003, ApJS, 148, 175
42. Spitzak, J. & Schneider, s.E. 1998, ApJS, 119, 159
43. Spitzer, L. 1978, Physical Processes in the Interstellar Medium, John Wiley & Sons: New York.)
44. Steidel, C.C. 1993, in The Environment and Evolution of Galaxies, eds. J. Shull & H.A. Thronson, Kluwer Academic Publ, p. 263
45. Taylor, A.R., Braun, R. 1998, Science with the SKA, see http://www.skatelescope.org/pages/science_gen.htm
46. Tozzi, P., *et al.* 2000, ApJ, 528, 597
47. Wolfe, A.M., *et al.* 1986, ApJS, 61, 249
48. Wolfire, M.G., *et al.* 2003, ApJ, 587, 278
49. Zwaan, M.A., *et al.* 1997, ApJ, 490, 173
50. Zwaan, M.A., & Briggs, F.H. 2000, ApJ, 530L, 61
51. Zwaan, M.A., 2001, MNRAS, 325, 1142
52. Zwaan, M.A., *et al.* 2003, AJ, 125, 2842

GRAVITATIONAL LENSING: COSMOLOGICAL MEASURES

R. L. WEBSTER AND C. M. TROTT

School of Physics, University of Melbourne, Parkville, Victoria, 3010, Australia
E-mail: rwebster@physics.unimelb.edu.au, ctrott@physics.unimelb.edu.au

For decades, gravitational lensing has been recognised as the most powerful method for measuring the mass of an astronomical object, in particular instances of near perfect alignment between background sources and foreground masses. Techniques to extend lensing methods to measure cosmological parameters are more recent. These lectures discuss the methodology of estimating the cosmological parameters, and present some of the best measurements to date.

1. Introduction

Gravitational lensing is the term used to describe the dynamical interaction between photons and the geometry of space-time. The physics of gravitational lensing is well understood, so that the observational consequences can be calculated and precisely modelled. Different observational outcomes depend on two primary variables: the cosmological model and the distribution of mass in the object nearest to the line-of-sight.

This review will begin by outlining the observational outcomes of gravitational lensing. Each of the subsequent sections will discuss specific experiments focussed on determining parameters of the cosmological model.

Figure 1 provides a sketch of the wavefront emanating from a source. Initially the wavefront is assumed to be spherical. As the wavefront passes near a massive object, the geometry of space-time is curved, and the wavefront is distorted. As it moves past the deflector, the wavefront is folded, so that an observer 'downstream' will see three different segments of the wavefront. For each segment, the observer will define the direction of the image as perpendicular to the wavefront, and the two orthogonal radii of its curvature will measure the magnification. If the observer is located in the region where the wavefront is folded, then multiple images will be observed. This is termed *strong lensing*. Observers outside this region, will still see observable effects, but these are termed *weak lensing*. It is clear from this geometry, that the observer will always see an odd number of images, unless the mass distribution is singular, as is the case if there is a supermassive black hole at the centre of the galaxy.

We will describe three different regimes, each of which is based on different astrophysics and requires different theoretical modelling.

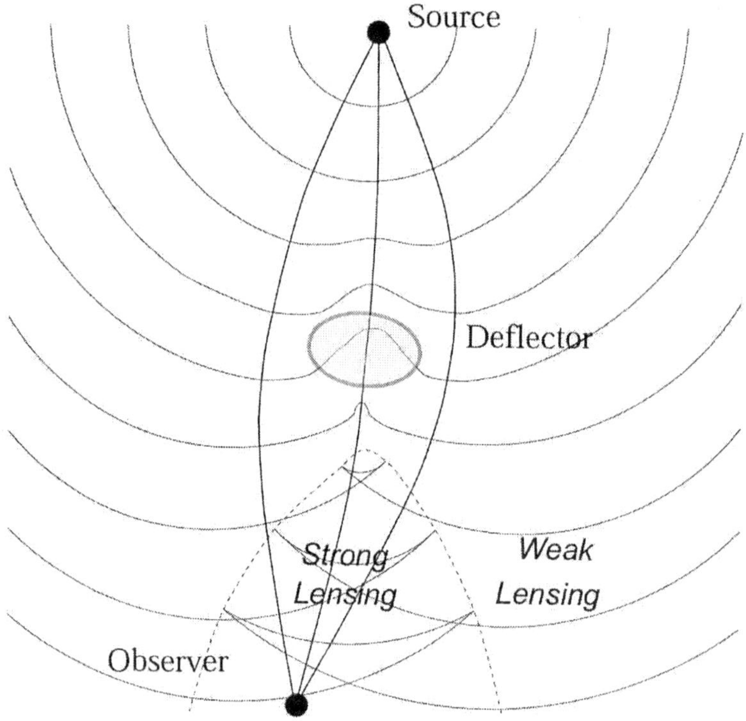

Figure 1. Sketch of the spherical wavefront from a distant source, passing a massive lens, as it travels towards an observer. The regions of strong and weak lensing are marked.

(1) Simple Lenses, which are strong gravitational lenses. In this case, the surface density of the deflector, Σ, is greater than the critical value Σ_{crit} defined in Eq. 12. We consider three different angular sizes: θ_S, the source size, θ_E, the radius of the Einstein ring defined later in this section, and θ_{Res} the resolution of the image, which depends on the seeing, the resolution of the telescope, etc.

If $\theta_{Res} > \theta_E > \theta_S$, then we see image magnification, but we cannot resolve the multiple images. An example is galactic μlensing[1] where we see background stars either in the bulge of our galaxy or in the nearby Large Magellanic Cloud, microlensed by a foreground star in our galaxy.

If $\theta_E > \theta_S > \theta_{Res}$, then we see an Einstein ring. Many examples of these are now known, both at optical and radio wavelengths, and are given on the CASTLES website: http://cfa-www.harvard.edu/glensdata/.

If $\theta_E > \theta_{Res} > \theta_S$, then we observe multiply-imaged quasars, and again excellent examples are given on the CASTLES website. In each of the three remaining inequalities, although strong lensing is occurring, the observational effects are very small, and currently unobservable.

(2) Simple Lenses, which are weak gravitational lenses. Generally this means that $\Sigma < \Sigma_{crit}$, although if the lensing mass is elliptical, it is possible for multiple images to form[2]. In the case of weak lensing by a simple lens, the background image is distorted and slightly magnified. If $\Sigma \lesssim 0.05\Sigma_{crit}$, the magnification is unobservable.

(3) Complex Lenses. In these cases, the effects of a distribution of strong lenses on the background source must be treated statistically. These can be considered as a caustic network in the source plane. An example is the μlensing of a multiply-imaged quasar by a foreground galaxy, or ensemble of stars. Q2237+0305 has been studied in detail by Wyithe and collaborators[3].

(4) Time Delay. If a background source is variable, then a delay is observed between features in the lightcurves due to the different path lengths over which each image is observed. The path length comprises two components: the geometric component and that due to gravitational potential. Examples of suitable background sources are quasars, particularly radio-loud quasars, Gamma Ray Bursts (GRBs) and binary stars.

In order to describe the observational effects of gravitational lensing, the basic astrophysics needs to be elucidated. In the following paragraphs, a brief outline of the main ideas is presented. A fuller discussion is provided in the textbook of Schneider and co-authors[4].

Nearly all observable instances of gravitational lensing are adequately described in the weak field limit of Einstein's theory of General Relativity. In this case, the Newtonian gravitational potential of the lens, $\Phi \ll c^2$. The geometry is defined in figure 2. In the basic configuration there are three important planes: the source plane, the deflector or lens plane and the plane of the observer.

The simplest lens is a point mass. The metric near this mass is described by

$$ds^2 = c^2\xi dt^2 - (\frac{dr^2}{\xi^2} + r^2 d\Omega^2) \tag{1}$$

$$\xi = 1 - \frac{r_s}{r} \tag{2}$$

$$d\Omega^2 = d\theta^2 + sin^2\theta d\phi^2 \tag{3}$$

$$r_s = \frac{2GM}{c^2} \tag{4}$$

where the last term is the Schwarzschild or gravitational radius of the point mass M. Standard textbooks on General Relativity, such as Hartle[5], provide a derivation of the bending angle in the photon path in the lens plane. In the weak field limit, only the linear term describing the *bending angle* α, needs to be considered, giving

$$\alpha = \frac{2r_s}{r_{min}} \tag{5}$$

where r_{min} is the minimum impact parameter of the photon. The verification of this result provides one of the classical tests of General Relativity. This has been

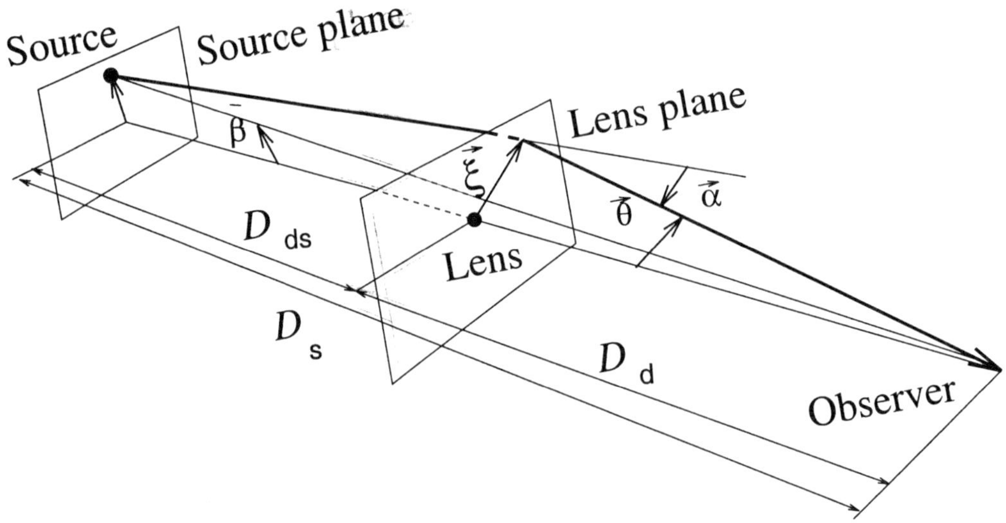

Figure 2. Sketch of source plane, lens plane and observer plane, with the photon path marked in bold. The angles described in the text are marked, as well as the relevant distances.

tested in the solar system to a relative accuracy of 0.001[6].

More complex lenses can be modelled as the sum of point masses, if the lens can be assumed to be thin, *i.e.* all the mass concentrated in the plane perpendicular to the line-of-sight to the source. In practice, if the lens is at a cosmological distance, this approximation will be true.

Many of the relationships used to measure observable quantities can be derived from one simple equation: *the lens equation.* Figure 2 illustrates the geometry of lensing by a single mass distribution. From this geometry, we can derive the lens equation:

$$\theta = \beta + \tilde{\alpha} \tag{6}$$

$$\tilde{\alpha} = \frac{D_{ds}}{D_d}\alpha \tag{7}$$

where the distance D_s is the angular diameter distance from the observer to the source, D_d is the angular diameter distance from the observer to the deflector and D_{ds} is the angular diameter distance from the deflector to the source. In this equation, θ is the observed angular position of the images and the bending angle α is obtained from modelling the mass distribution. The position angle of the unperturbed source β, must be the same for all images, and is a derived quantity. In gravitational lensing, the appropriate distances are affine distances, which are given by angular diameter distances. The angular diameter distance from z_1 to z_2 can be determined for all cosmologies, including those with a positive Λ term, by solving a differential equation:

$$D(z_1, z_2) = a(z_2) f(D_{comoving}) \tag{8}$$

$$D_{comoving} = \frac{c}{H_0} \int_{a(z_1)}^{a(z_2)} [a\Omega_m + a^2(1 - \Omega_m - \Omega_\Lambda) + a^4\Omega_\Lambda]^{-0.5} da \tag{9}$$

where a is the scale factor, f is the radial coordinate distance and Ω_m and Ω_Λ are the normalised density parameters due to matter and Λ.

If the source is on the observer-lens axis, and the lens is assumed to have a constant surface density, then the *critical surface density*, Σ_{crit}, can be derived:

$$\alpha = \frac{D_{ds}}{D_s} \frac{4G}{c^2 \xi} \pi \Sigma \xi^2 \tag{10}$$

$$\xi = D_d \theta \tag{11}$$

$$\Sigma_{crit} = \frac{c^2}{4\pi G} \frac{D_s}{D_d D_{ds}} \tag{12}$$

For cosmological distances, $\Sigma_{crit} \sim 1\,\mathrm{g\,cm^{-2}} \sim 10^{23.5}\,\mathrm{HI\,cm^{-2}}$.

If the source is on-axis, then the image is a ring called the *Einstein ring*. Its radius is:

$$\theta_E = \left(\frac{2r_s D_{ds}}{D_s D_d}\right)^{0.5} \tag{13}$$

For a galaxy mass, and a source at cosmological distances, this radius is about 1 arcsec. The Einstein ring maps the critical curve in the lens plane, which encloses a region where the average surface density is Σ_{crit}. This result can be used to determine the mass enclosed within the observed image configuration to within a few % (due to mass model variations) — $\sim$ 15% (due to variations in the cosmology)[7]. However the recent determination of the cosmological parameters by WMAP will reduce the total error on this mass estimate to a few percent.

The *time delay* between images is directly derivable from the lens equation. The bending angle is the gradient of the lensing potential, ψ, defined as

$$\psi = \frac{2}{c^2} \frac{D_d D_{ds}}{D_s} \int \Phi dl \tag{14}$$

and $\alpha = \nabla_\theta \psi$. Then the lens equation can be written as a gradient,

$$\vec{\nabla}_{\vec{\theta}} [\frac{1}{2}(\vec{\theta} - \vec{\beta})^2 - \psi] = 0. \tag{15}$$

The terms in square brackets are the geometric and gravitational components of the delay respectively. In order to see this equation as a time delay, it is rewritten as

$$t(\vec{\theta}) = \frac{1 + z_d}{c} \frac{D_d D_s}{D_{ds}} [\frac{1}{2}(\vec{\theta} - \vec{\beta})^2 - \psi] \tag{16}$$

$$= t_{geom} + t_{grav}, \tag{17}$$

where z_d is the lens redshift. Images are therefore located where $\vec{\nabla}_{\vec{\theta}} t = 0$, *i.e.* at the extrema of the two-dimensional time delay surface. These can be maxima, minima or saddlepoints in the time delay surface. The terms due to the geometric and the gravitational components are about the same size, and in cases where the source is approximately on-axis, almost cancel. For cosmological distances, the time delay is of the order of a year for galaxian masses, but it depends strongly on the impact parameter. The first image to vary, (*i.e.* the shortest path), will always be an image at a minimum with positive parity. This is the path which does not pass through a caustic.

The Equivalence Principle in General Relativity explicitly states that the bending angle does not depend on the wavelength of the photons. Thus one test of lensing is to observe similar behaviour at different wavelengths. However in the case of a non-uniform source, differential magnification can occur. An example is a quasar, where the source emits at different wavelengths on different scales.

Surface brightness is conserved by gravitational lensing. It is worth noting that different images of a source actually show images of the source from slightly different directions, and so will only be identical if the source is completely spherically symmetric. In particular, the images are not coherent.

The *magnification*, μ, of an image is given by the relative change in area between the source and the image, since surface brightness is conserved. For a symmetric lens, the magnification is:

$$\mu = |\frac{\delta\theta}{\delta\beta}| = \frac{\theta}{\beta}\frac{d\theta}{d\beta} \tag{18}$$

For a transparent, thin non-singular lens, one image is always brighter than the source would have been in the absence of the lens. Equations describing the observables for particular mass distributions can be found in Schneider *et al.* [4].

Gravitational lensing assumes that geometric optics is valid, but when

$$\lambda \sim \frac{GM}{c^2} \tag{19}$$

diffraction effects become important. This gives a maximum magnification of $\mu_{max} \sim r_s/\lambda$. In addition, polarisation is unaffected by gravitational lensing, but in the case of a strong gravitational field, the angle of polarisation will be rotated[8]. Even though surface brightness is conserved, it is possible to change the surface brightness of the CMB (for example) if a lens is moving transversely across the line-of-sight[9]:

$$\Delta T \sim 10\left(\frac{v_t}{1000\,\mathrm{km\,s^{-2}}}\right)K. \tag{20}$$

This effect is due to Special Relativity. More generally, transverse motion of the deflector will induce a wavelength change of

$$\frac{\Delta\lambda}{\lambda} \sim \frac{v_t}{c}\left(\frac{r_s}{\theta D_d}\right) \tag{21}$$

For quasar emission lines, for example, this shift is unlikely to be measurable.

2. H_0: Time Delays and Mass Distributions

Before the first gravitational lens was discovered, Refsdal[10] realised the potential of lensing to measure the Hubble Constant. This technique measures the delay in arrival times between lensed images, of a flux change in the source. In order to extract a value for H_0, the mass distribution of the lens, the angular diameter distances to the source and lens, and the image positions are required. Therefore, an accurate model of the lens and a good measurement of the time delay between images measures H_0, through the ratio of angular diameter distances. There have been many attempts to parameterise the lensing potential, in order to provide an accurate but simple model of the mass distribution. This simplifies the information required to measure H_0 and allows a statistical measure of H_0 using many multiply-imaged systems.

Fassnacht *et al.* [11] list eleven systems where time delays have been measured and a value for H_0 derived. For these systems, the value of H_0 is consistently lower than that measured by the WMAP experiment combined with previous data sets ($H_0 = 71^{+4}_{-3}\mathrm{km\,s^{-1}Mpc^{-1}}$)[12]. Examples of these systems are PKS1830-211, SBS1520+530 and B1608+656.

Winn *et al.* [13] have measured the time delay for the radio lens system PKS1830-211. Comprehensive lens models over the range of expected parameter values provide an estimate of $H_0 = 44 \pm 9$ $\mathrm{km\,s^{-1}Mpc^{-1}}$ using a singular isothermal ellipsoidal mass for the galaxy lens. The system SBS1520+530 comprises two lensing galaxies near the line-of-sight. For this system, Burud *et al.* [14] have used the pseudo-isothermal elliptical mass distributions (PIEMDs)[15] to derive a value of H_0 $= 51 \pm 9$ $\mathrm{km\,s^{-1}Mpc^{-1}}$. As they have monitoring data in only one waveband for the time delay measurement, they are unable to disentangle intrinsic fluctuations in the source from those due to microlensing in the foreground galaxy.

Fassnacht *et al.* [11] have studied the system B1608+656, where the two lensing galaxies are modelled as singular isothermal ellipsoids without external shear. For an $(\Omega_M, \Omega_\Lambda) = (0.3, 0.7)$ universe, they find $H_0 = 61$–65 $(\pm 1 \pm 2)$ $\mathrm{km\,s^{-1}Mpc^{-1}}$, where the values in brackets are the systematic uncertainties. They suggest the uncertainty due to the simplified lens model could be $\sim$15 $\mathrm{km\,s^{-1}Mpc^{-1}}$.

The last point is critical in understanding of the limitations of measuring H_0 using multiply-imaged sources. The lens model is by far the weakest link in the progression from time delay to H_0. To date, all lensing systems used to measure H_0 are at intermediate redshift, making high resolution imaging and modelling of the mass distribution difficult. Without a detailed knowledge of the mass distribution in the system, and galaxies near the line-of-sight which might significantly shear the lightcone, the determination H_0 is uncertain. The question can be reversed to ask whether a knowledge of the value of H_0 from WMAP can constrain the mass distribution in galaxies. Unfortunately, assuming H_0 does not provide a unique solution as there are too many parameters in a mass model of the lens.

Kochanek[16] has discussed the discrepancy between H_0 determined using standard mass models and the higher value of H_0 found by the HST Key Project[17] (and WMAP). Kochanek simplifies lenses by assuming the lensing potential can be expanded into multipoles, retaining the monopole and quadrupole terms. Combined with the slope of the potential in the annulus of mass between the lensed images, this provides a simple framework within which to determine H_0 from time delays and image positions.

Lewis & Ibata (2002)[18] suggest that an evolving equation of state for the universe (a quintessence model) can produce significant changes in the calculated value of H_0, possibly accounting for the differences between the published lens models and the WMAP results. For a sample of $\sim$100 lens systems with measured time delays the equation of state could be constrained, assuming that the mass model of the lens is known.

In summary, there are few well-studied lenses with uncomplicated mass distributions (no multiple lenses, no evidence of major disruption to the system) that can be used for a measurement of H_0. Lower redshift lenses, where more information is available to model mass distributions, will provide a better estimate of H_0. Each system has a large uncertainty, and so studying many systems may provide a statistical estimate of H_0, if the mass modelling does not bias the estimate of H_0. However, the inconsistencies between the values of H_0 determined from gravitational lensing and from the WMAP measurements suggest that fundamental aspects of galaxian mass distributions are yet to be understood.

3. $\Omega_{compact}$ – the Cosmological MACHO Experiment

Gravitational lensing of background sources depends only on the mass of the lens and its distribution, and not on other physical attributes of the lens, such as its luminosity. Thus gravitational lensing provides the most robust method for measuring dark matter distributions. The first set of measurements which will be described, are those which use the observable effects of a population of point mass lenses to determine $\Omega_{compact}$. For a point mass, where the bending angle is $\alpha = 2r_s/\xi$, the separation between the two images is $\Delta\theta = (\beta^2 + 4\theta_E^2)^{0.5}$, and the total magnification of the source is $\mu_{tot} = (u^2+2)/u(u^2+4)^{0.5}$ where $u = \beta\theta_E^{-1}$. Since the probability of lensing scales as β^2, if $\beta \sim \theta_E \sim \Delta\theta/2$ then the probability of lensing by a mass M scales like M. In addition, if $\beta \sim \theta_E$ then the total magnification is fixed: $\mu = 1.34$. This means that the probability of magnification by a factor ≥ 1.34 depends on the total mass in compact objects, and is independent of the distribution of masses.

In a seminal paper, Press and Gunn[19] showed that measurement of the frequency of an observable parameter (for example, multiple images) due to lensing by a compact object would allow the determination of $\Omega_{compact}$. They showed that in a

Table 1. Published limits on $\Omega_{compact}$.

Source	$\Omega_{compact}$	Mass range	Reference
VLA sources	< 0.4	$10^{11} - 10^{13}\,M_\odot$	Hewitt[21]
VLBI sources	< 0.4	$10^{7} - 10^{9}\,M_\odot$	Kassiola *et al.*.[22]
GRBs	≤ 0.2	$10^{-16} - 10^{-13}\,M_\odot$	Marani *et al.*[23]
GRBs	≤ 0.9	$10^{-12.5} - 10^{-9}\,M_\odot$	Marani *et al.*[23]
GRBs	$< 0.15(90\%)$	$10^{6.5}\,M_\odot$	Marani *et al.*[23]
QSO emission lines	< 0.2	$10^{-3} - 60\,M_\odot$	Dalcanton *et al.*[24]

flat universe, the optical depth to lensing by point masses was:

$$\begin{aligned} \tau &= 3\Omega_m\left(\frac{z+2+z(1+z)ln(1+z)}{z} - 4\right) \\ &= \Omega_m\frac{z^2}{4} \qquad \text{for } z \ll 1 \\ &= 0.3\Omega_m \qquad \text{for } z \sim 2 \end{aligned} \tag{22}$$

where Ω_m is the baryonic mass in critical units. Since that time, similar analyses have been performed using different populations as the background sources. If the source counts are steep, then magnification bias will be important (see Sec 5). Selected results are summarised in table 1 and a good early summary is given by Carr[20]. In the case of GRB detections, which have poor angular resolution, the differential time delay can be used to probe scales down to $\sim 10^3\,M_\odot$[25]. So far, about 1500 GRBs have been observed by BATSE, and there have been no convincing lenses discovered. As described in the previous section, these searches have assumed that the profiles of the two 'images' in the GRB are identical. However it is possible that the lensing signature may not be achromatic; as explained earlier, the two images actually image different parts of the source. If the source is beamed, for example, there may be measurable differences between images. Also, Williams and Wijers[26] have calculated the probability of μlensing GRBs and conclude that the effects could be significant. Other limits on $\Omega_{compact}$ use the differential magnification of GRBs measured by two or more satellites[23], the presence of 'spectral lines' in GRB profiles[23] and the redshift evolution in the ratio of continuum and emission line fluxes for QSOs[24].

4. Cosmological Parameters from Strong Lensing

For a well-defined sample of multiply-imaged sources at cosmological redshifts, lensing models provide a strong measure Λ, as the volume of space at high redshifts changes substantially as a function of Λ[27]. Based on a sample of six optical and eleven radio lens systems, Kochanek[28] obtained a value of $\Omega_\Lambda < 0.66(99\%)$ and a maximum likelihood value of $\Omega_\Lambda \sim 0$. He used both de Vaucouleurs profiles and singular isothermal spheres as mass models, finding the former to be inconsistent with observations.

In the most recent analysis in this field, Chae[29] has used the Cosmic Lens All Sky Survey (CLASS) of 8598 flat spectrum radio sources, to undertake a likelihood analysis of the probability distribution of finding thirteen multiply-imaged lenses in the SDSS sample. Chae's method, which is similar to previous work, builds a model for the lensing probability describing the galaxian mass distributions with the following functional forms: Schechter luminosity function, Tully-Fisher relation for late-type galaxies and Faber-Jackson for early-type galaxies. The galaxy mass distributions are modelled as singular isothermal spheres, whose velocity distributions can be either prolate or oblate.

Essentially, the differential probability that a source at z_s with flux F is lensed with image separation between $\Delta\theta$ and $\Delta\theta + d(\Delta\theta)$ due to galaxies at redshifts z to $z + \Delta z$ is:

$$\frac{d^2p(z, \Delta\theta; z_s, F)}{dzd(\Delta\theta)} \propto \sigma \times B(z_s, F) \times L(z, z_s, \Delta\theta) \tag{23}$$

where σ is the cross-section to lensing, $B(z_s, F)$ is the magnification bias and $L(z, z_s, \Delta\theta)$ is the effective luminosity function. This differential probability is then integrated over the line-of-sight and image separations to give the total expected probability given a particular cosmology. The galaxian parameters are fixed to the values determined in the SDSS[31] and 2dFGRS[30] samples. Unfortunately only 26 sources and 7 lenses in the CLASS sample have redshifts, so the redshift distribution is assumed from other studies. The statistical likelihood is then minimised to find the best fitting cosmological parameters given the number of lensed and unlensed galaxies found in the CLASS sample.

Chae determines the cosmological parameters under a number of different assumptions. Figure 3 shows plots of Ω_Λ as a function of Ω_m assuming the 2dFGRS luminosity function, both where the normalising velocity dispersions for the Faber-Jackson and Tully-Fisher relations are assumed, and where they are not. Table 2 provides a summary of Chae's most important results for $\Omega_m = 1 - \Omega_\lambda$ and for ω in the equation of state for the dark energy for a flat Λ-dominated universe. Chae claims that the small value of the mean image separation of the lens candidates of the CLASS sample gives a significantly higher value of Ω_Λ than Kochanek obtained. However the results are quite sensitive to the assumed slope of the faint end of the galaxy luminosity function.

The measurements of the cosmological parameters from latest strong lensing studies are concordant with the results from WMAP. However the results are strongly dependent on the choice of parameters for the galaxian mass models, and will not substantially improve until much larger samples of multiply-imaged sources, with known source and lens redshifts are discovered. The SDSS collaboration estimate they will find ~1000 lenses in their photometric sample, 100 of which will be above their spectroscopic limit[32]. Such a sample would provide an independent method for measuring Ω_Λ.

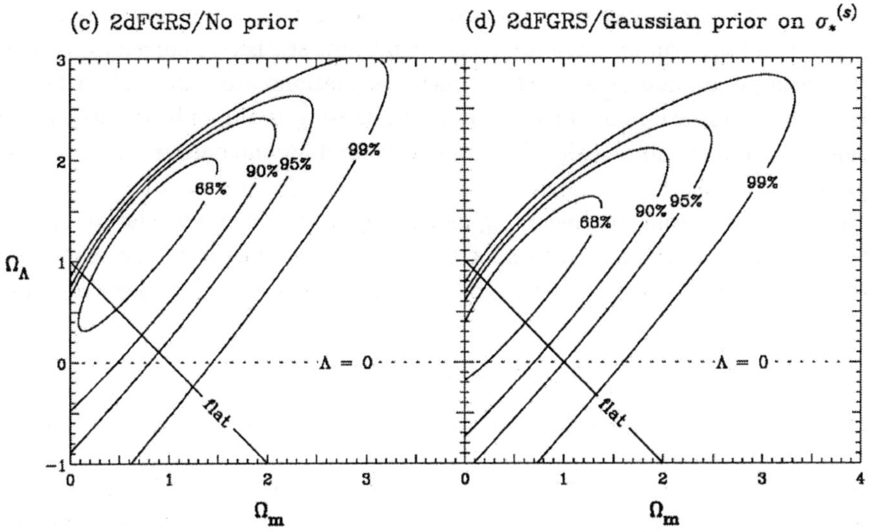

Figure 3. Ω_m, Λ plot from Chae showing likelihood contours for the cosmological parameters using the 2dFGRS luminosity function. Figure on the left (right) shows the regions without (with) the velocity dispersion assumed.

Table 2. 68% confidence limits on Ω_m and ω from Chae.

	Assuming Galaxy Parameters	Unconstrained
Ω_m	$0.40^{+0.28}_{-0.16}$	$0.31^{+0.27}_{-0.14}$
ω	$< -0.41^{+0.28}_{-0.16}$	$< -0.55^{+0.18}_{-0.11}$

5. The Bias Factor

When samples of gravitational lenses are selected from a magnitude-limited sample of sources, as is always the case, the magnification bias must be taken into account in determining the probability of gravitational lensing, and therefore the mass in lensing objects.

Suppose we observe sources to a flux limit f. If the sources in a particular direction are magnified by a factor μ, then our image will include sources with intrinsic magnitudes $> f/\mu$. However these sources are observed over a larger area of sky, by a factor μ. Thus a larger number of sources will be observed if the number counts of the background sources are steep enough. Suppose the intrinsic counts of sources with flux greater than f is n_0, and the observed counts greater than f is n. If $n_0(> f) \propto f^{-\alpha}$, then $n(> f) = \mu^{\alpha-1} n_0(> f)$. For bright quasars, $\alpha \sim 2.5$, and for faint quasars $\alpha \sim 0.65$, so a measurable magnification bias might be expected for bright quasars.

Observationally, magnification bias will induce an apparent correlation between the number or luminosity of a background source population and a foreground population. Many studies have produced significant statistical correlations, particularly on large angular scales, where mass correlations are expected to be weak. If the correlations are real, then gravitational lensing is the only sensible physical explanation, and the luminosity functions measured for the background sources are affected.

In the weak lensing or linear regime, the quasar-galaxy correlation ξ_{QG}, can be expressed as a simple function of the galaxy bias factor b, the slope of the background counts α and the cross-correlation between the magnification and the density contrast $\xi_{\mu\delta}$[33],

$$\xi_{QG}(\theta) = b(2.5\alpha - 1)\xi_{\mu\delta}(\theta) \tag{24}$$

$\xi_{\mu\delta}$ must be measured independently from the amplitude of the power spectrum of the density fluctuations.

Some of the uncertainty in this measurement can be reduced by using two different foreground lensing populations and measuring the bias factor as a function of scale, independently of other cosmological parameters[34]. A robust measure of ξ_{QG} taking account of non-linear biasing has yet to be made.

6. Cosmological Parameters from Weak Lensing

Weak lensing allows the determination of cosmological parameters by measuring of the distortion background sources. Masses near the line-of-sight magnify and distort the lightcone, changing the shape of the image on the sky. Statistical ensembles of background images are required to determine the lensing effects as the intrinsic shapes of individual galaxies are unknown.

6.1. *Theoretical Background*

The observed ellipticity ϵ_I, of the image can be related to the source ellipticity ϵ_S through the reduced complex shear[35],

$$\mathbf{g} = \frac{\gamma}{(1-\kappa)} \tag{25}$$

$$\epsilon_I = \frac{\epsilon_S + \mathbf{g}}{1 - \mathbf{g}^* \epsilon_S} \tag{26}$$

where γ is the complex shear, κ is the convergence of the lens and the latter equation is applicable for weakly lensed images. An individual galaxy cannot be deconvolved to find its true shape, but an average over many galaxies should produce a non-random signal of induced ellipticity. The image ellipticities are measured, averaged on a suitable angular scale and the mass distribution that has produced the mean ellipticities is constructed. The allowed cosmologies for constructed mass distribution are then determined. A range of statistical techniques are used to measure the cosmological parameters and are fully described in recent review papers[35,36].

6.2. *Parameter Dependencies*

Weak lensing depends on Ω_m, Λ, σ_8 and Γ. Ω_m and Λ define the length of the light path and the distribution of matter and therefore are both critical to the strength of the distortion. The normalisation of the power spectrum, σ_8, gives the overall strength of clustering and so it also normalises the strength of weak lensing. The shape parameter, Γ will be reflected in the polarisation correlation function and its relation to the power spectrum of density fluctuations.

Van Waerbeke *et al.* [37] show that the degeneracies in the measurement of the cosmological parameters using the CMB (most recently the WMAP experiment) are approximately orthogonal to the degeneracies in the weak lensing determinations. Thus weak lensing provides an alternative method which is independent of the Type Ia supernovae results.

6.3. *Measurement of Observational Parameters*

Weak lensing studies measure the distortion and magnification of resolved images due to large-scale structures near the line-of-sight. If all galaxies were intrinsically round and had sharp edges, this process would be straightforward. However, galaxies have different shapes, sizes and orientations on the sky. The ellipticity of an individual galaxy is measured using moments, since particularly at high redshift, galaxies are faint and poorly resolved. The effect of a PSF needs to be accurately removed from all images and this usually is accomplished by comparison with a nearby star on the CCD. Variations in the PSF across the CCD can reduce the effectiveness of this strategy, and the image residuals may have significant effects on the results if not properly included. Both fixed apertures and fixed isophotes are used to define boundaries of the galaxy images.

6.4. *Results*

The only way to determine the efficiency and possible utility of weak lensing in measuring the cosmological parameters is to simulate possible observations. Van Waerbeke *et al.* [37] have simulated maps of the sky in 5×5 and 10×10 sq. degree fields using a Gaussian random field source background and a foreground generated to represent the large scale structure for a given cosmology. Noise is then added to the simulation. The convergence map is then reconstructed using the technique of Bartelmann *et al.* [38]. This technique uses the χ^2 statistic to reconstruct the lensing potential from the measured reduced shear and magnification. They find a large (6σ) separation in the skewness measurement for open (Ω_m=0.3) and flat (Ω_m=1.0) universes, providing a useful discriminant independent of σ_8 and Γ. They also find that the power spectrum normalisation, σ_8, may be measurable to $\sim 2\%$, a smaller error than current CMB studies. The simulations assume that the redshifts of the source populations are precisely known. In practice, redshifts are likely to be photometric at best, increasing the errors on the parameter estimations. Most importantly, the simulations provide powerful arguments for extensive observational studies of weak lensing. Simulations by Bartelmann and Schneider[39] based on another statistical method, the mass aperture technique, find that Ω_m could be measured to $\approx 27\%$ and σ_8 to 8%, if an area of side length 8 degrees is imaged.

Barber[40] investigates the importance of known source redshifts on the determination of parameters using numerical simulation. He constructs a ΛCDM cosmology mass distribution and calculates the components of the lensing matrix to give a three dimensional shear field. Integration along the line-of-sight then provides the two dimensional shear. He finds the shear variance is well represented by a power law,

$$\langle\gamma^2\rangle(\theta, z_s) = 1.05 \times 10^{-3}\theta^{-1.12}z_s^{2.07}. \tag{27}$$

for $z_s < 1.6$ and 2 arcmin $\leq \theta \leq$ 32 arcmin, the angular scales likely to be probed by most observational studies. In particular, Barber claims that source redshifts differing by 0.1 can give errors in the parameters of $10 - 20\%$ on small scales.

Recently, Heavens[41] considered the measurement of w from weak lensing. He shows that full three dimensional information about the shear field can provide tight constraints on the value of w ($\approx 1\%$). CMB measurements do not constrain this parameter well ($w < -0.78$ at 95% confidence according to the recent WMAP results[12]). This may be an area where weak lensing can provide a stringent measurement.

Van Waerbeke *et al.* [42] have imaged 1.75 sq. degrees of sky at the CFHT. They measure a weak lensing signal, consistent with a ΛCDM cosmology, but do not have a large area to statistically measure values for the cosmological parameters. This survey is expanding and four colour photometric redshifts will be included in the analysis, providing better parameter estimation in the next few years.

The problems of anisotropic PSFs, source redshifts, cosmic variance and intrinsic galaxy alignments ensure that the measurement of cosmological parameters using weak lensing remains an observational challenge. In the coming years, larger deeper surveys such as VISTA, SDSS and CFHT will greatly reduce the errors in the measurement of the cosmological parameters. Since the degeneracies are orthogonal to the CMB measurements, the investment of substantial effort in this technique is warranted.

7. Conclusions

Gravitational lensing determinations of the parameters for the cosmological model provide robust and independent measurements of Ω_m, Ω_Λ, $\Omega_{compact}$, σ_8, H_0, b, ω and Γ. Each method has observational or modelling limitations at the present time, but the potential for either an improvement over existing WMAP measurements, or confirmation of alternative methods is sufficient to warrant a substantial investment in the required observational programs.

References

1. B. Paczynski, *Ap.J.* **304**, 1 (1986)
2. K. Subramanian and S.A. Cowling, *M.N.R.A.S.* **219**, 333 (1986)
3. J.S.B. Wyithe, R.L. Webster, E.L. Turner and D.J. Mortlock, *M.N.R.A.S.* **315**, 62 (2000)
4. P. Schneider, J. Ehlers and E.E Falco, *Gravitational Lenses*, Springer-Verlag, Berlin (1992)
5. J.B. Hartle, *Gravity, An Introduction to Einstein's General Relativity*, Addison Wesley, San Francisco, (2003)
6. D.E. Lebach *et al.*, *Phys.Rev.Lett.* **75**, 1439L (1995)
7. C.S. Kochanek, *Ap.J.* **373**, 354 (1991)
8. S. Pinault, *M.N.R.A.S.* **179**, 691 (1977)
9. M. Birkenshaw, in *Lecture in Physics, No 330, Gravitational Lenses,* eds. J.N. Moran *et al.*, Springer-Verlag, Berlin (1989), p59.
10. S. Refsdal, *M.N.R.A.S.* **128**, 307 (1964)
11. C.D. Fassnacht *et al.*,*Ap.J.* **581**, 823 (2002)
12. C.L. Bennett *et al.*, *Ap.J.S.* **148**, 1 (2003)
13. J.N Winn *et al.*, *Ap.J.* **575**, 103 (2002)
14. I. Burud *et al.*, *A.& A.* **391**, 481 (2002)
15. C. Faure *et al.*, *A.& A.* **386**, 69 (2002)
16. C.S. Kochanek, *Ap.J.* **583**, 49 (2003)
17. W.L. Freedman *et al.*, *Ap.J.* **553**, 47 (2001)
18. G.F. Lewis and R.A. Ibata, *M.N.R.A.S.* **337**, 26 (2002)
19. W.H. Press and J.E. Gunn, *Ap.J.* **185**, 397 (1973)
20. B.J.Carr, *Ann.Rev.Astron.Astrophys.* **32**, 531 (1994)
21. J.N. Hewett *et al.*, in *Lecture in Physics, No 330, Gravitational Lenses,* eds. J.N. Moran *et al.*, Springer-Verlag, Berlin (1989), p147.
22. A. Kassiola, I. Kovner and R.D. Blandford, *Ap.J.* **381**, 6 (1991)
23. G.F. Marani *et al.*, *Ap.J.* **512**, L13 (1999)
24. J.J. Dalcanton *et al.*, *Ap.J.* **424**, 550 (1994)
25. O.M. Blaes and R.L.Webster, *Ap.J.L.* **284**, 1 (1992)
26. L.L.R. Williams and R.A.M.J. Wijers *M.N.R.A.S.* **286**, L11 (1997)
27. E.L. Turner, *Ap.J.L.* **365**, 43 (1990)
28. C.S. Kochanek, *Ap.J.* **466**, 638 (1996)
29. K.-H. Chae, *M.N.R.A.S.* **346**, 746 (2003)
30. P. Norberg *et al.*, *M.N.R.A.S.* **328**, 64 (2001)
31. M.R. Blanton *et al.*, *A.J.* **121**, 2358 (2001)
32. G.T. Richards *et al.*, *BAAS Meeting 194* (1999)
33. M. Bartelmann, *A.& A.* **298**, 661 (1995)
34. L. van Waerbeke, *A.& A.* **334**, 1 (1998)
35. Y. Mellier, *Ann.Rev.Astron.Astrophys.* **37**, 127 (1999)
36. M. Bartelmann and P. Schneider, *Ph.R.* **340**, 291 (2001)
37. L. van Waerbeke, F. Bernardeau and Y. Mellier, *A.& A.* **342**, 15 (1999)
38. M. Bartelmann *et al.*, *Ap.J.* **464**, 115 (1996)
39. M. Bartelmann and P. Schneider, *A.& A.* **345**, 17 (1999)
40. A.J. Barber, *A.J.* **335**, 909 (2002)
41. A. Heavens, *M.N.R.A.S.* **343**, 1327 (2003)
42. L. van Waerbeke, *et al.*, *A.& A.* **358**, 30 (2000)

PARTICLE PHYSICS AND COSMOLOGY

JOHN ELLIS

Theoretical Physics Division, CERN, CH- 1211 Geneva 23, Switzerland
E-mail: John.Ellis@cern.ch

In the first Lecture, the Big Bang and the Standard Model of particle physics are introduced, as well as the structure of the latter and open issues beyond it. Neutrino physics is discussed in the second Lecture, with emphasis on models for neutrino masses and oscillations. The third Lecture is devoted to supersymmetry, including the prospects for discovering it at accelerators or as cold dark matter. Inflation is reviewed from the viewpoint of particle physics in the fourth Lecture, including simple models with a single scalar inflaton field: the possibility that this might be a sneutrino is proposed. Finally, the fifth Lecture is devoted to topics further beyond the Standard Model, such as grand unification, baryo- and leptogenesis — that might be due to sneutrino inflaton decays — and ultra-high-energy cosmic rays — that might be due to the decays of metastable superheavy dark matter particles.

1. Introduction to the Standard Models

1.1. *The Big Bang and Particle Physics*

The Universe is currently expanding almost homogeneously and isotropically, as discovered by Hubble, and the radiation it contains is cooling as it expands adiabatically:

$$a \times T \simeq \text{Constant}, \tag{1}$$

where a is the scale factor of the Universe and T is the temperature. There are two important pieces of evidence that the scale factor of the Universe was once much smaller than it is today, and correspondingly that its temperature was much higher. One is the *Cosmic Microwave Background* [1], which bathes us in photons with a density

$$n_\gamma \simeq 400 \text{ cm}^{-3}, \tag{2}$$

with an effective temperature $T \simeq 2.7$ K. These photons were released when electrons and nuclei combined to form atoms, when the Universe was some 3000 times hotter and the scale factor correspondingly 3000 times smaller than it is today. The second is the agreement of the *Abundances of Light Elements* [2], in particular those of ^{4}He, Deuterium and ^{6}Li, with calculations of cosmological nucelosynthesis. For these elements to have been produced by nuclear fusion, the Universe must once have been some 10^9 times hotter and smaller than it is today.

During this epoch of the history of the Universe, its energy density would have been dominated by relativistic particles such as photons and neutrinos, in which case the age t of the Universe is given approximately by

$$t \propto a^2 \propto \frac{1}{T^2}. \tag{3}$$

The constant of proportionality between time and temperature is such that $t \simeq 1$ second when the temperature $T \simeq 1$ MeV, near the start of cosmological nucleosynthesis. Since typical particle energies in a thermal plasma are $\mathcal{O}(T)$, and the Boltzmann distribution guarantees large densities of particles weighing $\mathcal{O}(T)$, the history of the earlier Universe when $T > \mathcal{O}(1)$ MeV was dominated by elementary particles weighing an MeV or more [3].

The landmarks in the history of the Universe during its first second presumably included the epoch when protons and neutrons were created out of quarks, when $T \sim 200$ MeV and $t \sim 10^{-5}$ s. Prior to that, there was an epoch when the symmetry between weak and electromagnetic interactions was broken, when $T \sim 100$ GeV and $t \sim 10^{-10}$ s. Laboratory experiments with accelerators have already explored physics at energies $E \lesssim 100$ GeV, and the energy range $E \lesssim 1000$ GeV, corresponding to the history of the Universe when $t \gtrsim 10^{-12}$ s, will be explored at CERN's LHC accelerator that is scheduled to start operation in 2007 [4]. Our ideas about physics at earlier epochs are necessarily more speculative, but one possibility is that there was an inflationary epoch when the age of the Universe was somewhere between 10^{-40} and 10^{-30} s.

We return later to possible experimental probes of the physics of these early epochs, but first we review the Standard Model of particle physics, which underlies our description of the Universe since it was 10^{-10} s old.

1.2. *Summary of the Standard Model of Particle Physics*

The Standard Model of particle physics has been established by a series of experiments and theoretical developments over the past century [5], including:

- 1897 – The discovery of the electron;
- 1910 – The discovery of the nucleus;
- 1930 – The nucleus found to be made of protons and neutrons; neutrino postulated;
- 1936 – The muon discovered;
- 1947 – Pion and strange particles discovered;
- 1950s – Many strongly-interacting particles discovered;
- 1964 – Quarks proposed;
- 1967 – The Standard Model proposed;
- 1973 – Neutral weak interactions discovered;
- 1974 – The charm quark discovered;
- 1975 – The τ lepton discovered;

- 1977 – The bottom quark discovered;
- 1979 – The gluon discovered;
- 1983 – The intermediate $W^{\pm}, Z^0$ bosons discovered;
- 1989 – Three neutrino species counted;
- 1994 – The top quark discovered;
- 1998 – Neutrino oscillations discovered.

All the above historical steps, apart from the last (which was made with neutrinos from astrophysical sources), fit within the Standard Model, and the Standard Model continues to survive all experimental tests at accelerators.

The Standard Model contains the following set of spin-1/2 matter particles:

$$\text{Leptons}: \begin{pmatrix} \nu_e \\ e \end{pmatrix}, \begin{pmatrix} \nu_\mu \\ \mu \end{pmatrix}, \begin{pmatrix} \nu_\tau \\ \tau \end{pmatrix} \tag{4}$$

$$\text{Quarks}: \begin{pmatrix} u \\ d \end{pmatrix}, \begin{pmatrix} c \\ s \end{pmatrix}, \begin{pmatrix} b \\ t \end{pmatrix} \tag{5}$$

We know from experiments at CERN's LEP accelerator in 1989 that there can only be three neutrinos [6]:

$$N_\nu = 2.9841 \pm 0.0083, \tag{6}$$

which is a couple of standard deviations below 3, but that cannot be considered a significant discrepancy. I had always hoped that N_ν might turn out to be non-integer: $N_\nu = \pi$ would have been good, and $N_\nu = e$ would have been even better, but this was not to be! The constraint (6) is also important for possible physics beyond the Standard Model, such as supersymmetry as we discuss later. The measurement (6) implies, by extension, that there can only be three charged leptons and hence no more quarks, by analogy and in order to preserve the calculability of the Standard Model [7].

The forces between these matter particles are carried by spin-1 bosons: electromagnetism by the familiar massless photon γ, the weak interactions by the massive intermediate $W^{\pm}$ and Z^0 bosons that weigh $\simeq 80, 91$ GeV, respectively, and the strong interactions by the massless gluon. *Among the key objectives of particle physics are attempts to unify these different interactions, and to explain the very different masses of the various matter particles and spin-1 bosons.*

Since the Standard Model is the rock on which our quest for new physics must be built, we now review its basic features [5] and examine whether its successes offer any hint of the direction in which to search for new physics. Let us first recall the structure of the charged-current weak interactions, which have the current-current form:

$$\frac{1}{4}\mathcal{L}_{cc} = \frac{G_F}{\sqrt{2}} \; J^+_\mu \; J^{-\mu}, \tag{7}$$

where the charged currents violate parity maximally:

$$J^+_\mu \; = \; \Sigma_{\ell=e,\mu,\tau} \bar{\ell}\gamma_\mu(1-\gamma_5)\nu_\ell \; + \; \text{similarly for quarks.} \tag{8}$$

The charged current (8) can be interpreted as a generator of a weak SU(2) isospin symmetry acting on the matter-particle doublets in (5). The matter fermions with left-handed helicities are doublets of this weak SU(2), whereas the right-handed matter fermions are singlets. It was suggested already in the 1930s, and with more conviction in the 1960s, that the structure (8) could most naturally be obtained by exchanging massive $W^{\pm}$ vector bosons with coupling g and mass m_W:

$$\frac{G_F}{\sqrt{2}} \equiv \frac{g^2}{8m_W^2}. \tag{9}$$

In 1973, neutral weak interactions with an analogous current-current structure were discovered at CERN:

$$\frac{1}{4}\mathcal{L}_{NC} = \frac{G_F^{NC}}{\sqrt{2}} \; J_\mu^0 \; J^{\mu 0}, \tag{10}$$

and it was natural to suggest that these might also be carried by massive neutral vector bosons Z^0.

The $W^{\pm}$ and Z^0 bosons were discovered at CERN in 1983, so let us now review the theory of them, as well as the Higgs mechanism of spontaneous symmetry breaking by which we believe they acquire masses [8]. The vector bosons are described by the Lagrangian

$$\mathcal{L} = -\frac{1}{4} \; G_{\mu\nu}^i G^{i\mu\nu} - \frac{1}{4} F_{\mu\nu} F^{\mu\nu} \tag{11}$$

where $G_{\mu\nu}^i \equiv \partial_\mu W_\nu^i - \partial_\nu W_\mu^i + ig\epsilon_{ijk} W_\mu^j W_\nu^k$ is the field strength for the SU(2) vector boson W_μ^i, and $F_{\mu\nu} \equiv \partial_\mu W_\nu^i - \partial_\nu W_\mu^i$ is the field strength for a U(1) vector boson B_μ that is needed when we incorporate electromagnetism. The Lagrangian (11) contains bilinear terms that yield the boson propagators, and also trilinear and quartic vector-boson interactions.

The vector bosons couple to quarks and leptons via

$$\mathcal{L}_F = -\sum_f i \;\; \left[\bar{f}_L \gamma^\mu D_\mu f_L + \bar{f}_R \gamma^\mu D_\mu f_R \right] \tag{12}$$

where the D_μ are covariant derivatives:

$$D_\mu \equiv \partial_\mu - i \; g \; \sigma_i \; W_\mu^i - i \; g' \; Y \; B_\mu \tag{13}$$

The SU(2) piece appears only for the left-handed fermions f_L, whereas the U(1) vector boson B_μ couples to both left- and right-handed compnents, via their respective hypercharges Y.

The origin of all the masses in the Standard Model is postulated to be a weak doublet of scalar Higgs fields, whose kinetic term in the Lagrangian is

$$\mathcal{L}_\phi = -|D_\mu \phi|^2 \tag{14}$$

and which has the magic potential:

$$\mathcal{L}_V = -V(\phi) : V(\phi) = -\mu^2 \phi^\dagger \phi + \frac{\lambda}{2} (\phi^\dagger \phi)^2 \tag{15}$$

Because of the negative sign for the quadratic term in (15), the symmetric solution $<0|\phi|0>=0$ is unstable, and if $\lambda > 0$ the favoured solution has a non-zero vacuum expectation value which we may write in the form:

$$<0|\phi|0>=<0|\phi^\dagger|0>=v\begin{pmatrix}0\\ \frac{1}{\sqrt{2}}\end{pmatrix} : v^2=\frac{\mu^2}{2\lambda} \tag{16}$$

corresponding to spontaneous breakdown of the electroweak symmetry.

Expanding around the vacuum: $\phi=<0|\phi|0>+\hat{\phi}$, the kinetic term (14) for the Higgs field yields mass terms for the vector bosons:

$$\mathcal{L}_\phi \ni -\frac{g^2v^2}{2}\ W_\mu^+\ W^{\mu-} - g'^2\ \frac{v^2}{2}\ B_\mu\ B^\mu + g\ g'v^2\ B_\mu\ W^{\mu 3} - g^2\ \frac{v^2}{2}\ W_\mu^3\ W^{\mu 3} \tag{17}$$

corresponding to masses

$$m_{W^\pm}=\frac{gv}{2} \tag{18}$$

for the charged vector bosons. The neutral vector bosons (W_μ^3, B_μ) have a 2×2 mass-squared matrix:

$$\begin{pmatrix}\frac{g^2}{2} & \frac{-gg'}{2}\\ \frac{-gg'}{2} & \frac{g'^2}{2}\end{pmatrix} v^2 \tag{19}$$

This is easily diagonalized to yield the mass eigenstates:

$$Z_\mu=\frac{gW_\mu^3-g'B_\mu}{\sqrt{g^2+g'^2}}\ :\ m_Z=\frac{1}{2}\sqrt{g^2+g'^2}v\ ;\ A_\mu=\frac{g'W_\mu^3+gB_\mu}{\sqrt{g^2+g'^2}}\ :\ m_A=0 \tag{20}$$

that we identify with the massive Z^0 and massless γ, respectively. It is useful to introduce the electroweak mixing angle θ_W defined by

$$\sin\theta_W=\frac{g'}{\sqrt{g^2+g'^2}} \tag{21}$$

in terms of the weak SU(2) coupling g and the weak U(1) coupling g'. Many other quantities can be expressed in terms of $\sin\theta_W$ (21): for example, $m_W^2/m_Z^2=\cos^2\theta_W$.

With these boson masses, one indeed obtains charged-current interactions of the current-current form (8) shown above, and the neutral currents take the form:

$$J_\mu^0 \equiv J_\mu^3 - \sin^2\theta_W\ J_\mu^{em}\ ,\ \ G_F^{NC}\equiv\frac{g^2+g'^2}{8m_Z^2} \tag{22}$$

The ratio of neutral- and charged-current interaction strengths is often expressed as

$$\rho=\frac{G_F^{NC}}{G_F}=\frac{m_W^2}{m_Z^2\cos^2\theta_W} \tag{23}$$

which takes the value unity in the Standard Model, apart from quantum corrections (loop effects).

The previous field-theoretical discussion of the Higgs mechanism can be rephrased in more physical language. It is well known that a massless vector boson such as the photon γ or gluon g has just two polarization states: $\lambda = \pm 1$. However, a massive vector boson such as the ρ has three polarization states: $\lambda = 0, \pm 1$. This third polarization state is provided by a spin-0 field. In order to make $m_{W^\pm, Z^0} \neq 0$, this should have non-zero electroweak isospin $I \neq 0$, and the simplest possibility is a complex isodoublet (ϕ^+, ϕ^0), as assumed above. This has four degrees of freedom, three of which are eaten by the $W^\pm$ amd Z^0 as their third polarization states, leaving us with one physical Higgs boson H. Once the vacuum expectation value $|\langle 0|\phi|0\rangle| = v/\sqrt{2}\ :\ \ v = \mu/\sqrt{2\lambda}$ is fixed, the mass of the remaining physical Higgs boson is given by

$$m_H^2 = 2\mu^2 = 4\lambda v^2, \tag{24}$$

which is a free parameter in the Standard Model.

1.3. *Precision Tests of the Standard Model*

The quantity that was measured most accurately at LEP was the mass of the Z^0 boson [6]:

$$m_Z \;=\; 91,187.5 \pm 2.1 \text{ MeV}, \tag{25}$$

as seen in Fig. 1. Strikingly, m_Z is now known more accurately than the muon decay constant! Attaining this precision required understanding astrophysical effects — those of terrestrial tides on the LEP beam energy, which were $\mathcal{O}(10)$ MeV, as well as meteorological — when it rained, the water expanded the rock in which LEP was buried, again changing the beam energy, and seasonal variations in the level of water in Lake Geneva also caused the rock around LEP to expand and contract — as well as electrical — stray currents from the nearby electric train line affected the LEP magnets [9].

LEP experiments also made precision measurements of many properties of the Z^0 boson [6], such as the total cross section:

$$\sigma \;=\; \frac{12\pi}{m_Z^2}\frac{\Gamma_{ee}\Gamma_{had}}{\Gamma_Z^2}, \tag{26}$$

where $\Gamma_Z(\Gamma_{ee}, \Gamma_{had})$ is the total Z^0 decay rate (rate for decays into e^+e^-, hadrons). Eq. (26) is the classical (tree-level) expression, which is reduced by about 30 % by radiative corrections. The total decay rate is given by:

$$\Gamma_Z \;=\; \Gamma_{ee} + \Gamma_{\mu\mu} + \Gamma_{\tau\tau} + N_\nu \Gamma_{\nu\nu} + \Gamma_{had}, \tag{27}$$

where we expect $\Gamma_{ee} = \Gamma_{\mu\mu} = \Gamma_{\tau\tau}$ because of lepton universality, which has been verified experimentally, as seen in Fig. 2 [6]. Other partial decay rates have been

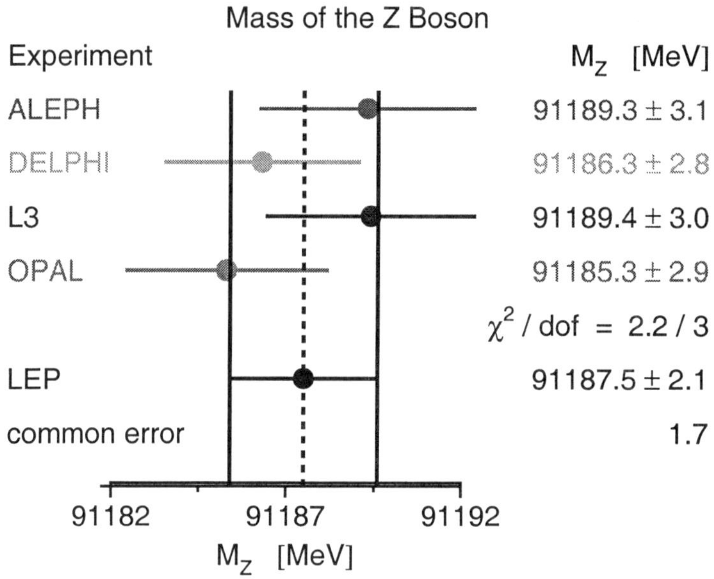

Figure 1. The mass of the Z^0 vector boson is one of the parameters of the Standard Model that has been measured most accurately [6].

measured via the branching ratios

$$R_{b,c} \equiv \frac{\Gamma_{\bar{b}b,\bar{c}c}}{\Gamma_{had}}, \tag{28}$$

as seen in Fig. 3.

Also measured have been various forward-backward asymmetries $A_{\ell,q}$ in the production of leptons and quarks, as well as the polarization of τ leptons produced in Z^0 decay, as also seen in Fig. 3. Various other measurements are also shown there, including the mass and decay rate of the $W^\pm$, the mass of the top quark, and low-energy neutral-current measurements in ν-nucleon scattering and parity violation in atomic Cesium. The Standard Model is quite compatible with all these measurements, although some of them may differ by a couple of standard deviations: if they did not, we should be suspicious! Overall, the electroweak measurements tell us that [6]:

$$\sin^2\theta_W = 0.23148 \pm 0.00017, \tag{29}$$

providing us with a strong hint for grand unification, as we see later.

1.4. *The Search for the Higgs Boson*

The precision electroweak measurements at LEP and elsewhere are sensitive to radiative corrections via quantum loop diagrams, in particular those involving particles such as the top quark and the Higgs boson that are too heavy to be observed directly at LEP [10,11]. Many of the electroweak observables mentioned above exhibit

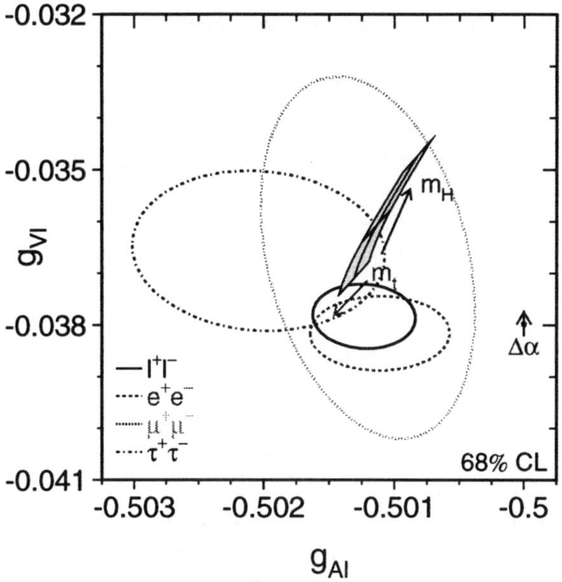

Figure 2. Precision measurements of the properties of the charged leptons e, μ and τ indicate that they have universal couplings to the weak vector bosons [6], whose value favours a relatively light Higgs boson.

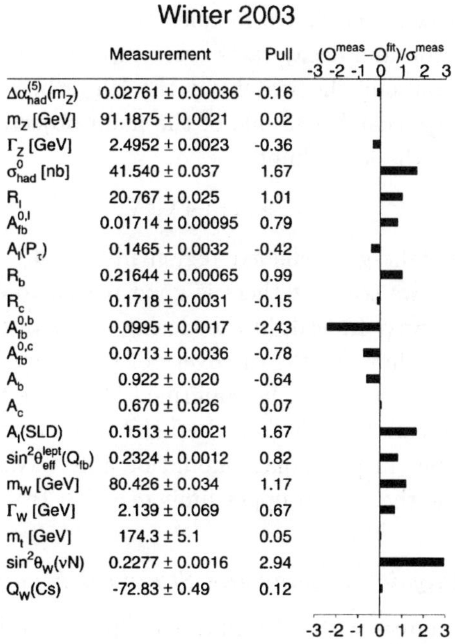

Figure 3. Precision electroweak measurements and the pulls they exert in a global fit [6].

quadratic sensitivity to the mass of the top quark:

$$\Delta \propto G_F m_t^2. \tag{30}$$

The measurements of these electroweak observables enabled the mass of the top quark to be predicted before it was discovered, and the measured value:

$$m_t = 174.3 \pm 5.1 \text{ GeV} \tag{31}$$

agrees quite well with the prediction

$$m_t = 177.5 \pm 9.3 \text{ GeV} \tag{32}$$

derived from precision electroweak data [6]. Electroweak observables are also sensitive logarithmically to the mass of the Higgs boson:

$$\Delta \propto \left(\frac{\alpha}{\pi}\right) \ln\left(\frac{m_H^2}{m_Z^2}\right), \tag{33}$$

so their measurements can also be used to predict the mass of the Higgs boson. This prediction can be made more definite by combining the precision electroweak data with the measurement (31) of the mass of the top quark. Making due allowance for theoretical uncertainties in the Standard Model calculations, as seen in Fig. 4, one may estimate that [6]:

$$m_H = 91^{+58}_{-37} \text{ GeV}, \tag{34}$$

whereas m_H is not known from first principles in the Standard Model.

The Higgs production and decay rates are completely fixed as functions of the unknown mass m_H, enabling the search for the Higgs boson to be planned as a function of m_H [12]. This search was one of the main objectives of experiments at LEP, which established the lower limit:

$$m_H > 114.4 \text{ GeV}, \tag{35}$$

that is shown as the light gray shaded region in Fig. 4. Combining this limit with the estimate (34), we see that there is good reason to expect that the Higgs boson may not be far away. Indeed, in the closing weeks of the LEP experimental programme, there was a hint for the discovery of the Higgs boson at LEP with a mass ~ 115 GeV, but this could not be confirmed [13]. In the future, experiments at the Fermilab Tevatron collider and then the LHC will continue the search for the Higgs boson. The latter, in particular, should be able to discover it whatever its mass may be, up to the theoretical upper limit $m_H \lesssim 1$ TeV [4].

1.5. *Roadmap to Physics Beyond the Standard Model*

The Standard Model agrees with all confirmed experimental data from accelerators, but is theoretically very unsatisfactory [14,15]. It does not explain the particle quantum numbers, such as the electric charge Q, weak isospin I, hypercharge Y and colour, and contains at least 19 arbitrary parameters. These include three

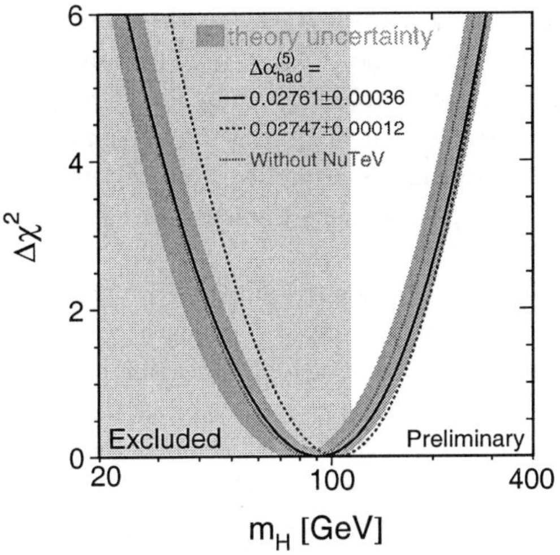

Figure 4. Estimate of the mass of the Higgs boson obtained from precision electroweak measurements. The mid-gray band indicates theoretical uncertainties, and the different curves demonstrate the effects of different plausible estimates of the renormalization of the fine-structure constant at the Z^0 peak [6].

independent vector-boson couplings and a possible CP-violating strong-interaction parameter, six quark and three charged-lepton masses, three generalized Cabibbo weak mixing angles and the CP-violating Kobayashi-Maskawa phase, as well as two independent masses for weak bosons.

The Big Issues in physics beyond the Standard Model are conveniently grouped into three categories [14,15]. These include the problem of **Mass**: what is the origin of particle masses, are they due to a Higgs boson, and, if so, why are the masses so small; **Unification**: is there a simple group framework for unifying all the particle interactions, a so-called Grand Unified Theory (GUT); and **Flavour**: why are there so many different types of quarks and leptons and why do their weak interactions mix in the peculiar way observed? Solutions to all these problems should eventually be incorporated in a Theory of Everything (TOE) that also includes gravity, reconciles it with quantum mechanics, explains the origin of space-time and why it has four dimensions, makes coffee, etc. String theory, perhaps in its current incarnation of M theory, is the best (only?) candidate we have for such a TOE [16], but we do not yet understand it well enough to make clear experimental predictions.

As if the above 19 parameters were insufficient to appall you, at least nine more parameters must be introduced to accommodate the neutrino oscillations discussed in the next Lecture: 3 neutrino masses, 3 real mixing angles, and 3 CP-violating phases, of which one is in principle observable in neutrino-oscillation experiments and the other two in neutrinoless double-beta decay experiments. In fact even the simplest models for neutrino masses involve 9 further parameters, as discussed later.

Moreover, there are many other cosmological parameters that we should also seek to explain. Gravity is characterized by at least two parameters, the Newton constant G_N and the cosmological vacuum energy. We may also want to construct a field-theoretical model for inflation, and we certainly need to explain the baryon asymmetry of the Universe. So there is plenty of scope for physics beyond the Standard Model.

The first clear evidence for physics beyond the Standard Model of particle physics has been provided by neutrino physics, which is also of great interest for cosmology, so this is the subject of Lecture 2. Since there are plenty of good reasons to study supersymmetry [15], including the possibility that it provides the cold dark matter, this is the subject of Lecture 3. Inflation is the subject of Lecture 4, and various further topics such as GUTs, baryo/leptogenesis and ultra-high-energy cosmic rays are discussed in Lecture 5. As we shall see later, neutrino physics may be the key to both inflation and baryogenesis.

2. Neutrino Physics

2.1. *Neutrino Masses?*

There is no good reason why either the total lepton number L or the individual lepton flavours $L_{e,\mu,\tau}$ should be conserved. Theorists have learnt that the only conserved quantum numbers are those associated with exact local symmetries, just as the conservation of electromagnetic charge is associated with local U(1) invariance. On the other hand, there is no exact local symmetry associated with any of the lepton numbers, so we may expect non-zero neutrino masses.

However, so far we have only upper experimental limits on neutrino masses [17]. From measurements of the end-point in Tritium β decay, we know that:

$$m_{\nu_e} \lesssim 2.5 \text{ eV}, \tag{36}$$

which might be improved down to about 0.5 eV with the proposed KATRIN experiment [18]. From measurements of $\pi \to \mu\nu$ decay, we know that:

$$m_{\nu_\mu} < 190 \text{ KeV}, \tag{37}$$

and there are prospects to improve this limit by a factor ~ 20. Finally, from measurements of $\tau \to n\pi\nu$ decay, we know that:

$$m_{\nu_\tau} < 18.2 \text{ MeV}, \tag{38}$$

and there are prospects to improve this limit to ~ 5 MeV.

Astrophysical upper limits on neutrino masses are stronger than these laboratory limits. The 2dF data were used to infer an upper limit on the sum of the neutrino masses of 1.8 eV [19], which has recently been improved using WMAP data to [20]

$$\Sigma_{\nu_i} m_{\nu_i} < 0.7 \text{ eV}, \tag{39}$$

as seen in Fig. 5. This impressive upper limit is substantially better than even the most stringent direct laboratory upper limit on an individual neutrino mass.

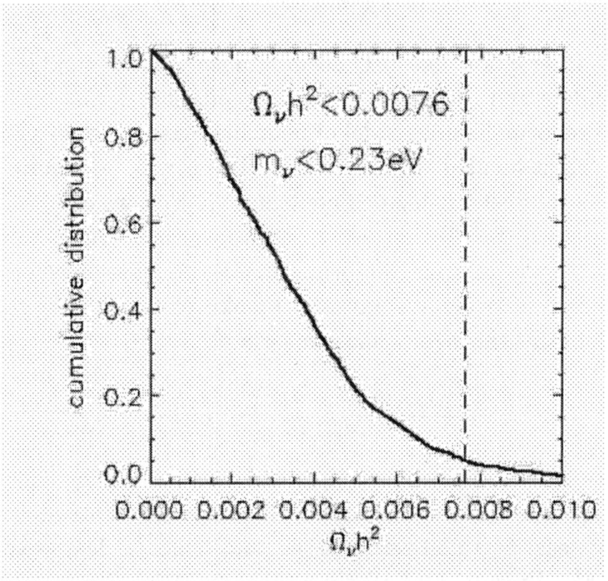

Figure 5. Likelihood function for the sum of neutrino masses provided by WMAP [20]: the quoted upper limit applies if the 3 light neutrino species are degenerate.

Another interesting laboratory limit on neutrino masses comes from searches for neutrinoless double-β decay, which constrain the sum of the neutrinos' Majorana masses weighted by their couplings to electrons [21]:

$$\langle m_\nu \rangle_e \equiv |\Sigma_{\nu_i} m_{\nu_i} U_{ei}^2| \lesssim 0.35 \text{ eV} \tag{40}$$

which might be improved to ~ 0.01 eV in a future round of experiments.

Neutrinos have been seen to oscillate between their different flavours [22,23], showing that the separate lepton flavours $L_{e,\mu,\tau}$ are indeed not conserved, though the conservation of total lepton number L is still an open question. The observation of such oscillations strongly suggests that the neutrinos have different masses.

2.2. *Models of Neutrino Masses and Mixing*

The conservation of lepton number is an accidental symmetry of the renormalizable terms in the Standard Model Lagrangian. However, one could easily add to the Standard Model non-renormalizable terms that would generate neutrino masses, even without introducing any new fields. For example, a non-renormalizable term of the form [24]

$$\frac{1}{M} \nu H \cdot \nu H, \tag{41}$$

where M is some large mass beyond the scale of the Standard Model, would generate a neutrino mass term:

$$m_\nu \nu \cdot \nu : \; m_\nu \; = \; \frac{\langle 0|H|0\rangle^2}{M}. \tag{42}$$

However, a new interaction like (41) seems unlikely to be fundamental, and one should like to understand the origin of the large mass scale M.

The minimal renormalizable model of neutrino masses requires the introduction of weak-singlet 'right-handed' neutrinos N. These will in general couple to the conventional weak-doublet left-handed neutrinos via Yukawa couplings Y_ν that yield Dirac masses $m_D = Y_\nu \langle 0|H|0\rangle \sim m_W$. In addition, these 'right-handed' neutrinos N can couple to themselves via Majorana masses M that may be $\gg m_W$, since they do not require electroweak symmetry breaking. Combining the two types of mass term, one obtains the seesaw mass matrix [25]:

$$(\nu_L, N) \begin{pmatrix} 0 & M_D \\ M_D^T & M \end{pmatrix} \begin{pmatrix} \nu_L \\ N \end{pmatrix}, \tag{43}$$

where each of the entries should be understood as a matrix in generation space.

In order to provide the two measured differences in neutrino masses-squared, there must be at least two non-zero masses, and hence at least two heavy singlet neutrinos N_i [26,27]. Presumably, all three light neutrino masses are non-zero, in which case there must be at least three N_i. This is indeed what happens in simple GUT models such as SO(10), but some models [28] have more singlet neutrinos [29]. In this Lecture, for simplicity we consider just three N_i.

The effective mass matrix for light neutrinos in the seesaw model may be written as:

$$\mathcal{M}_\nu \; = \; Y_\nu^T \frac{1}{M} Y_\nu v^2, \tag{44}$$

where we have used the relation $m_D = Y_\nu v$ with $v \equiv \langle 0|H|0\rangle$. Taking $m_D \sim m_q$ or m_ℓ and requiring light neutrino masses $\sim 10^{-1}$ to 10^{-3} eV, we find that heavy singlet neutrinos weighing $\sim 10^{10}$ to 10^{15} GeV seem to be favoured.

It is convenient to work in the field basis where the charged-lepton masses $m_{\ell^\pm}$ and the heavy singlet-neutrino mases M are real and diagonal. The seesaw neutrino mass matrix $\mathcal{M}_\nu$ (44) may then be diagonalized by a unitary transformation U:

$$U^T \mathcal{M}_\nu U \; = \; \mathcal{M}_\nu^d. \tag{45}$$

This diagonalization is reminiscent of that required for the quark mass matrices in the Standard Model. In that case, it is well known that one can redefine the phases of the quark fields [30] so that the mixing matrix U_{CKM} has just one CP-violating phase [31]. However, in the neutrino case, there are fewer independent field phases, and one is left with 3 physical CP-violating parameters:

$$U \; = \; \tilde{P}_2 V P_0 : \; P_0 \equiv \mathrm{Diag}\left(e^{i\phi_1}, e^{i\phi_2}, 1\right). \tag{46}$$

Here $\tilde{P}_2 = \mathrm{Diag}\left(e^{i\alpha_1}, e^{i\alpha_2}, e^{i\alpha_3}\right)$ contains three phases that can be removed by phase rotations and are unobservable in light-neutrino physics, though they do play a rôle at high energies, as discussed in Lecture 5, V is the light-neutrino mixing matrix first considered by Maki, Nakagawa and Sakata (MNS) [32], and P_0 contains 2 CP-violating phases $\phi_{1,2}$ that are observable at low energies. The MNS matrix describes neutrino oscillations

$$V = \begin{pmatrix} c_{12} & s_{12} & 0 \\ -s_{12} & c_{12} & 0 \\ 0 & 0 & 1 \end{pmatrix} \begin{pmatrix} 1 & 0 & 0 \\ 0 & c_{23} & s_{23} \\ 0 & -s_{23} & c_{23} \end{pmatrix} \begin{pmatrix} c_{13} & 0 & s_{13} \\ 0 & 1 & 0 \\ -s_{13}e^{-i\delta} & 0 & c_{13}e^{-i\delta} \end{pmatrix}. \quad (47)$$

The three real mixing angles $\theta_{12,23,13}$ in (47) are analogous to the Euler angles that are familiar from the classic rotations of rigid mechanical bodies. The phase δ is a specific quantum effect that is also observable in neutrino oscillations, and violates CP, as we discuss below. The other CP-violating phases $\phi_{1,2}$ are in principle observable in neutrinoless double-β decay (40).

2.3. *Neutrino Oscillations*

In quantum physics, particles such as neutrinos propagate as complex waves. Different mass eigenstates m_i travelling with the same momenta p oscillate with different frequencies:

$$e^{iE_it} : \ E_i^2 = p^2 + m_i^2. \quad (48)$$

Now consider what happens if one produces a neutrino beam of one given flavour, corresponding to some specific combination of mass eigenstates. After propagating some distance, the different mass eigenstates in the beam will acquire different phase weightings (48), so that the neutrinos in the beam will be detected as a mixture of different neutrino flavours. These oscillations will be proportional to the mixing $\sin^2 2\theta$ between the different flavours, and also to the differences in masses-squared Δm_{ij}^2 between the different mass eigenstates.

The first of the mixing angles in (47) to be discovered was θ_{23}, in atmospheric neutrino experiments. Whereas the numbers of downward-going atmospheric ν_μ were found to agree with Standard Model predictions, a deficit of upward-going ν_μ was observed, as seen in Fig. 6. The data from the Super-Kamiokande experiment, in particular [22], favour near-maximal mixing of atmospheric neutrinos:

$$\theta_{23} \sim 45^o, \ \Delta m_{23}^2 \sim 2.4 \times 10^{-3} \ \mathrm{eV}^2. \quad (49)$$

Recently, the K2K experiment using a beam of neutrinos produced by an accelerator has found results consistent with (49) [33]. It seems that the atmospheric ν_μ probably oscillate primarily into ν_τ, though this has yet to be established.

More recently, the oscillation interpretation of the long-standing solar-neutrino deficit has been established, in particular by the SNO experiment. Solar neutrino experiments are sensitive to the mixing angle θ_{12} in (47). The recent data

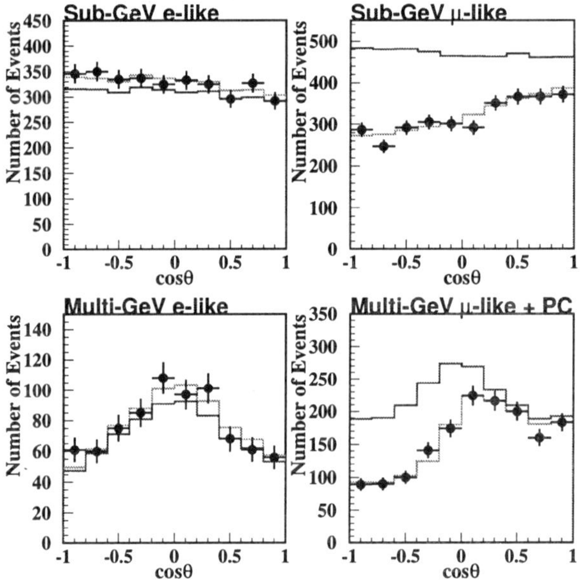

Figure 6. The zenith angle distributions of atmospheric neutrinos exhibit a deficit of downward-moving ν_μ, which is due to neutrino oscillations [22].

from SNO [23] and Super-Kamiokande [34] prefer quite strongly the large-mixing-angle (LMA) solution to the solar neutrino problem with

$$\theta_{12} \sim 30^o, \ \Delta m^2_{12} \sim 6 \times 10^{-5} \text{ eV}^2, \tag{50}$$

though they have been unable to exclude completely the LOW solution with lower δm^2. However, the KamLAND experiment on reactors produced by nuclear power reactors has recently found a deficit of ν_e that is highly compatible with the LMA solution to the solar neutrino problem [35], as seen in Fig. 7, and excludes any other solution.

Using the range of θ_{12} allowed by the solar and KamLAND data, one can establish a correlation between the relic neutrino density $\Omega_\nu h^2$ and the neutrinoless double-β decay observable $\langle m_\nu \rangle_e$, as seen in Fig. 8 [37]. Pre-WMAP, the experimental limit on $\langle m_\nu \rangle_e$ could be used to set the bound

$$10^{-3} \lesssim \Omega_\nu h^2 \lesssim 10^{-1}. \tag{51}$$

Alternatively, now that WMAP has set a tighter upper bound $\Omega_\nu h^2 < 0.0076$ (39) [20], one can use this correlation to set an upper bound:

$$< m_\nu >_e \lesssim 0.1 \text{ eV}, \tag{52}$$

which is difficult to reconcile with the neutrinoless double-β decay signal reported in [21].

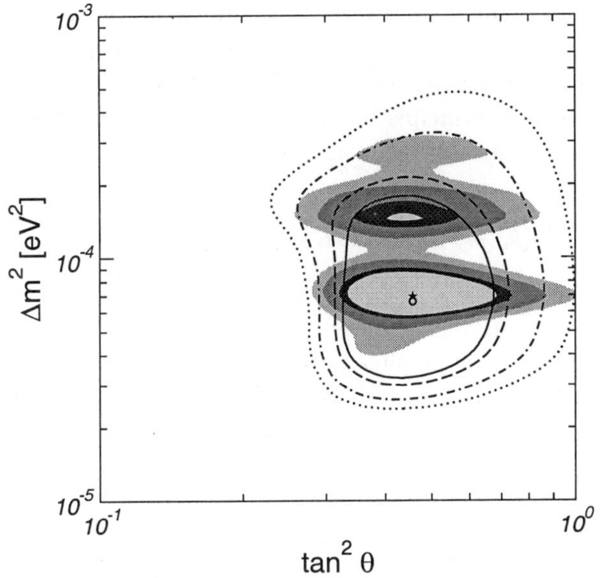

Figure 7. The KamLAND experiment (shadings) finds [35] a deficit of reactor neutrinos that is consistent with the LMA neutrino oscillation parameters previously estimated (ovals) on the basis of solar neutrino experiments [36].

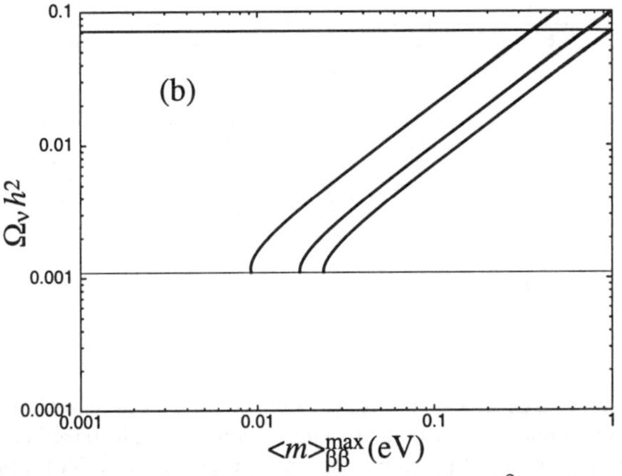

Figure 8. The correlation between the relic density of neutrinos $\Omega_\nu h^2$ and the neutrinoless double-β decay observable: the different lines indicate the ranges allowed by neutrino oscillation experiments [37].

The third mixing angle θ_{13} in (47) is basically unknown, with experiments such as Chooz [38] and Super-Kamiokande only establishing upper limits. *A fortiori*, we have no experimental information on the CP-violating phase δ.

The phase δ could in principle be measured by comparing the oscillation probabilities for neutrinos and antineutrinos and computing the CP-violating asymmetry [39]:

$$\begin{aligned} P(\nu_e \to \nu_\mu) - P(\bar{\nu}_e \to \bar{\nu}_\mu) &= 16 s_{12} c_{12} s_{13} c_{13}^2 s_{23} c_{23} \sin\delta \\ &\quad \sin\left(\frac{\Delta m_{12}^2}{4E} L\right) \sin\left(\frac{\Delta m_{13}^2}{4E} L\right) \sin\left(\frac{\Delta m_{23}^2}{4E} L\right), \end{aligned} \tag{53}$$

as seen in Fig. 9 [40]. This is possible only if Δm_{12}^2 and s_{12} are large enough — as now suggested by the success of the LMA solution to the solar neutrino problem, and if s_{13} is large enough — which remains an open question.

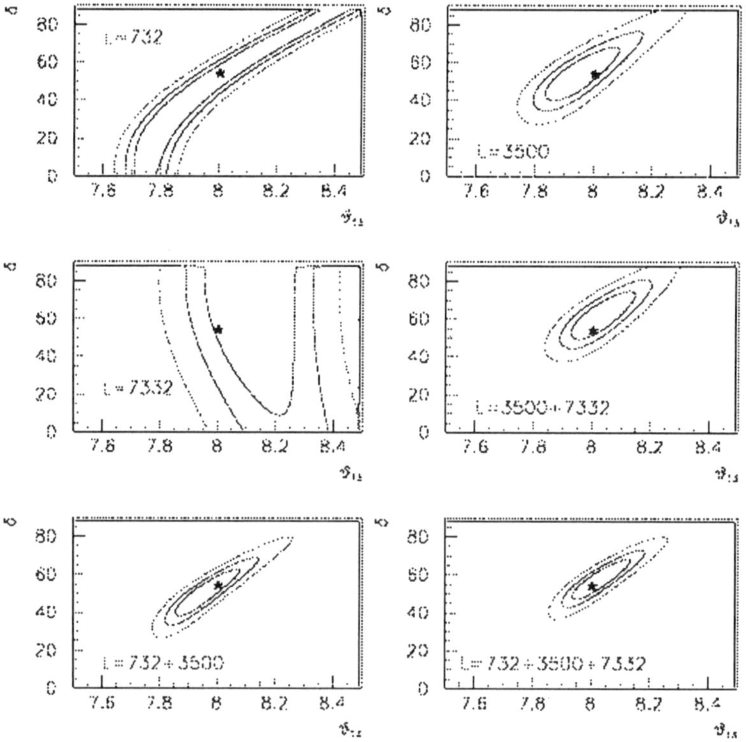

Figure 9. Possible measurements of θ_{13} and δ that could be made with a neutrino factory, using a neutrino energy threshold of about 10 GeV. Using a single baseline correlations are very strong, but can be largely reduced by combining information from different baselines and detector techniques [40], enabling the CP-violating phase δ to be extracted.

A number of long-baseline neutrino experiments using beams from accelerators are now being prepared in the United States, Europe and Japan, with the objectives of measuring more accurately the atmospheric neutrino oscillation parameters, $\Delta m^2_{23}, \theta_{23}$ and θ_{13}, and demonstrating the production of ν_τ in a ν_μ beam. Beyond these, ideas are being proposed for intense 'super-beams' of low-energy neutrinos, produced by high-intensity, low-energy accelerators such as the SPL [41] proposed at CERN. A subsequent step could be a storage ring for unstable ions, whose decays would produce a 'β beam' of pure ν_e or $\bar{\nu}_e$ neutrinos. These experiments might be able to measure δ via CP and/or T violation in neutrino oscillations [42]. A final step could be a full-fledged neutrino factory based on a muon storage ring, which would produce pure ν_μ and $\bar{\nu}_e$ (or ν_e and $\bar{\nu}_\mu$ beams and provide a greatly enhanced capability to search for or measure δ via CP violation in neutrino oscillations [43].

We have seen above that the effective low-energy mass matrix for the light neutrinos contains 9 parameters, 3 mass eigenvalues, 3 real mixing angles and 3 CP-violating phases. However, these are not all the parameters in the minimal seesaw model. As shown in Fig. 10, this model has a total of 18 parameters [44,45]. The additional 9 parameters comprise the 3 masses of the heavy singlet 'right-handed' neutrinos M_i, 3 more real mixing angles and 3 more CP-violating phases. As illustrated in Fig. 10, many of these may be observable via renormalization in supersymmetric models [46,45,47,48], which may generate observable rates for flavour-changing lepton decays such as $\mu \to e\gamma, \tau \to \mu\gamma$ and $\tau \to e\gamma$, and CP-violating observables such as electric dipole moments for the electron and muon. Some of these extra parameters may also have controlled the generation of matter in the Universe via leptogenesis [49], as discussed in Lecture 5.

3. Supersymmetry

3.1. *Why?*

The main theoretical reason to expect supersymmetry at an accessible energy scale is provided by the *hierarchy problem* [51]: why is $m_W \ll m_P$, or equivalently why is $G_F \sim 1/m_W^2 \gg G_N = 1/m_P^2$? Another equivalent question is why the Coulomb potential in an atom is so much greater than the Newton potential: $e^2 \gg G_N m^2 = m^2/m_P^2$, where m is a typical particle mass?

Your first thought might simply be to set $m_P \gg m_W$ by hand, and forget about the problem. Life is not so simple, because quantum corrections to m_H and hence m_W are quadratically divergent in the Standard Model:

$$\delta m^2_{H,W} \simeq \mathcal{O}(\frac{\alpha}{\pi})\Lambda^2, \tag{54}$$

which is $\gg m_W^2$ if the cutoff Λ, which represents the scale where new physics beyond the Standard Model appears, is comparable to the GUT or Planck scale. For example, if the Standard Model were to hold unscathed all the way up the

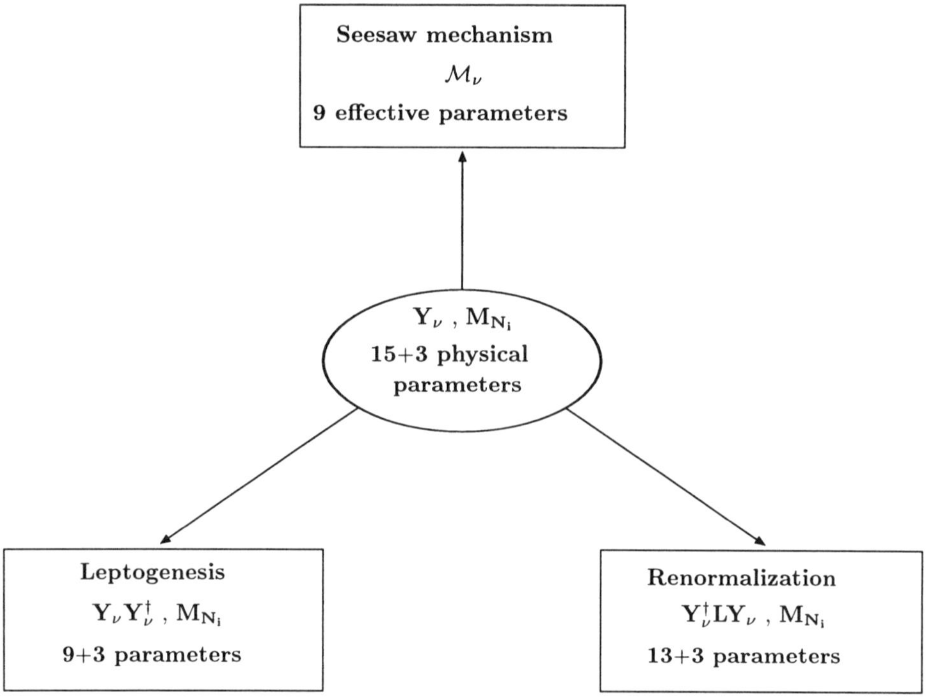

Figure 10. Roadmap for the physical observables derived from Y_ν and N_i [50].

Planck mass $m_P \sim 10^{19}$ GeV, the radiative correction (54) would be 36 orders of magnitude greater than the physical values of $m^2_{H,W}$!

In principle, this is not a problem from the mathematical point of view of renormalization theory. All one has to do is postulate a tree-level value of m^2_H that is (very nearly) equal and opposite to the 'correction' (54), and the correct physical value may be obtained by a delicate cancellation. However, this fine tuning strikes many physicists as rather unnatural: they would prefer a mechanism that keeps the 'correction' (54) comparable at most to the physical value [51].

This is possible in a supersymmetric theory, in which there are equal numbers of bosons and fermions with identical couplings. Since bosonic and fermionic loops have opposite signs, the residual one-loop correction is of the form

$$\delta m^2_{H,W} \simeq \mathcal{O}(\frac{\alpha}{\pi})(m^2_B - m^2_F), \qquad (55)$$

which is $\lesssim m_{H,W}^2$ and hence naturally small if the supersymmetric partner bosons B and fermions F have similar masses:

$$|m_B^2 - m_F^2| \lesssim 1 \text{ TeV}^2. \tag{56}$$

This is the best motivation we have for finding supersymmetry at relatively low energies [51]. In addition to this first supersymmetric miracle of removing (55) the quadratic divergence (54), many logarithmic divergences are also absent in a supersymmetric theory [52], a property that also plays a rôle in the construction of supersymmetric GUTs [14].

Supersymmetry had been around for some time before its utility for stabilizing the hierarchy of mass scales was realized. Some theorists had liked it because it offered the possibility of unifying fermionic matter particles with bosonic force-carrying particles. Some had liked it because it reduced the number of infinities found when calculating quantum corrections — indeed, theories with enough supersymmetry can even be completely finite [52]. Theorists also liked the possibility of unifying Higgs bosons with matter particles, though the first ideas for doing this did not work out very well [53]. Another aspect of supersymmetry, that made some theorists think that its appearance should be inevitable, was that it was the last possible symmetry of field theory not yet known to be exploited by Nature [54]. Yet another asset was the observation that making supersymmetry a local symmetry, like the Standard Model, necessarily introduced gravity, offering the prospect of unifying *all* the particle interactions. Moreover, supersymmetry seems to be an essential requirement for the consistency of string theory, which is the best candidate we have for a Theory of Everything, including gravity. However, none of these 'beautiful' arguments gave a clue about the scale of supersymmetric particle masses: this was first provided by the hierarchy argument outlined above.

Could any of the known particles in the Standard Model be paired up in supermultiplets? Unfortunately, none of the known fermions q, ℓ can be paired with any of the 'known' bosons $\gamma, W^{\pm}Z^0, g, H$, because their internal quantum numbers do not match [53]. For example, quarks q sit in triplet representations of colour, whereas the known bosons are either singlets or octets of colour. Then again, leptons ℓ have non-zero lepton number $L = 1$, whereas the known bosons have $L = 0$. Thus, the only possibility seems to be to introduce new supersymmetric partners (spartners) for all the known particles, as seen in the Table below: quark $\rightarrow$ squark, lepton $\rightarrow$ slepton, photon $\rightarrow$ photino, Z $\rightarrow$ Zino, W $\rightarrow$ Wino, gluon $\rightarrow$ gluino, Higgs $\rightarrow$ Higgsino. The best that one can say for supersymmetry is that it economizes on principle, not on particles!

Particle	Spin	Spartner	Spin
quark: q	$\frac{1}{2}$	squark: $\tilde{q}$	0
lepton: ℓ	$\frac{1}{2}$	slepton: $\tilde{\ell}$	0
photon: γ	1	photino: $\tilde{\gamma}$	$\frac{1}{2}$
W	1	wino: $\tilde{W}$	$\frac{1}{2}$
Z	1	zino: $\tilde{Z}$	$\frac{1}{2}$
Higgs: H	0	higgsino: $\tilde{H}$	$\frac{1}{2}$

The minimal supersymmetric extension of the Standard Model (MSSM) [55] has the same vector interactions as the Standard Model, and the particle masses arise in much the same way. However, in addition to the Standard Model particles and their supersymmetric partners in the Table, the minimal supersymmetric extension of the Standard Model (MSSM), requires two Higgs doublets $H, \bar{H}$ with opposite hypercharges in order to give masses to all the matter fermions, whereas one Higgs doublet would have sufficed in the Standard Model. The two Higgs doublets couple via an extra coupling called μ, and it should also be noted that the ratio of Higgs vacuum expectation values

$$\tan\beta \equiv \frac{\langle \bar{H} \rangle}{\langle H \rangle} \tag{57}$$

is undetermined and should be treated as a free parameter.

3.2. *Hints of Supersymmetry*

There are some phenomenological hints that supersymmetry may, indeed, appear at the TeV scale. One is provided by the strengths of the different Standard Model interactions, as measured at LEP [56]. These may be extrapolated to high energy scales including calculable renormalization effects [57], to see whether they unify as predicted in a GUT. The answer is no, if supersymmetry is not included in the calculations. In that case, GUTs would require a ratio of the electromagnetic and weak coupling strengths, parametrized by $\sin^2\theta_W$, different from what is observed (29), if they are to unify with the strong interactions. On the other hand, as seen in Fig. 11, minimal supersymmetric GUTs predict just the correct ratio for the weak and electromagnetic interaction strengths, *i.e.*, value for $\sin^2\theta_W$ (29).

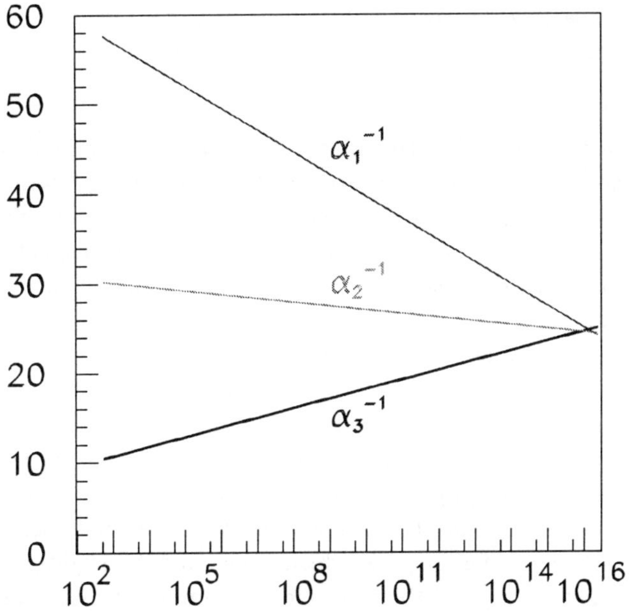

Figure 11. The measurements of the gauge coupling strengths at LEP, including $\sin^2 \theta_W$ (29), evolve to a unified value if supersymmetry is included [56].

A second hint is the fact that precision electroweak data prefer a relatively light Higgs boson weighing less than about 200 GeV [6]. This is perfectly consistent with calculations in the minimal supersymmetric extension of the Standard Model (MSSM), in which the lightest Higgs boson weighs less than about 130 GeV [58].

A third hint is provided by the astrophysical necessity of cold dark matter. This could be provided by a neutral, weakly-interacting particle weighing less than about 1 TeV, such as the lightest supersymmetric particle (LSP) χ [59]. This is expected to be stable in the MSSM, and hence should be present in the Universe today as a cosmological relic from the Big Bang [60,59]. Its stability arises because there is a multiplicatively-conserved quantum number called R parity, that takes the values +1 for all conventional particles and -1 for all sparticles [53]. The conservation of R parity can be related to that of baryon number B and lepton number L, since

$$R = (-1)^{3B+L+2S} \tag{58}$$

where S is the spin. There are three important consequences of R conservation:

(1) sparticles are always produced in pairs, *e.g.*, $\bar{p}p \to \tilde{q}\tilde{g}X$, $e^+e^- \to \tilde{\mu} + \tilde{\mu}^-$,

(2) heavier sparticles decay to lighter ones, *e.g.*, $\tilde{q} \to q\tilde{g}, \tilde{\mu} \to \mu\tilde{\gamma}$, and
(3) the lightest sparticle (LSP) is stable, because it has no legal decay mode.

This last feature constrains strongly the possible nature of the lightest supersymmetric sparticle [59]. If it had either electric charge or strong interactions, it would surely have dissipated its energy and condensed into galactic disks along with conventional matter. There it would surely have bound electromagnetically or via the strong interactions to conventional nuclei, forming anomalous heavy isotopes that should have been detected.

A priori, the LSP might have been a sneutrino partner of one of the 3 light neutrinos, but this possibility has been excluded by a combination of the LEP neutrino counting and direct searches for cold dark matter. Thus, the LSP is often thought to be the lightest neutralino χ of spin 1/2, which naturally has a relic density of interest to astrophysicists and cosmologists: $\Omega_\chi h^2 = \mathcal{O}(0.1)$ [59].

Finally, a fourth hint may be coming from the measured value of the muon's anomalous magnetic moment, $g_\mu - 2$, which seems to differ slightly from the Standard Model prediction [61,62]. If there is indeed a significant discrepancy, this would require new physics at the TeV scale or below, which could easily be provided by supersymmetry, as we see later.

3.3. *Constraints on Supersymmetric Models*

Important experimental constraints on supersymmetric models have been provided by the unsuccessful direct searches at LEP and the Tevatron collider. When compiling these, the supersymmetry-breaking masses of the different unseen scalar particles are often assumed to have a universal value m_0 at some GUT input scale, and likewise the fermionic partners of the vector bosons are also commonly assumed to have universal fermionic masses $m_{1/2}$ at the GUT scale — the so-called constrained MSSM or CMSSM.

The allowed domains in some of the $(m_{1/2}, m_0)$ planes for different values of $\tan\beta$ and the sign of μ are shown in Fig. 12. The various panels of this figure feature the limit $m_{\chi^\pm} \gtrsim 104$ GeV provided by chargino searches at LEP [63]. The LEP neutrino counting and other measurements have also constrained the possibilities for light neutralinos, and LEP has also provided lower limits on slepton masses, of which the strongest is $m_{\tilde{e}} \gtrsim 99$ GeV [64], as illustrated in panel (a) of Fig. 12. The most important constraints on the supersymmetric partners of the u, d, s, c, b squarks and on the gluinos are provided by the FNAL Tevatron collider: for equal masses $m_{\tilde{q}} = m_{\tilde{g}} \gtrsim 300$ GeV. In the case of the $\tilde{t}$, LEP provides the most stringent limit when $m_{\tilde{t}} - m_\chi$ is small, and the Tevatron for larger $m_{\tilde{t}} - m_\chi$ [63].

Another important constraint in Fig. 12 is provided by the LEP lower limit on the Higgs mass: $m_H > 114.4$ GeV [13]. Since m_h is sensitive to sparticle masses,

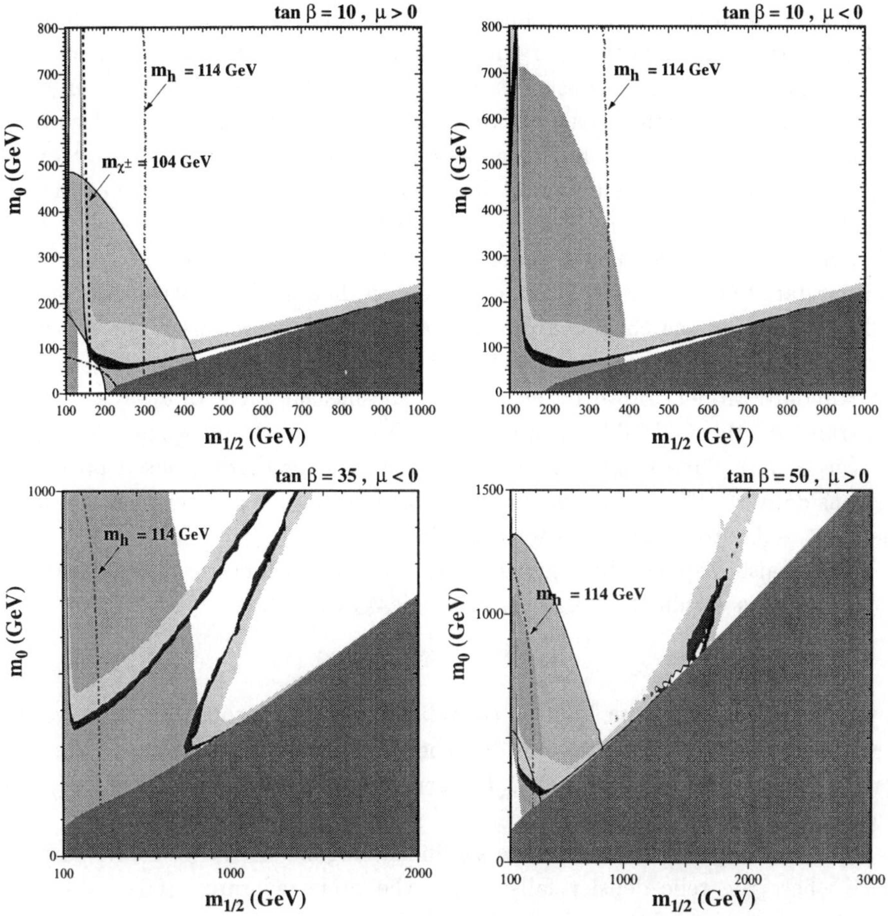

Figure 12. Compilations of phenomenological constraints on the CMSSM for (a) $\tan\beta = 10, \mu > 0$, (b) $\tan\beta = 10, \mu < 0$, (c) $\tan\beta = 35, \mu < 0$ and (d) $\tan\beta = 50, \mu > 0$ [65]. The near-vertical lines are the LEP limits $m_{\chi^\pm} = 104$ GeV (dashed) [63], shown in (a) only, and $m_h = 114$ GeV (dot-dash) [13]. Also, in the lower left corner of (a), we show the $m_{\tilde{e}} = 99$ GeV contour [64]. The large dark shaded regions are excluded because the LSP is charged. The light shaded areas have $0.1 \leq \Omega_\chi h^2 \leq 0.3$, and the smaller dark shaded regions have $0.094 \leq \Omega_\chi h^2 \leq 0.129$, as favoured by WMAP [65]. The medium shaded regions that are most prominent in panels (b) and (c) are excluded by $b \to s\gamma$ [66]. The mid-light shaded regions in panels (a) and (d) show the $\pm 2\,\sigma$ ranges of $g_\mu - 2$ [61].

particularly $m_{\tilde{t}}$, via loop corrections:

$$\delta m_h^2 \propto \frac{m_t^4}{m_W^2} \ln\left(\frac{m_{\tilde{t}}^2}{m_t^2}\right) + \ldots \tag{59}$$

the Higgs limit also imposes important constraints on the soft supersymmetry-breaking CMSSM parameters, principally $m_{1/2}$ [67] as displayed in Fig. 12.

Also shown in Fig. 12 is the constraint imposed by measurements of $b \to s\gamma$ [66].

These agree with the Standard Model, and therefore provide bounds on supersymmetric particles, such as the chargino and charged Higgs masses, in particular.

The final experimental constraint we consider is that due to the measurement of the anomalous magnetic moment of the muon. Following its first result last year [68], the BNL E821 experiment has recently reported a new measurement [61] of $a_\mu \equiv \frac{1}{2}(g_\mu - 2)$, which deviates by about 2 standard deviations from the best available Standard Model predictions based on low-energy $e^+e^- \to$ hadrons data [62]. On the other hand, the discrepancy is more like 0.9 standard deviations if one uses $\tau \to$ hadrons data to calculate the Standard Model prediction. Faced with this confusion, and remembering the chequered history of previous theoretical calculations [69], it is reasonable to defer judgement whether there is a significant discrepancy with the Standard Model. However, either way, the measurement of a_μ is a significant constraint on the CMSSM, favouring $\mu > 0$ in general, and a specific region of the $(m_{1/2}, m_0)$ plane if one accepts the theoretical prediction based on $e^+e^- \to$ hadrons data [70]. The regions preferred by the current $g - 2$ experimental data and the $e^+e^- \to$ hadrons data are shown in Fig. 12.

Fig. 12 also displays the regions where the supersymmetric relic density $\rho_\chi = \Omega_\chi \rho_{critical}$ falls within the range preferred by WMAP [20]:

$$0.094 < \Omega_\chi h^2 < 0.129 \tag{60}$$

at the 2-σ level. The upper limit on the relic density is rigorous, but the lower limit in (60) is optional, since there could be other important contributions to the overall matter density. Smaller values of $\Omega_\chi h^2$ correspond to smaller values of $(m_{1/2}, m_0)$, in general.

We see in Fig. 12 that there are significant regions of the CMSSM parameter space where the relic density falls within the preferred range (60). What goes into the calculation of the relic density? It is controlled by the annihilation cross section [59]:

$$\rho_\chi = m_\chi n_\chi \,, \quad n_\chi \sim \frac{1}{\sigma_{ann}(\chi\chi \to \ldots)} \,, \tag{61}$$

where the typical annihilation cross section $\sigma_{ann} \sim 1/m_\chi^2$. For this reason, the relic density typically increases with the relic mass, and this combined with the upper bound in (60) then leads to the common expectation that $m_\chi \lesssim \mathcal{O}(1)$ GeV.

However, there are various ways in which the generic upper bound on m_χ can be increased along filaments in the $(m_{1/2}, m_0)$ plane. For example, if the next-to-lightest sparticle (NLSP) is not much heavier than χ: $\Delta m / m_\chi \lesssim 0.1$, the relic density may be suppressed by coannihilation: $\sigma(\chi + \text{NLSP} \to \ldots)$ [71]. In this way, the allowed CMSSM region may acquire a 'tail' extending to larger sparticle masses. An example of this possibility is the case where the NLSP is the lighter stau: $\tilde{\tau}_1$ and $m_{\tilde{\tau}_1} \sim m_\chi$, as seen in Figs. 12(a) and (b) [72].

Another mechanism for extending the allowed CMSSM region to large m_χ is rapid annihilation via a direct-channel pole when $m_\chi \sim \frac{1}{2} m_{Higgs}$ [73,74]. This may yield a 'funnel' extending to large $m_{1/2}$ and m_0 at large $\tan\beta$, as seen in panels (c) and (d) of Fig. 12 [74]. Yet another allowed region at large $m_{1/2}$ and m_0 is the 'focus-point' region [75], which is adjacent to the boundary of the region where electroweak symmetry breaking is possible. The lightest supersymmetric particle is relatively light in this region.

3.4. *Benchmark Supersymmetric Scenarios*

As seen in Fig. 12, all the experimental, cosmological and theoretical constraints on the MSSM are mutually compatible. As an aid to understanding better the physics capabilities of the LHC and various other accelerators, as well as non-accelerator experiments, a set of benchmark supersymmetric scenarios have been proposed [76]. Their distribution in the $(m_{1/2}, m_0)$ plane is sketched in Fig. 13. These benchmark scenarios are compatible with all the accelerator constraints mentioned above, including the LEP searches and $b \to s\gamma$, and yield relic densities of LSPs in the range suggested by cosmology and astrophysics. The benchmarks are not intended to sample 'fairly' the allowed parameter space, but rather to illustrate the range of possibilities currently allowed.

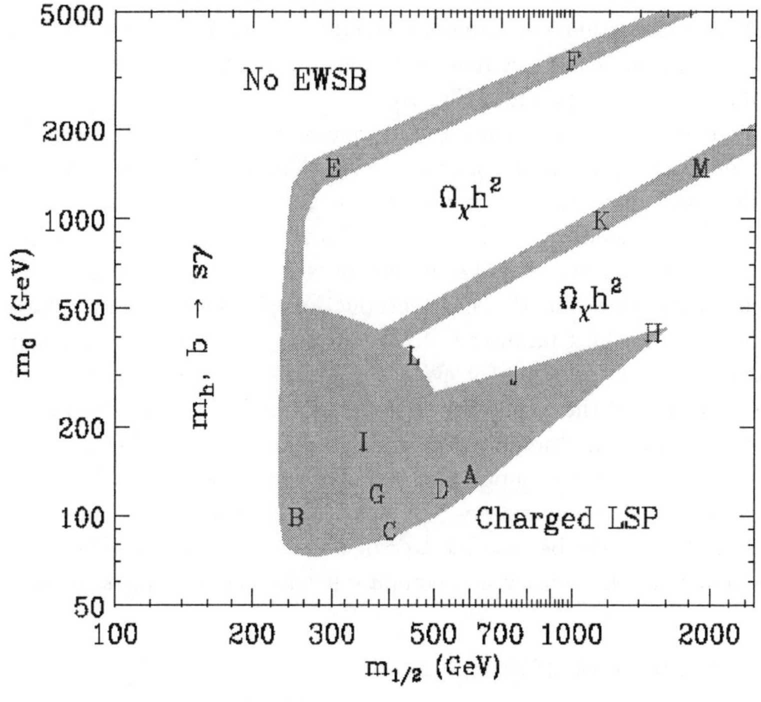

Figure 13. Sketch of the locations of the benchmark points proposed in [76] in the region of the $(m_{1/2}, m_0)$ plane where $\Omega_\chi h^2$ falls within the range preferred by cosmology (shaded). Note that the filaments of the allowed parameter space extending to large $m_{1/2}$ and/or m_0 are sampled.

In addition to a number of benchmark points falling in the 'bulk' region of parameter space at relatively low values of the supersymmetric particle masses, as see in Fig. 13, we also proposed [76] some points out along the 'tails' of parameter space extending out to larger masses. These clearly require some degree of fine-tuning to obtain the required relic density [77] and/or the correct $W^{\pm}$ mass [78], and some are also disfavoured by the supersymmetric interpretation of the $g_{\mu} - 2$ anomaly, but all are logically consistent possibilities.

3.5. *Prospects for Discovering Supersymmetry at Accelerators*

In the CMSSM discussed here, there are just a few prospects for discovering supersymmetry at the FNAL *Tevatron collider* [76], but these could be increased in other supersymmetric models [79]. On the other hand, there are good prospects for discovering supersymmetry at the *LHC*, and Fig. 14 shows its physics reach for observing pairs of supersymmetric particles. The signature for supersymmetry — multiple jets (and/or leptons) with a large amount of missing energy — is quite distinctive, as seen in Fig. 15 [80,81]. Therefore, the detection of the supersymmetric partners of quarks and gluons at the LHC is expected to be quite easy if they weigh less than about 2.5 TeV [82]. Moreover, in many scenarios one should be able to observe their cascade decays into lighter supersymmetric particles. As seen in Fig. 16, large fractions of the supersymmetric spectrum should be seen in most of the benchmark scenarios, although there are a couple where only the lightest supersymmetric Higgs boson would be seen [76], as seen in Fig. 16.

Electron-positron colliders provide very clean experimental environments, with egalitarian production of all the new particles that are kinematically accessible, including those that have only weak interactions, and hence are potentially complementary to the LHC, as illustrated in Fig. 16. Moreover, polarized beams provide a useful analysis tool, and $e\gamma$, $\gamma\gamma$ and e^-e^- colliders are readily available at relatively low marginal costs. However, the direct production of supersymmetric particles at such a collider cannot be guaranteed [84]. We do not yet know what the supersymmetric threshold energy may be (or even if there is one!). We may well not know before the operation of the LHC, although $g_{\mu} - 2$ might provide an indication [70], if the uncertainties in the Standard Model calculation can be reduced. However, if an e^+e^- collider is above the supersymmetric threshold, it will be able to measure very accurately the sparticle masses. By combining its measurements with those made at the LHC, it may be possible to calculate accurately from first principles the supersymmetric relic density and compare it with the astrophysical value.

3.6. *Searches for Dark Matter Particles*

In the above discussion, we have paid particular attention to the region of parameter space where the lightest supersymmetric particle could constitute the cold dark matter in the Universe [59]. How easy would this be to detect?

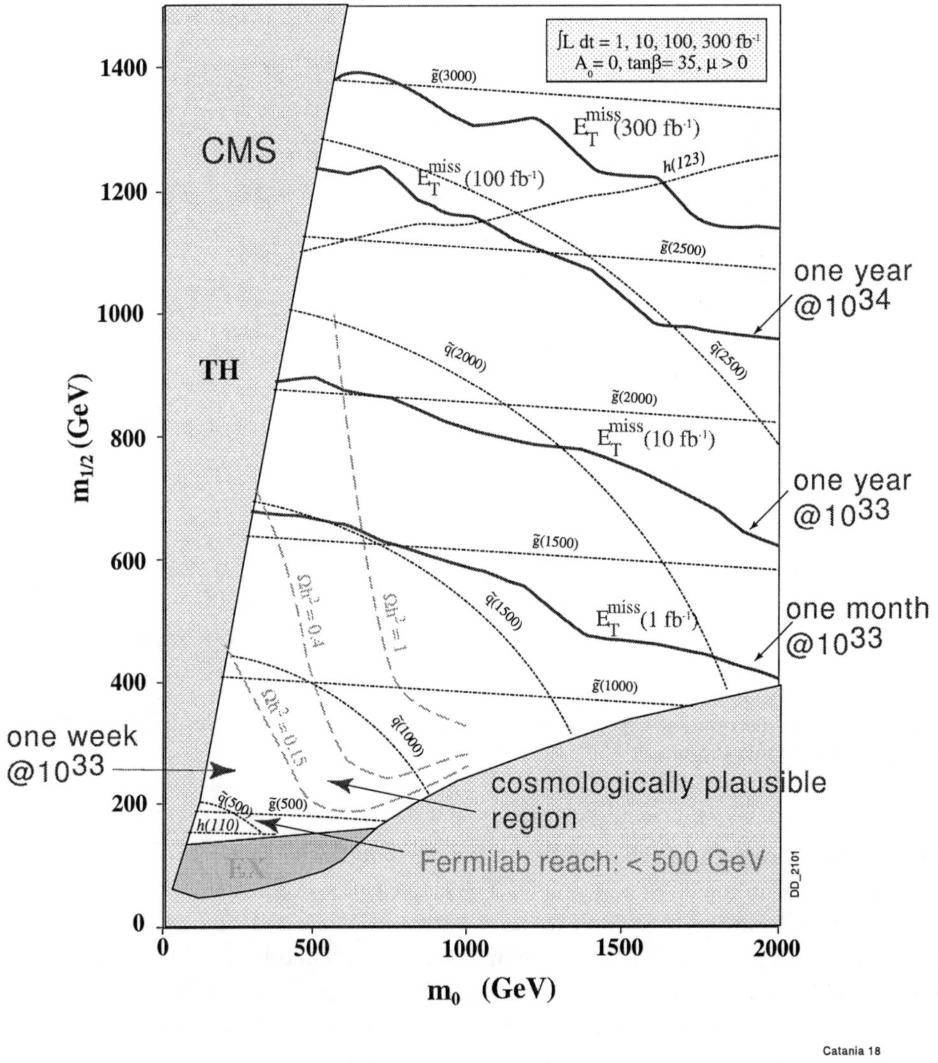

Figure 14. The regions of the $(m_0, m_{1/2})$ plane that can be explored by the LHC with various integrated luminosities [82], using the missing energy + jets signature [81].

• One strategy is to look for relic annihilations in the *galactic halo*, which might produce detectable antiprotons or positrons in the cosmic rays [85]. Unfortunately, the rates for their production are not very promising in the benchmark scenarios we studied [86].

• Alternatively, one might look for annihilations in the core of our galaxy, which might produce detectable *gamma rays*. As seen in the left panel of Fig. 17, this may

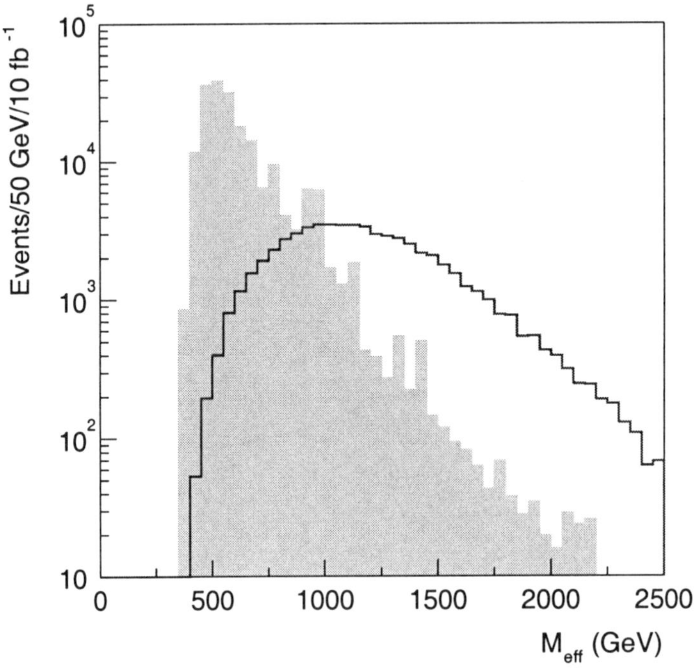

Figure 15. The distribution expected at the LHC in the variable $M_{\rm eff}$ that combines the jet energies with the missing energy [83,80,81].

be possible in certain benchmark scenarios [86], though the rate is rather uncertain because of the unknown enhancement of relic particles in our galactic core.

• A third strategy is to look for *annihilations inside the Sun or Earth*, where the local density of relic particles is enhanced in a calculable way by scattering off matter, which causes them to lose energy and become gravitationally bound [87]. The signature would then be energetic neutrinos that might produce detectable muons. Several underwater and ice experiments are underway or planned to look for this signature, and this strategy looks promising for several benchmark scenarios, as seen in the right panel of Fig. 17 [86]. It will be interesting to have such neutrino telescopes in different hemispheres, which will be able to scan different regions of the sky for astrophysical high-energy neutrino sources.

• The most satisfactory way to look for supersymmetric relic particles is directly via their *elastic scattering on nuclei* in a low-background laboratory experiment [88]. There are two types of scattering matrix elements, spin-independent — which are normally dominant for heavier nuclei, and spin-dependent — which could be interesting for lighter elements such as fluorine. The best experimental sensitivities so far are for spin-independent scattering, and one experiment has claimed a positive

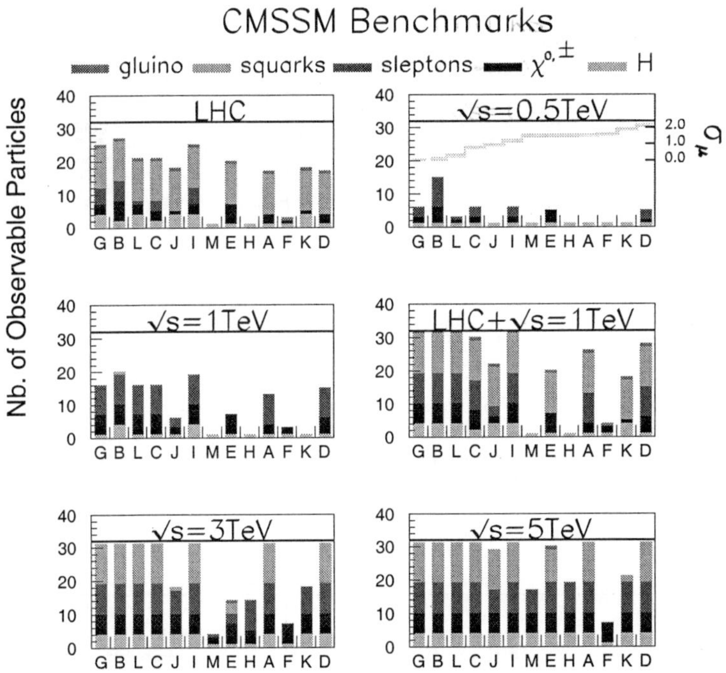

Figure 16. The numbers of different sparticles expected to be observable at the LHC and/or linear e^+e^- colliders with various energies, in each of the proposed benchmark scenarios [76], ordered by their difference from the present central experimental value of $g_\mu - 2$ [61].

signal [89]. However, this has not been confirmed by a number of other experiments [90]. In the benchmark scenarios the rates are considerably below the present experimental sensitivities [86], but there are prospects for improving the sensitivity into the interesting range, as seen in Fig. 18.

4. Inflation

4.1. *Motivations*

One of the main motivations for inflation [95] is the *horizon* or *homogeneity* problem: why are distant parts of the Universe so similar:

$$\left(\frac{\delta T}{T}\right)_{CMB} \sim 10^{-5}\,? \tag{62}$$

In conventional Big Bang cosmology, the largest patch of the CMB sky which could have been causally connected, *i.e.*, across which a signal could have travelled at the speed of light since the initial singularity, is about 2 degrees. So how did

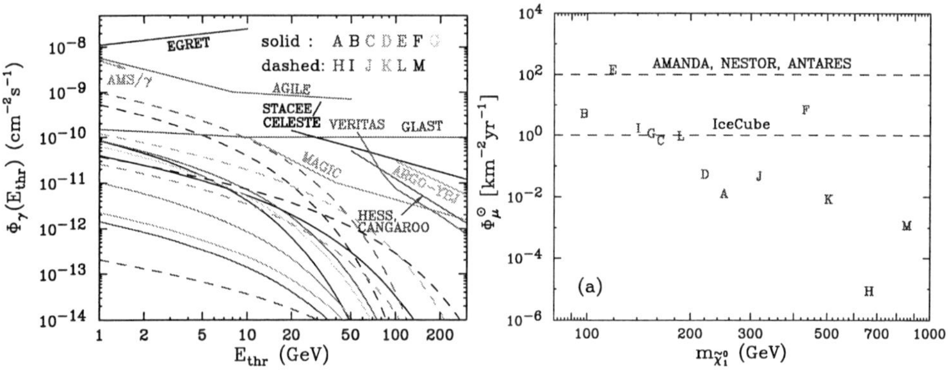

Figure 17. Left panel: Spectra of photons from the annihilations of dark matter particles in the core of our galaxy, in different benchmark supersymmetric models [86]. Right panel: Signals for muons produced by energetic neutrinos originating from annihilations of dark matter particles in the core of the Sun, in the same benchmark supersymmetric models [86].

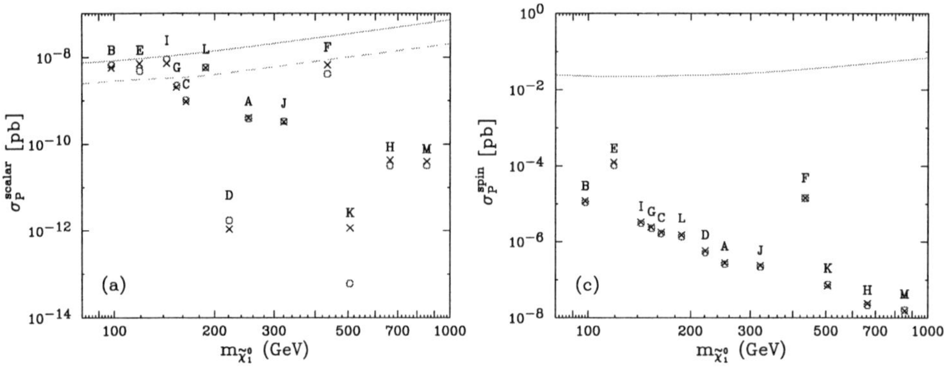

Figure 18. Left panel: elastic spin-independent scattering of supersymmetric relics on protons calculated in benchmark scenarios [86], compared with the projected sensitivities for CDMS II [91] and CRESST [92] (solid) and GENIUS [93] (dashed). The predictions of the `SSARD` code (crosses) and `Neutdriver`[94] (circles) for neutralino-nucleon scattering are compared [86]. The labels A, B, ...,L correspond to the benchmark points as shown in Fig. 13. Right panel: prospects for detecting elastic spin-dependent scattering in the benchmark scenarios, which are less bright [86].

opposite parts of the Universe, 180 degrees apart, 'know' how to coordinate their temperatures and densities?

Another problem of conventional Big bang cosmology is the *size* or *age* problem. The Hubble expansion rate in conventional Big bang cosmology is given by:

$$H^2 \equiv \left(\frac{\dot{a}}{a}\right)^2 = \frac{8\pi G_N \rho}{3} - \frac{k}{a^2}, \tag{63}$$

where $k = 0$ or ± 1 is the curvature. The only dimensionful coefficient in (63) is the Newton constant, $G_N \equiv 1/M_P^2 : M_P \simeq 1.2 \times 10^{19}$ GeV. A generic solution of (63) would have a characteristic scale size $a \sim \ell_P \equiv 1/M_P \sim 10^{-33}$ s and live to the ripe old age of $t \sim t_P \equiv \ell_P/c \simeq 10^{-43}$ s. Why is our Universe so long-lived and big? Clearly, we live in an atypical solution of (63)!

A related issue is the *flatness* problem. Defining, as usual

$$\Omega \equiv \frac{\rho}{\rho_c} : \rho_c \equiv \frac{3H^2}{8\pi G_N}, \tag{64}$$

we have

$$\Omega(t) = \frac{1}{1 - \frac{(k/a^2)}{(8\pi G_N \rho/3)}}. \tag{65}$$

Since $\rho \sim a^{-4}$ during the radiation-dominated era and $\sim a^{-3}$ during the matter-dominated era, it is clear from (65) that $\Omega(t) \to 0$ rapidly: for Ω to be $\mathcal{O}(1)$ as it is today, $|\Omega - 1|$ must have been $\mathcal{O}(10^{-60})$ at the Planck epoch when $t_P \sim 10^{-43}$ s. The density of the very early Universe must have been very finely tuned in order for its geometry to be almost flat today.

Then there is the *entropy* problem: why are there so many particles in the visible Universe: $S \sim 10^{90}$? A 'typical' Universe would have contained $\mathcal{O}(1)$ particles in its size $\sim \ell_P^3$.

All these particles have diluted what might have been the primordial density of *unwanted massive particles* such as magnetic monopoles and gravitinos. Where did they go?

The basic idea of inflation [96] is that, at some early epoch in the history of the Universe, its energy density may have been dominated by an almost constant term:

$$\left(\frac{\dot{a}}{a}\right)^2 = \frac{8\pi G_N \rho}{3} - \frac{k}{a^2} : \rho = V, \tag{66}$$

leading to a phase of almost de Sitter expansion. It is easy to see that the second (curvature) term in (66) rapidly becomes negligible, and that

$$a \simeq a_I e^{Ht} : H = \sqrt{\frac{8\pi G_N}{3} V} \tag{67}$$

during this inflationary expansion.

It is then apparent that the *horizon* would have expanded (near-) exponentially, so that the entire visible Universe might have been within our pre-inflationary horizon. This would have enabled initial homogeneity to have been established. The trick is not somehow to impose connections beyond the horizon, but rather to make the horizon much larger than naively expected in conventional Big Bang cosmology:

$$a_H \simeq a_I e^{H\tau} \gg c\tau, \tag{68}$$

where $H\tau$ is the number of e-foldings during inflation. It is also apparent that the $-\frac{k}{a^2}$ term in (66) becomes negligible, so that the Universe is almost *flat* with $\Omega_{tot} \simeq 1$. However, as we see later, perturbations during inflation generate a small deviation from unity: $|\Omega_{tot} - 1| \simeq 10^{-5}$. Following inflation, the conversion of the inflationary vacuum energy into particles reheats the Universe, filling it with the required *entropy*. Finally, the closest pre-inflationary *monopole* or *gravitino* is pushed away, further than the origin of the CMB, by the exponential expansion of the Universe.

From the point of view of general relativity, the (near-) constant inflationary vacuum energy is equivalent to a cosmological constant Λ:

$$R_{\mu\nu} - \frac{1}{2}g_{\mu\nu}R = 8\pi G_N T_{\mu\nu} + \Lambda g_{\mu\nu}. \tag{69}$$

We may compare the right-hand side of (69) with the energy-momentum tensor of a standard fluid:

$$T_{\mu\nu} = -pg_{\mu\nu} + (\rho + p)U_\mu U_\nu \tag{70}$$

where $U_\mu = (1,0,0,0)$ is the four-momentum vector for a comoving fluid. We can therefore write

$$R_{\mu\nu} - \frac{1}{2}g_{\mu\nu}R = 8\pi G_N T^\Lambda_{\mu\nu}, \tag{71}$$

where

$$\rho_\Lambda \equiv \frac{\Lambda}{8\pi G_N} \equiv -p_\Lambda. \tag{72}$$

Thus, we see that inflation has negative pressure. The value of the cosmological constant today, as suggested by recent observations [97,98], is *many* orders of magnitude smaller than would have been required during inflation: $\rho_\Lambda \sim 10^{-50}$ GeV^4 compared with the density $V \sim 10^{+65}$ GeV^4 required during inflation, as we see later.

Such a small value of the cosmological energy density is also *much smaller* than many contributions to it from identifiable physics sources: $\rho(QCD) \sim 10^{-4}$ GeV^4, $\rho(Electroweak) \sim 10^9$ GeV^4, $\rho(GUT) \sim 10^{64}(?)$ GeV^4 and $\rho(QuantumGravity) \sim 10^{74}(?)$ GeV^4. Particle physics offers no reason to expect the present-day vacuum energy to lie within the range suggested by cosmology, and raises the question why it is not many orders of magnitude larger.

4.2. *Some Inflationary Models*

The first inflationary potential V to be proposed was one with a 'double-dip' structure à la Higgs [96]. The *old inflation* idea was that the Universe would have started in the false vacuum with $V \neq 0$, where it would have undergone many e-foldings of de Sitter expansion. Then, the Universe was supposed to have tunnelled through the potential barrier to the true vacuum with $V \simeq 0$, and subsequently thermalized. The inflation required before this tunnelling was

$$H\tau \gtrsim 60: \ H = \sqrt{\frac{\Lambda}{3}}, \ \Lambda = 8\pi G_N V. \tag{73}$$

The problem with this old inflationary scenario was that the phase transition to the new vacuum would never have been completed. The Universe would look like a 'Swiss cheese' in which the bubbles of true vacuum would be expanding as $t^{1/2}$ or $t^{2/3}$, while the 'cheese' between them would still have been expanding exponentially as e^{Ht}. Thus, the fraction of space in the false vacuum would be

$$f \sim exp\left[Ht\left(3 - \frac{4\pi}{3}\frac{\Gamma}{H^4}\right)\right], \tag{74}$$

where Γ is the bubble nucleation rate per unit four-volume. The fraction $f \to 0$ only if $\Gamma/H^4 \simeq \mathcal{O}(1)$, but in this case there would not have been sufficient e-foldings for adequate inflation.

One of the fixes for this problem trades under the name of *new inflation* [99]. The idea is that the near-exponential expansion of the Universe took place in a flat region of the potential $V(\phi)$ that is not separated from the true vacuum by any barrier. It might have been reached after a first-order transition of the type postulated in old inflation, in which case one can regard our Universe as part of a bubble that expanded near-exponentially inside the 'cheese' of old vacuum, and there could be regions beyond our bubble that are still expanding (near-) exponentially. For the Universe to roll eventually downhill into the true vacuum, $V(\phi)$ could not quite be constant, and hence the Hubble expansion rate H during inflation was also not constant during new inflation.

An example of such a scenario is *chaotic inflation* [100], according to which there is no 'bump' in the effective potential $V(\phi)$, and hence no phase transition between old and new vacua. Instead, any given region of the Universe is assumed to start with some random value of the inflaton field ϕ and hence the potential $V(\phi)$, which decreases monotonically to zero. If the initial value of $V(\phi)$ is large enough, and the potential flat enough, (our part of) the Universe will undergo sufficient expansion.

Another fix for old inflation trades under the name of *extended inflation* [101]. Here the idea is that the tunnelling rate Γ depends on some other scalar field χ that varies while the inflaton ϕ is still stuck in the old vacuum. If $\Gamma(\chi)$ is initially small, but χ then changes so that $\Gamma(\chi)$ becomes large, the problem of completing the transition in the 'Swiss cheese' Universe is solved.

All these variants of inflation rely on some type of elementary scalar inflaton field. Therefore, the discovery of a Higgs boson would be a psychological boost for inflation, even though the electroweak Higgs boson cannot be responsible for it directly. Moreover, just as supersymmetry is well suited for stabilizing the mass scale of the electroweak Higgs boson, it may also be needed to keep the inflationary potential under control [102]. Later in this Lecture, I discuss a specific supersymmetric inflationary model.

4.3. *Density Perturbations*

The above description is quite classical. In fact, one should expect quantum fluctuations in the initial value of the inflaton field ϕ, which would cause the roll-over into the true vacuum to take place inhomogeneously, and different parts of the Universe to expand differently. As we discuss below in more detail, these quantum fluctuations would give rise to a Gaussian random field of perturbations with similar magnitudes on different scale sizes, just as the astrophysicists have long wanted. The magnitudes of these perturbations would be linked to the value of the effective potential during inflation, and would be visible in the CMB as adiabatic temperature fluctuations:

$$\frac{\delta T}{T} \sim \frac{\delta \rho}{\rho} \sim \mu^2 G_N, \tag{75}$$

where $\mu \equiv V^{1/4}$ is a typical vacuum energy scale during inflation. As we discuss later in more detail, consistency with the CMB data from COBE *et al.*, that find $\delta T/T \simeq 10^{-5}$, is obtained if

$$\mu \simeq 10^{16} \text{ GeV}, \tag{76}$$

comparable with the GUT scale.

Each density perturbation can be regarded as an embryonic potential well, into which non-relativistic cold dark matter particles may fall, increasing the local contrast in the mass-energy density. On the other hand, relativistic hot dark matter particles will escape from small-scale density perturbations, modifying their rate of growth. This also depends on the expansion rate of the Universe and hence the cosmological constant. Present-day data are able to distinguish the effects of different categories of dark matter. In particular, as we already discussed, the WMAP and other data tell us that the density of hot dark matter neutrinos is relatively small [20]:

$$\Omega_\nu h^2 < 0.0076, \tag{77}$$

whereas the density of cold dark matter is relatively large [20]:

$$\Omega_{CDM} h^2 = 0.1126^{+0.0081}_{-0.0091}, \tag{78}$$

and the cosmological constant is even larger: $\Omega_\Lambda \simeq 0.73$.

The cold dark matter amplifies primordial perturbations already while the conventional baryonic matter is coupled to radiation before (re)combination. Once this epoch is passed and the CMB decouples from the conventional baryonic matter, the baryons become free to fall into the 'holes' prepared for them by the cold dark matter that has fallen into the overdense primordial perturbations. In this way, structures in the Universe, such as galaxies and their clusters, may be formed earlier than they would have appeared in the absence of cold dark matter.

All this theory is predicated on the presence of primordial perturbations laid down by inflation [103], which we now explore in more detail.

There are in fact two types of perturbations, namely density fluctuations and gravity waves. To describe the first, we consider the density field $\rho(\mathbf{x})$ and its perturbations $\delta(\mathbf{x}) \equiv (\rho(\mathbf{x}) - <\rho>)/<\rho>$, which we can decompose into Fourier modes:

$$\delta(\mathbf{x}) = \int d^3\mathbf{x}\delta_{\mathbf{k}}e^{-i\mathbf{k}\cdot\mathbf{x}}. \tag{79}$$

The density perturbation on a given scale λ is then given by

$$\left(\frac{\delta\rho}{\rho}\right)^2_\lambda = \left(\frac{k^3|\delta_{\mathbf{k}}|^2}{2\pi^2}\right)_{k^{-1}=\lambda}, \tag{80}$$

whose evolution depends on the ratio λ/a_H, where $a_H \equiv c \cdot t$ is the naive horizon size.

The evolution of small-scale perturbations with $\lambda/a_H < 1$ depends on the astrophysical dynamics, such as the equation of state, dissipation, the Jeans instability, etc.:

$$\ddot{\delta}_{\mathbf{k}} + 2H\dot{\delta}_{\mathbf{k}} + v_s^2\frac{k^2}{a^2}\delta_{\mathbf{k}} = 4\pi G_N <\rho> \delta_{\mathbf{k}}, \tag{81}$$

where v_s is the sound speed: $v_s^2 = dp/d\rho$. If the wave number k is larger than the characteristic Jeans value

$$k_J = \sqrt{\frac{4\pi G_N a^2 <\rho>}{v_s^2}}, \tag{82}$$

the density perturbation $\delta_{\mathbf{k}}$ oscillates, whereas it grows if $k < k_J$. Cold dark matter effectively provides $v_s \rightarrow 0$, in which case $k_J \rightarrow \infty$ and perturbations with all wave numbers grow.

In order to describe the evolution of large-scale perturbations with $\lambda/a_H > 1$, we use the gauge-invariant ratio $\delta\rho/\rho + p$, which remains constant outside the horizon a_H. Hence, the value when such a density perturbation comes back within the horizon is identical with its value when it was inflated beyond the horizon. During inflation, one had $\rho + p \simeq <\dot{\phi}^2>$, and

$$\delta\rho = \delta\phi \times \frac{\partial V}{\partial\phi} = \delta\phi \times V'(\phi). \tag{83}$$

During roll-over, one has $\ddot{\phi} + 3H\dot{\phi} + V'(\phi) = 0$, and, if the roll-over is slow, one has

$$\dot{\phi} \simeq -\frac{V'(\phi)}{3H}, \tag{84}$$

where the Hubble expansion rate

$$H^2 = \frac{8\pi}{3}G_N\left(V(\phi) + \frac{1}{2}\dot{\phi}^2\right) \simeq \frac{8\pi G_N}{3}V(\phi). \tag{85}$$

The quantum fluctuations of the inflaton field in de Sitter space are given by:

$$\delta\phi \simeq \frac{H}{2\pi}, \tag{86}$$

so initially

$$\frac{\delta\rho}{\rho+p} \simeq \frac{\delta\phi\cdot V'}{\dot{\phi}^2} \simeq \frac{H^3\cdot V'}{(V')^2} \simeq \frac{V^{3/2}}{V'}. \tag{87}$$

This is therefore also the value when the perturbation comes back within the horizon:

$$\left(\frac{\delta\rho}{\rho}\right)_{\lambda=a_H} \equiv A_S(\phi) = \frac{\sqrt{2}\kappa^2}{8\pi^{3/2}}\frac{H^2}{|H'|} : \kappa^2 \equiv 8\pi G_N, \tag{88}$$

assuming that $\rho \gg p$ at this epoch.

Gravity-wave perturbations obey an equation analogous to (81):

$$\ddot{h}_{\mathbf{k}}^{1,2} + 2H\dot{h}_{\mathbf{k}}^{1,2} + v_s^2\frac{k^2}{a^2}h_{\mathbf{k}}^{1,2} = 0, \tag{89}$$

for each of the two graviton polarization states $h_{\mu\nu}^{1,2}$, where

$$g_{\mu\nu} = g_{\mu\nu}^{FRW} + h_{\mu\nu}. \tag{90}$$

The $h_{\mathbf{k}}^{1,2}$ also remain unchanged outside the horizon a_H, and have initial values

$$h_{\mathbf{k}}^{1,2} \simeq \frac{H}{2\pi}, \tag{91}$$

yielding

$$k^3|h_{\mathbf{k}}|_{\lambda=a_H} \equiv A_T(\phi) = \frac{\kappa}{4\pi^{3/2}}H : H \simeq \sqrt{\frac{8\pi G_N}{3}V}. \tag{92}$$

Comparing (88, 92), we see that

$$r \equiv \frac{A_S(\phi)}{A_T(\phi)} = \frac{\kappa}{\sqrt{2}}\frac{H}{|H'|} : \kappa^2 = 8\pi G_N. \tag{93}$$

Hence, if the roll-over is very slow, so that $|H'|$ is very small, the density waves dominate over the tensor gravity waves. However, in the real world, also the gravity waves may be observable, furnishing a possible signature of inflation [104].

4.4. *Inflation in Scalar Field Theories*

Let now consider in more detail chaotic inflation in a generic scalar field theory [104], described by a Lagrangian

$$\mathcal{L}(\phi) = \frac{1}{2}\partial^\mu\phi\partial_\mu\phi - V(\phi), \tag{94}$$

where the first term yields the kinetic energy of the inflaton field ϕ and the second term is the inflaton potential. One may treat the inflaton field as a fluid with density

$$\rho = \frac{1}{2}\dot{\phi}^2 + V(\phi), \tag{95}$$

and pressure

$$p = \frac{1}{2}\dot{\phi}^2 - V(\phi). \tag{96}$$

Inserting these expressions into the standard FRW equations, we find that the Hubble expansion rate is given by

$$H^2 = \frac{8\pi}{3\pi^2}\left[\frac{1}{2}\dot{\phi}^2 + V(\phi)\right], \tag{97}$$

as discussed above, the deceleration rate is given by

$$\left(\frac{\ddot{a}}{a}\right) = \frac{8\pi}{3\pi^2}\left[V(\phi) - \frac{1}{2}\dot{\phi}^2\right], \tag{98}$$

and the equation of motion of the inflaton field is

$$\ddot{\phi} + 3H\dot{\phi} + V'(\phi) = 0. \tag{99}$$

The first term in (99) is assumed to be negligible, in which case the equation of motion is dominated by the second (Hubble drag) term, and one has

$$\dot{\phi} \simeq -\frac{V'}{3H}, \tag{100}$$

as assumed above. In this slow-roll approximation, when the kinetic term in (97) is negligible, and the Hubble expansion rate is dominated by the potential term:

$$H \simeq \sqrt{\frac{1}{3M_P^2}V(\phi)}. \tag{101}$$

where $M_P \equiv 1/\sqrt{8\pi G_N} \simeq 2.4\times 10^{18}$ GeV. It is convenient to introduce the following slow-roll parameters:

$$\epsilon \equiv \frac{1}{2}M_P^2\left(\frac{V'}{V}\right)^2, \ \eta \equiv M_P^2\left(\frac{V''}{V}\right), \ \xi \equiv M_P^4\left(\frac{VV'''}{V^2}\right). \tag{102}$$

Various observable quantities can then be expressed in terms of ϵ, η and ξ, including the spectral index for scalar density perturbations:

$$n_s = 1 - 6\epsilon + 2\eta, \tag{103}$$

the ratio of scalar and tensor perturbations at the quadrupole scale:

$$r \equiv \frac{A_T}{A_S} = 16\epsilon, \tag{104}$$

the spectral index of the tensor perturbations:

$$n_T = -2\epsilon, \tag{105}$$

and the running parameter for the scalar spectral index:

$$\frac{dn_s}{d\ln k} = \frac{2}{3}\left[(n_s - 1)^2 - 4\eta^2\right] + 2\xi. \tag{106}$$

The amount e^N by which the Universe expanded during inflation is also controlled by the slow-roll parameter ϵ:

$$e^N : N = \int H dt = \frac{2\sqrt{\pi}}{m_P}\int_{\phi_{initial}}^{\phi_{final}} \frac{d\phi}{\sqrt{\epsilon(\phi)}}. \tag{107}$$

In order to explain the size of a feature in the observed Universe, one needs:

$$N = 62 - \ln\frac{k}{a_0 H_0} - \ln\frac{10^{16}\text{GeV}}{V_k^{1/4}} + \frac{1}{4}\ln\frac{V_k}{V_e} - \frac{1}{3}\ln\frac{V_e^{1/4}}{\rho_{reheating}^{1/4}}, \qquad (108)$$

where k characterizes the size of the feature, V_k is the magnitude of the inflaton potential when the feature left the horizon, V_e is the magnitude of the inflaton potential at the end of inflation, and $\rho_{reheating}$ is the density of the Universe immediately following reheating after inflation.

As an example of the above general slow-roll theory, let us consider chaotic inflation [100] with a $V = \frac{1}{2}m^2\phi^2$ potential [a], and compare its predictions with the WMAP data [20]. In this model, the conventional slow-roll inflationary parameters are

$$\epsilon = \frac{2M_P^2}{\phi_I^2}, \ \eta = \frac{2M_P^2}{\phi_I^2}, \ \xi = 0, \qquad (109)$$

where ϕ_I denotes the *a priori* unknown inflaton field value during inflation at a typical CMB scale k. The overall scale of the inflationary potential is normalized by the WMAP data on density fluctuations:

$$\Delta_R^2 = \frac{V}{24\pi^2 M_P^2 \epsilon} = 2.95 \times 10^{-9} A \quad : \quad A = 0.77 \pm 0.07, \qquad (110)$$

yielding

$$V^{\frac{1}{4}} = M_P^4 \sqrt{\epsilon \times 24\pi^2 \times 2.27 \times 10^{-9}} = 0.027 M_P \times \epsilon^{\frac{1}{4}}, \qquad (111)$$

corresponding to

$$m^{\frac{1}{2}}\phi_I = 0.038 \times M_P^{\frac{3}{2}} \qquad (112)$$

in any simple chaotic ϕ^2 inflationary model. The above expression (108) for the number of e-foldings after the generation of the CMB density fluctuations observed by COBE could be as low as $N \simeq 50$ for a reheating temperature T_{RH} as low as 10^6 GeV. In the ϕ^2 inflationary model, this value of N would imply

$$N = \frac{1}{4}\frac{\phi_I^2}{M_P^2} \simeq 50, \qquad (113)$$

corresponding to

$$\phi_I^2 \simeq 200 \times M_P^2. \qquad (114)$$

Inserting this requirement into the WMAP normalization condition (111), we find the following required mass for any quadratic inflaton:

$$m \simeq 1.8 \times 10^{13} \text{ GeV}. \qquad (115)$$

---

[a] This is motivated by the sneutrino inflation model [105] discussed later.

This is comfortably within the range of heavy singlet (s)neutrino masses usually considered, namely $m_N \sim 10^{10}$ to 10^{15} GeV, motivating the sneutrino inflation model [105] discussed below.

Is this simple ϕ^2 model compatible with the WMAP data? It predicts the following values for the primary CMB observables [105]: the scalar spectral index

$$n_s = 1 - \frac{8M_P^2}{\phi_I^2} \simeq 0.96, \tag{116}$$

the tensor-to scalar ratio

$$r = \frac{32M_P^2}{\phi_I^2} \simeq 0.16, \tag{117}$$

and the running parameter for the scalar spectral index:

$$\frac{dn_s}{d\ln k} = \frac{32M_P^4}{\phi_I^4} \simeq 8 \times 10^{-4}. \tag{118}$$

The value of n_s extracted from WMAP data depends whether, for example, one combines them with other CMB and/or large-scale structure data. However, the ϕ^2 model value $n_s \simeq 0.96$ appears to be compatible with the data at the 1-σ level. The ϕ^2 model value $r \simeq 0.16$ for the relative tensor strength is also compatible with the WMAP data. In fact, we note that the favoured individual values for n_s, r and $dn_s/d\ln k$ reported in an independent analysis [106] *all coincide* with the ϕ^2 model values, within the latter's errors!

One of the most interesting features of the WMAP analysis is the possibility that $dn_s/d\ln k$ might differ from zero. The ϕ^2 model value $dn_s/d\ln k \simeq 8 \times 10^{-4}$ derived above is negligible compared with the WMAP preferred value and its uncertainties. However, $dn_s/d\ln k = 0$ appears to be compatible with the WMAP analysis at the 2-σ level or better, so we do not regard this as a death-knell for the ϕ^2 model.

4.5. *Could the Inflaton be a Sneutrino?*

This 'old' idea [107] has recently been resurrected [105]. We recall that seesaw models [25] of neutrino masses involve three heavy singlet right-handed neutrinos weighing around 10^{10} to 10^{15} GeV, which certainly includes the preferred inflaton mass found above (115). Moreover, supersymmetry requires each of the heavy neutrinos to be accompanied by scalar sneutrino partners. In addition, singlet (s)neutrinos have no interactions with vector bosons, so their effective potential may be as flat as one could wish. Moreover, supersymmetry safeguards the flatness of this potential against radiative corrections. Thus, singlet sneutrinos have no problem in meeting the slow-roll requirements of inflation.

On the other hand, their Yukawa interactions Y_D are eminently suitable for converting the inflaton energy density into particles via $N \to H + \ell$ decays and their supersymmetric variants. Since the magnitudes of these Yukawa interactions are not completely determined, there is flexibility in the reheating temperature after

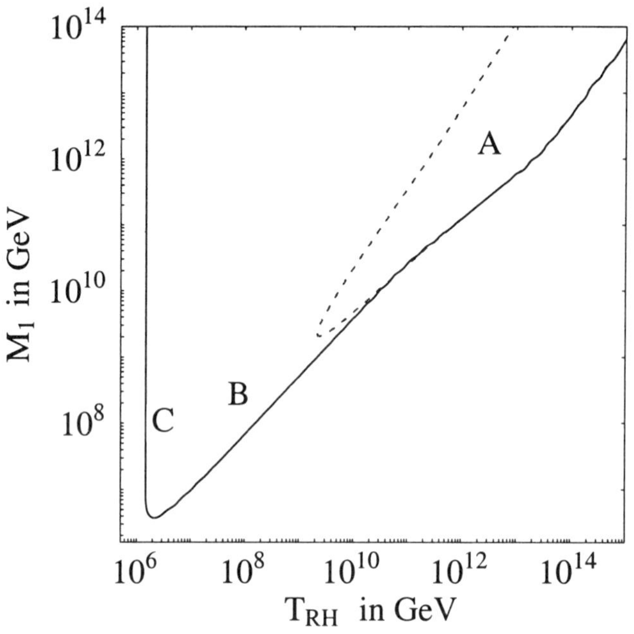

Figure 19. The solid curve bounds the region allowed for leptogenesis in the (T_{RH}, M_{N_1}) plane, assuming a baryon-to-entropy ratio $Y_B > 7.8 \times 10^{-11}$ and the maximal CP asymmetry $\epsilon_1^{max}(M_{N_1})$. In the area bounded by the dashed curve leptogenesis is entirely thermal [105].

inflation, as we see in Fig. 19 [105]. Thus the answer to the question in the title of this Section seems to be 'yes', so far.

5. Further Beyond

Some key cosmological and astrophysical problems may be resolved only by appeal to particle physics beyond the ideas we have discussed so far. One of the greatest successes of Big Bang cosmology has been an explanation of the observed abundances of light elements, ascribed to cosmological nucleosynthesis when the temperature $T \sim 1$ to 0.1 MeV. This requires a small baryon-to-entropy ratio $n_B/s \simeq 10^{-10}$. How did this small baryon density originate?

Looking back to the previous quark epoch, there must have been a small excess of quarks over antiquarks. All the antiquarks would then have annihilated with quarks when the temperature of the Universe was ~ 200 MeV, producing radiation and leaving the small excess of quarks to survive to form baryons. So how did the small excess of quarks originate?

Sakharov [108] pointed out that microphysics, in the form of particle interactions, could generate a small excess of quarks if the following three conditions were satisfied:

• *The interactions of matter and antimatter particles should differ*, in the sense that both charge conjugation C and its combination CP with mirror reflection should be broken, as discovered in the weak interactions.

• *There should exist interactions capable of changing the net quark number.* Such interactions do exist in the Standard Model, mediated by unstable field configurations called sphalerons. They have not been observed at low temperatures, where they would be mediated by heavy states called sphalerons and are expected to be very weak, but they are thought to have been important when the temperature of the Universe was $\gtrsim 100$ GeV. Alternatively, one may appeal to interactions in Grand Unified Theories (GUTs) that are thought to change quarks into leptons and *vice versa* when their energies $\sim 10^{15}$ GeV.

• *There should have been a breakdown of thermal equilibrium.* This could have occurred during a phase transition in the early Universe, for example during the electroweak phase transition when $T \sim 100$ GeV, during inflation, or during a GUT phase transition when $T \sim 10^{15}$ GeV.

The great hope in the business of cosmological baryogenesis is to find a connection with physics accessible to accelerator experiments, and some examples will be mentioned later in this Lecture.

Another example of observable phenomena related to GUT physics may be ultra-high-energy cosmic rays (UHECRs) [109], which have energies $\gtrsim 10^{11}$ GeV. The UHECRs might either have originated from some astrophysical source, such as an active galactic nuclei (AGNs) or gamma-ray bursters (GRBs), or they might be due to the decays of metastable GUT-scale particles, a possibility discussed in the last part of this Lecture.

5.1. *Grand Unified Theories*

The philosophy of grand unification [110] is to seek a simple group that includes the untidy separate interactions of the Standard Model, QCD and the electroweak sector. The hope is that this Grand Unification can be achieved while neglecting gravity, at least as a first approximation. If the grand unification scale turns out to be significantly less than the Planck mass, this is not obviously a false hope. The Grand Unification scale is indeed expected to be exponentially large:

$$\frac{m_{GUT}}{m_W} = \exp\left(\mathcal{O}\left(\frac{1}{\alpha_{em}}\right)\right) \tag{119}$$

and typical estimates are that $m_{GUT} = \mathcal{O}(10^{16}$ GeV). Such a calculation involves an extrapolation of known physics by many orders of magnitude further than, *e.g.*, the extrapolation that Newton made from the apple to the Solar System.

If the grand unification scale is indeed so large, most tests of it are likely to be indirect, such as relations between Standard Model vector couplings and between particle masses. Any new interactions, such as those that might cause protons to decay or give masses to neutrinos, are likely to be very strongly suppressed.

To examine the indirect GUT predictions for the Standard Model vector interactions in more detail, one needs to study their variations with the energy scale [57], which are described by the following two-loop renormalization equations:

$$Q \frac{\partial \alpha_i(Q)}{\partial Q} = -\frac{1}{2\pi} \left(b_i + \frac{b_{ij}}{4\pi} \alpha_j(Q) \right) [\alpha_i(Q)]^2 \tag{120}$$

where the b_i receive the one-loop contributions

$$b_i = \begin{pmatrix} 0 \\ -\frac{22}{3} \\ -11 \end{pmatrix} + N_g \begin{pmatrix} \frac{4}{3} \\ \frac{4}{3} \\ \frac{4}{3} \end{pmatrix} + N_H \begin{pmatrix} \frac{1}{10} \\ \frac{1}{6} \\ 0 \end{pmatrix} \tag{121}$$

from vector bosons, N_g matter generations and N_H Higgs doublets, respectively, and at two loops

$$b_{ij} = \begin{pmatrix} 0 & 0 & 0 \\ 0 & -\frac{136}{3} & 0 \\ 0 & 0 & -102 \end{pmatrix} + N_g \begin{pmatrix} \frac{19}{15} & \frac{3}{5} & \frac{44}{15} \\ \frac{1}{5} & \frac{49}{3} & 4 \\ \frac{4}{30} & \frac{3}{2} & \frac{76}{3} \end{pmatrix} + N_H \begin{pmatrix} \frac{9}{50} & \frac{9}{10} & 0 \\ \frac{3}{10} & \frac{13}{6} & 0 \\ 0 & 0 & 0 \end{pmatrix}. \tag{122}$$

These coefficients are all independent of any specific GUT model, depending only on the light particles contributing to the renormalization.

Including supersymmetric particles as in the MSSM, one finds [111]

$$b_i = \begin{pmatrix} 0 \\ -6 \\ -9 \end{pmatrix} + N_g \begin{pmatrix} 2 \\ 2 \\ 2 \end{pmatrix} + N_H \begin{pmatrix} \frac{3}{10} \\ \frac{1}{2} \\ 0 \end{pmatrix} \tag{123}$$

and

$$b_{ij} = \begin{pmatrix} 0 & 0 & 0 \\ 0 & -24 & 0 \\ 0 & 0 & -54 \end{pmatrix} + N_g \begin{pmatrix} \frac{38}{15} & \frac{6}{5} & \frac{88}{15} \\ \frac{2}{5} & 14 & 8 \\ \frac{11}{5} & 3 & \frac{68}{3} \end{pmatrix} + N_H \begin{pmatrix} \frac{9}{50} & \frac{9}{10} & 0 \\ \frac{3}{10} & \frac{7}{2} & 0 \\ 0 & 0 & 0 \end{pmatrix}, \tag{124}$$

again independent of any specific supersymmetric GUT.

Calculations with these equations show that non-supersymmetric models are not consistent with the measurements of the Standard Model interactions at LEP and elsewhere. However, although extrapolating the experimental determinations of the interaction strengths using the non-supersymmetric renormalization-group equations (121), (122) does not lead to a common value at any renormalization scale, we saw in Fig. 11 that extrapolation using the supersymmetric equations (123), (124) **does** lead to possible unification at $m_{GUT} \sim 10^{16}$ GeV [56].

The simplest GUT model is based on the group SU(5) [110], whose most useful representations are the complex vector $\underline{5}$ representation denoted by F_α, its conjugate $\underline{\bar{5}}$ denoted by $\bar{F}^\alpha$, the complex two-index antisymmetric tensor $\underline{10}$ representation $T_{[\alpha\beta]}$, and the adjoint $\underline{24}$ representation A^α_β. The latter is used to accommodate the vector bosons of SU(5):

$$\begin{pmatrix} & & & \vdots & \bar{X} & \bar{Y} \\ & g_{1,\ldots,8} & & \vdots & \bar{X} & \bar{Y} \\ & & & \vdots & \bar{X} & \bar{Y} \\ \cdots & \cdots & \cdots & \cdots & \cdots & \cdots \\ X & X & X & \vdots & & \\ & & & \vdots & W_{1,2,3} & \\ Y & Y & Y & \vdots & & \end{pmatrix} \tag{125}$$

where the $g_{1,\ldots,8}$ are the gluons of QCD, the $W_{1,2,3}$ are weak bosons, and the (X, Y) are new vector bosons, whose interactions we discuss in the next section.

The quarks and leptons of each generation are accommodated in $\underline{\bar{5}}$ and $\underline{10}$ representations of SU(5):

$$\bar{F} = \begin{pmatrix} d^c_R \\ d^c_Y \\ d^c_B \\ \cdots \\ -e^- \\ \nu_e \end{pmatrix}_L , \quad T = \begin{pmatrix} 0 & u^c_B & -u^c_Y & \vdots & -u_R & -d_R \\ -u^c_B & 0 & u^c_R & \vdots & -u_Y & -d_Y \\ u^c_Y & -u^c_R & 0 & \vdots & -u_B & -d_B \\ \cdots & \cdots & \cdots & \cdots & \cdots & \cdots \\ u_R & u_Y & u_B & \vdots & 0 & -e^c \\ d_R & d_Y & d_B & \vdots & e^c & 0 \end{pmatrix}_L \tag{126}$$

The particle assignments are unique up to the effects of mixing between generations, which we do not discuss in detail here [112].

5.2. *Baryon Decay and Baryogenesis*

Baryon instability is to be expected on general grounds, since there is no exact symmetry to guarantee that baryon number B is conserved, just as we discussed previously for lepton number. Indeed, baryon decay is a generic prediction of GUTs, which we illustrate with the simplest SU(5) model. We see in (125) that there are two species of vector bosons in SU(5) that couple the colour indices (1,2,3) to the electroweak indices (4,5), called X and Y. As we can see from the matter representations (126), these may enable two quarks or a quark and lepton to annihilate. Combining these possibilities leads to interactions with $\Delta B = \Delta L = 1$. The forms of effective four-fermion interactions mediated by the exchanges of massive Z and

Y bosons, respectively, are [113]:

$$\left(\epsilon_{ijk} u_{R_k} \gamma_\mu u_{L_j}\right) \frac{g_X^2}{8m_X^2} \quad (2e_R \, \gamma^\mu \, d_{L_i} + e_L \, \gamma^\mu \, d_{R_i}) \ ,$$

$$\left(\epsilon_{ijk} u_{R_k} \gamma_\mu d_{L_j}\right) \frac{g_Y^2}{8m_X^2} \quad (\nu_L \, \gamma^\mu \, d_{R_i}), \tag{127}$$

up to generation mixing factors.

Since the couplings $g_X = g_Y$ in an SU(5) GUT, and $m_X \simeq m_Y$, we expect that

$$G_X \equiv \frac{g_X^2}{8m_X^2} \simeq G_Y \equiv \frac{g_Y^2}{8m_Y^2}. \tag{128}$$

It is clear from (127) that the baryon decay amplitude $A \propto G_X$, and hence the baryon $B \to \ell +$ meson decay rate

$$\Gamma_B = cG_X^2 m_p^5, \tag{129}$$

where the factor of m_p^5 comes from dimensional analysis, and c is a coefficient that depends on the GUT model and the non-perturbative properties of the baryon and meson.

The decay rate (129) corresponds to a proton lifetime

$$\tau_p = \frac{1}{c} \frac{m_X^4}{m_p^5}. \tag{130}$$

It is clear from (130) that the proton lifetime is very sensitive to m_X, which must therefore be calculated very precisely. In minimal SU(5), the best estimate was

$$m_X \simeq (1 \text{ to } 2) \times 10^{15} \times \Lambda_{QCD} \tag{131}$$

where Λ_{QCD} is the characteristic QCD scale. Making an analysis of the generation mixing factors [112], one finds that the preferred proton (and bound neutron) decay modes in minimal SU(5) are

$$p \to e^+\pi^0 \ , \quad e^+\omega \ , \quad \bar{\nu}\pi^+ \ , \quad \mu^+ K^0 \ , \quad \ldots$$

$$n \to e^+\pi^- \ , \quad e^+\rho^- \ , \quad \bar{\nu}\pi^0 \ , \quad \ldots \tag{132}$$

and the best numerical estimate of the lifetime is

$$\tau(p \to e^+\pi^0) \simeq 2 \times 10^{31\pm1} \times \left(\frac{\Lambda_{QCD}}{400 \text{ MeV}}\right)^4 \quad y \tag{133}$$

This is in *prima facie* conflict with the latest experimental lower limit

$$\tau(p \to e^+\pi^0) > 1.6 \times 10^{33} \ y \tag{134}$$

from super-Kamiokande [114].

We saw earlier that supersymmetric GUTs, including SU(5), fare better with coupling unification. They also predict a larger GUT scale [111]:

$$m_X \simeq 10^{16} \text{ GeV}, \tag{135}$$

so that $\tau(p \to e^+\pi^0)$ is considerably longer than the experimental lower limit. However, this is not the dominant proton decay mode in supersymmetric SU(5) [115]. In this model, there are important $\Delta B = \Delta L = 1$ interactions mediated by the exchange of colour-triplet Higgsinos $\tilde{H}_3$, dressed by gaugino exchange [116]:

$$G_X \to \mathcal{O}\left(\frac{\lambda^2 g^2}{16\pi^2}\right) \frac{1}{m_{\tilde{H}_3}\tilde{m}} \tag{136}$$

where λ is a Yukawa coupling. Taking into account colour factors and the increase in λ for more massive particles, it was found [115] that decays into neutrinos and strange particles should dominate:

$$p \to \bar{\nu}K^+ \ , \quad n \to \bar{\nu}K^0 \ , \quad \ldots \tag{137}$$

Because there is only one factor of a heavy mass $m_{\tilde{H}_3}$ in the denominator of (136), these decay modes are expected to dominate over $p \to e^+\pi^0$, etc., in minimal supersymmetric SU(5). Calculating carefully the other factors in (136) [115], it seems that the modes (137) may now be close to exclusion at rates compatible with this model. The current experimental limit is $\tau(p \to \bar{\nu}K^+) > 6.7 \times 10^{32}y$. However, there are other GUT models [28] that remain compatible with the baryon decay limits.

The presence of baryon-number-violating interactions opens the way to cosmological baryogenesis via the out-of-equilibrium decays of GUT bosons [117]:

$$X \to q + \bar{\ell} \quad vs \quad \bar{X} \to \bar{q} + \ell. \tag{138}$$

In the presence of C and CP violation, the branching ratios for $X \to q + \bar{\ell}$ and $\bar{X} \to \bar{q} + \ell$ may differ. Such a difference may in principle be generated by quantum (loop) corrections to the leading-order interactions of GUT bosons. This effect is too small in the minimal SU(5) GUT described above [118], but could be larger in some more complicated GUT. One snag is that, with GUT bosons as heavy as suggested above, the CP-violating decay asymmetry may tend to get washed out by thermal effects. This difficulty may in principle be avoided by appealing to the decays of GUT Higgs bosons, which might weigh $\ll 10^{15}$ GeV, though this possibility is not strongly motivated.

Although neutrino masses might arise without a GUT framework, they appear very naturally in most GUTs, and this framework helps motivate the mass scale $\sim 10^{10}$ to 10^{15} GeV required for the heavy singlet neutrinos. Their decays provide an alternative mechanism for generating the baryon asymmetry of the Universe, namely leptogenesis [49]. In the presence of C and CP violation, the branching ratios for $N \to$ Higgs + ℓ may differ from that for $N \to$ Higgs + $\bar{\ell}$, producing a net lepton asymmetry. The likely masses for heavy singlet neutrinos could be significantly lower than the GUT scale, so it may be easier to avoid thermal washout effects. However, you may ask what is the point of generating a lepton asymmetry, since we want a quark asymmetry? The answer is provided by the weak sphaleron interactions that are present in the Standard Model, and would have converted part of the lepton asymmetry into the desired quark asymmetry. We now discuss how this scenario might have operated in the minimal seesaw model for neutrino masses discussed in Lecture 2.

5.3. *Leptogenesis in the Seesaw Model*

As mentioned in the second Lecture, the minimal seesaw neutrino model contains 18 parameters [44], of which only 9 are observable in low-energy neutrino interactions: 3 light neutrino masses, 3 real mixing angles $\theta_{12,23,31}$, the oscillation phase δ and the 2 Majorana phases $\phi_{1,2}$.

To see how the extra 9 parameters appear [45], we reconsider the full lepton sector, assuming that we have diagonalized the charged-lepton mass matrix:

$$(Y_\ell)_{ij} \;=\; Y^d_{\ell_i}\delta_{ij}, \tag{139}$$

as well as that of the heavy singlet neutrinos:

$$M_{ij} \;= M^d_i\delta_{ij}. \tag{140}$$

We can then parametrize the neutrino Dirac coupling matrix Y_ν in terms of its real and diagonal eigenvalues and unitary rotation matrices:

$$Y_\nu \;=\; Z^* Y^d_{\nu_k} X^\dagger, \tag{141}$$

where X has 3 mixing angles and one CP-violating phase, just like the CKM matrix, and we can write Z in the form

$$Z \;=\; P_1 \bar{Z} P_2, \tag{142}$$

where $\bar{Z}$ also resembles the CKM matrix, with 3 mixing angles and one CP-violating phase, and the diagonal matrices $P_{1,2}$ each have two CP-violating phases:

$$P_{1,2} \;=\; \mathrm{Diag}\left(e^{i\theta_{1,3}}, e^{i\theta_{2,4}}, 1\right). \tag{143}$$

In this parametrization, we see explicitly that the neutrino sector has 18 parameters [44]: the 3 heavy-neutrino mass eigenvalues M^d_i, the 3 real eigenvalues of $Y^D_{\nu_i}$, the $6 = 3 + 3$ real mixing angles in X and $\bar{Z}$, and the $6 = 1 + 5$ CP-violating phases in X and $\bar{Z}$ [45].

The total decay rate of a heavy neutrino N_i may be written in the form

$$\Gamma_i \;=\; \frac{1}{8\pi}\left(Y_\nu Y_\nu^\dagger\right)_{ii} M_i. \tag{144}$$

One-loop CP-violating diagrams involving the exchange of heavy neutrino N_j would generate an asymmetry in N_i decay of the form:

$$\epsilon_{ij} \;=\; \frac{1}{8\pi}\frac{1}{\left(Y_\nu Y_\nu^\dagger\right)_{ii}}\mathrm{Im}\left(\left(Y_\nu Y_\nu^\dagger\right)_{ij}\right)^2 f\left(\frac{M_j}{M_i}\right), \tag{145}$$

where $f(M_j/M_i)$ is a known kinematic function.

Thus we see that leptogenesis [49] is proportional to the product

$$Y_\nu Y_\nu^\dagger \; = \; P_1^* \bar{Z}^* \left(Y_\nu^d\right)^2 \bar{Z}^T P_1, \tag{146}$$

which depends on 13 of the real parameters and 3 CP-violating phases. As mentioned in Lecture 2, the extra seesaw parameters also contribute to the renormalization of soft supersymmetry-breaking masses, in leading order via the combination

$$Y_\nu^\dagger Y_\nu \; = \; X \left(Y_\nu^d\right)^2 X^\dagger, \tag{147}$$

which depends on just 1 CP-violating phase, with two more phases appearing in higher orders, when one allows the heavy singlet neutrinos to be non-degenerate [47].

In order to see how the low-energy sector is embedded in the full parametrization of the seesaw model, and hence its (lack of) relation to leptogenesis [50], we first recall that the 3 phases in $\tilde{P}_2$ (46) become observable when one also considers high-energy quantities. Next, we introduce a complex orthogonal matrix

$$R \; \equiv \; \sqrt{M^d}^{-1} Y_\nu U \sqrt{\mathcal{M}^d}^{-1} \left[v \sin\beta\right], \tag{148}$$

which has 3 real mixing angles and 3 phases: $R^T R = 1$. These 6 additional parameters may be used to characterize Y_ν, by inverting

$$Y_\nu \; = \; \frac{\sqrt{M^d} R \sqrt{\mathcal{M}^d} U^\dagger}{\left[v \sin\beta\right]}, \tag{149}$$

giving us the grand total of $18 = 9+3+6$ parameters [45]. The leptogenesis observable (146) may now be written in the form

$$Y_\nu Y_\nu^\dagger \; = \; \frac{\sqrt{M^d} R \mathcal{M}_\nu^d R^\dagger \sqrt{M^d}}{\left[v^2 \sin^2\beta\right]}, \tag{150}$$

which depends on the 3 phases in R, but *not* the 3 low-energy phases $\delta, \phi_{1,2}$, *nor* the 3 real MNS mixing angles [45]!

The basic reason for this is that one makes a unitary sum over all the light lepton species in evaluating the asymmetry ϵ_{ij}. It is easy to derive a compact expression for ϵ_{ij} in terms of the heavy neutrino masses and the complex orthogonal matrix R:

$$\epsilon_{ij} \; = \; \frac{1}{8\pi} M_j f\left(\frac{M_j}{M_i}\right) \frac{\mathrm{Im}\left(\left(R \mathcal{M}_\nu^d R^\dagger\right)_{ij}\right)^2}{\left(R \mathcal{M}_\nu^d R^\dagger\right)_{ii}}, \tag{151}$$

which depends explicitly on the extra phases in R. How can we measure them?

In general, one may formulate the following strategy for calculating leptogenesis in terms of laboratory observables [45,50]:

- Measure the neutrino oscillation phase δ and the Majorana phases $\phi_{1,2}$,
- Measure observables related to the renormalization of soft supersymmetry-breaking parameters, that are functions of $\delta, \phi_{1,2}$ and the leptogenesis phases,

- Extract the effects of the known values of δ and $\phi_{1,2}$, and isolate the leptogenesis parameters.

In the absence of complete information on the first two steps above, we are currently at the stage of preliminary explorations of the multi-dimensional parameter space. As seen in Fig. 20, the amount of the leptogenesis asymmetry is explicitly independent of δ [50]. However, in order to make more definite predictions, one must make extra hypotheses.

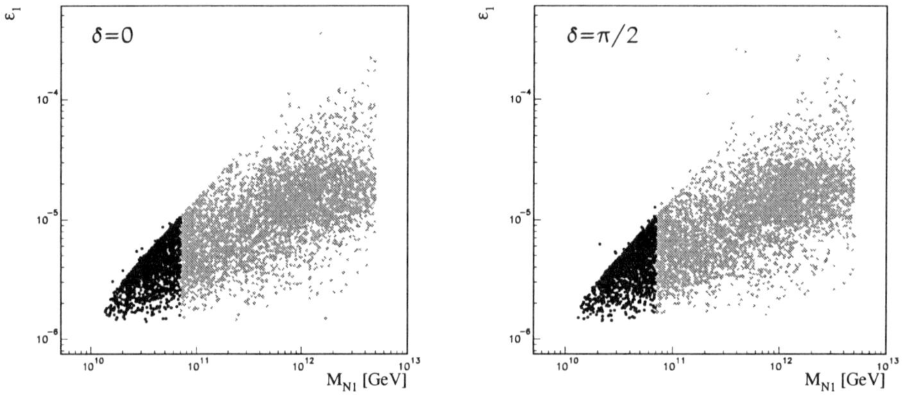

Figure 20. Comparison of the CP-violating asymmetries in the decays of heavy singlet neutrinos giving rise to the cosmological baryon asymmetry via leptogenesis (left panel) without and (right panel) with maximal CP violation in neutrino oscillations [50]. They are indistinguishable.

One possibility is that the inflaton might be a heavy singlet sneutrino, as discussed in the previous Lecture [105]. As shown there, this hypothesis would require a mass $\simeq 1.8 \times 10^{13}$ GeV for the lightest sneutrino, which is well within the range favoured by seesaw models. As also discussed in the previous Lecture, this sneutrino inflaton model predicts values of the spectral index of scalar perturbations, the fraction of tensor perturbations and other CMB observables that are consistent with the WMAP data. The sneutrino inflaton model is quite compatible with a low reheating temperature, as seen in Fig. 19. Moreover, because of this and the other constraints on the seesaw model parameters in this model, it makes predictions for the branching ratio for $\mu \to e\gamma$ that are more precise than in the generic seesaw model. As seen in Fig. 21, it predicts that this decay should appear within a couple of orders of magnitude of the present experimental upper limit [105].

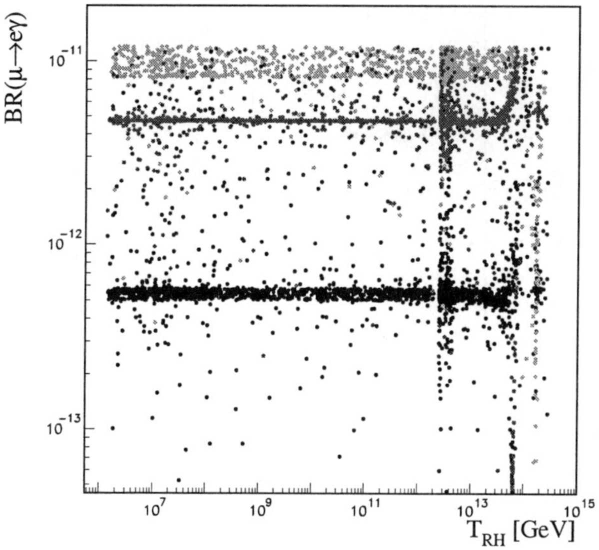

Figure 21. Calculations of BR($\mu \to e\gamma$) in the sneutrino inflation model. The lower locus of points corresponds to $\sin\theta_{13} = 0.0$, $M_2 = 10^{14}$ GeV, and 5×10^{14} GeV $< M_3 < 5 \times 10^{15}$ GeV. The middle locus of points corresponds to $\sin\theta_{13} = 0.0$, $M_2 = 5 \times 10^{14}$ GeV and $M_3 = 5 \times 10^{15}$ GeV, while the upper set of points correspond to $\sin\theta_{13} = 0.1$, $M_2 = 10^{14}$ GeV and $M_3 = 5 \times 10^{14}$ GeV [105]. We assume for illustration that $(m_{1/2}, m_0) = (800, 170)$ GeV and $\tan\beta = 10$.

5.4. *Ultra-High-Energy Cosmic Rays*

The flux of cosmic rays falls approximately as E^{-3} from $E \sim 1$ GeV, through $E \sim 10^6$ GeV where there is a small change in slope called the 'knee', continuing to about 10^{10} GeV, the 'ankle'. Beyond about 5×10^{10} GeV, as seen in Fig. 22, one expects a cutoff [119] due to the photopion reaction $p + \gamma_{CMB} \to \Delta^+ \to p + \pi^0, n + \pi^+$, for all primary cosmic rays that originate from more than about 50 Mpc away. However, some experiments report cosmic-ray events with higher energies of 10^{11} GeV or more [109]. If this excess flux beyond the GKZ cutoff is confirmed, conventional physics would require it to originate from distances $\lesssim 100$ Mpc, in which case one would expect to see some discrete sources. Analogous cutoffs are expected for primary cosmic-ray photons or nuclei, as also seen in Fig. 22.

There are two general categories of sources considered for such ultra-high-energy cosmic rays (UHECRs): *bottom-up* and *top-down* scenarios [109].

Astrophysical sources capable of accelerating high-energy cosmic rays in some bottom-up scenario must be larger than the gyromagnetic radius R corresponding

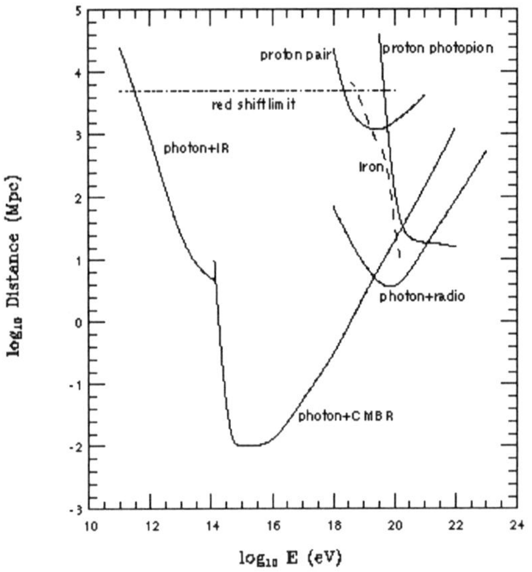

Figure 22. Energetic particles propagating through the Universe scatter on relic photons, imposing a cutoff on the maximum distance over which they can propagate [119].

to their internal magnetic field B:

$$R \sim \left(\frac{100}{Z}\right) \left(\frac{E}{10^{11}\text{GeV}}\right) \left(\frac{\mu G}{B}\right) \text{ kpc}, \tag{152}$$

where Z is the atomic number of the cosmic ray particle. Candidate astrophysical sources include gamma-ray bursters (GRBs) and active galactic nuclei (AGNs).

If UHECRs are produced by such localized sources, one would expect to see a clustering in their arrival directions. Such clustering has been claimed in both the AGASA and Yakutsk data [120], but I personally do not find the evidence overwhelming. A correlation has also been claimed with BL Lac objects [121], which are AGNs emitting relativistic jets pointing towards us, but this is also a claim that I should like to see confirmed by more data, as will be provided soon by the HiRes and Auger experiments.

Favoured top-down scenarios involve physics at the GUT scale $\gtrsim 10^{15}$ GeV that produces UHECRs with energies $\sim 10^{12}$ GeV via some 'trickle-down' mechanism. Suggestions have included topological structures, such as cosmic strings, that are present in some GUTs and would radiate energetic particles, and the decays of metastable superheavy relic particles.

In the latter case, one would expect most of the observed UHECRs to come from the decays of relics in our own galactic halo. In this case, one would expect the UHECRs to exhibit an anisotropy correlated with the orientation of the galaxy [122]. The present data are insufficient to confirm or exclude an isotropy of the magnitude predicted in different halo models, but the Auger experiment should be able to decide the issue. One might naively expect that superheavy relic particles would be spread smoothly through the halo, and hence that they would not cause clustering in the UHECRs. However, this is not necessarily the case, as many cold dark matter models predict clumps within the halo [123], which could contribute a clustered component on top of an apparently smooth background.

How might suitable metastable superheavy relic particles arise [124]? The proton is a prototype for a metastable particle. As discussed earlier in this Lecture, we know that its lifetime must exceed about 10^{33} y or so, much longer than if it decayed via conventional weak interactions. On the other hand, there is no known exact symmetry principle capable of preventing the proton from decaying. Therefore, we believe that it is only metastable, decaying very slowly via some higher-dimensional non-renormalizable interaction that violates baryon number. For example, as we saw earlier, in many GUT models there is a dimension-6 $qqq\ell$ interaction with a coefficient $\propto 1/M^2$, where M is some superheavy mass scale. This would yield a decay amplitude $A \sim 1/M^2$, and hence a long lifetime $\tau \sim \frac{M^4}{m_p^5}$.

We must work harder in the case of a superheavy relic weighing $\gtrsim 10^{12}$ GeV, but the principle is the same. For an interaction of dimension $4 + n$, we expect

$$\tau \sim \frac{M^{2n}}{m_{\mathrm{relic}}^{2n+1}} \tag{153}$$

This could yield a lifetime greater than the age of the Universe, even for $m_{\mathrm{relic}} \sim 10^{12}$ GeV, if M and/or n are large enough, for example if $M \sim 10^{17}$ GeV and $n \geq 9$ [125].

Phenomenological constraints on such metastable relic particles were considered some time ago for reasons other than explaining UHECRs [126]. Constraints from the abundances of light elements, from the CMB and from the high-energy ν flux have been considered. They provide no obstacle to postulating a superheavy relic particle with $\Omega h^2 \sim 0.1$ if $\tau \gtrsim 10^{15}$ y. Hence, metastable superheavy relic particles could in principle constitute most of the cold dark matter.

Possible theoretical candidates within a general framework of string and/or M theory have been considered [124,125]. These models have the generic feature that, in addition to the interactions of the Standard Model, there are others that act on a different set of 'hidden' matter particles, which communicate with the Standard Model only via higher-order interactions scaled by some inverse power of a large mass scale M. Just as the strong nuclear interactions bind quarks to form metastable massive particles, the protons, so some 'hidden-sector' interactions might become strong at some higher energy scale, and form analogous, but supermassive, metastable particles. Just like the proton, these massive 'cryptons' generally decay through high-dimension interactions into multiple quarks and leptons. The energetic quarks hadronize via QCD in a way that can be modelled using information from Z^0 decays at LEP. Several simulations have shown that the resulting spectrum of UHECRs is compatible with the available data, whether supersymmetry is included in the jet fragmentation process, or not, as shown in Fig. 23 [127].

A crucial issue is whether there is a mechanism that might produce a relic

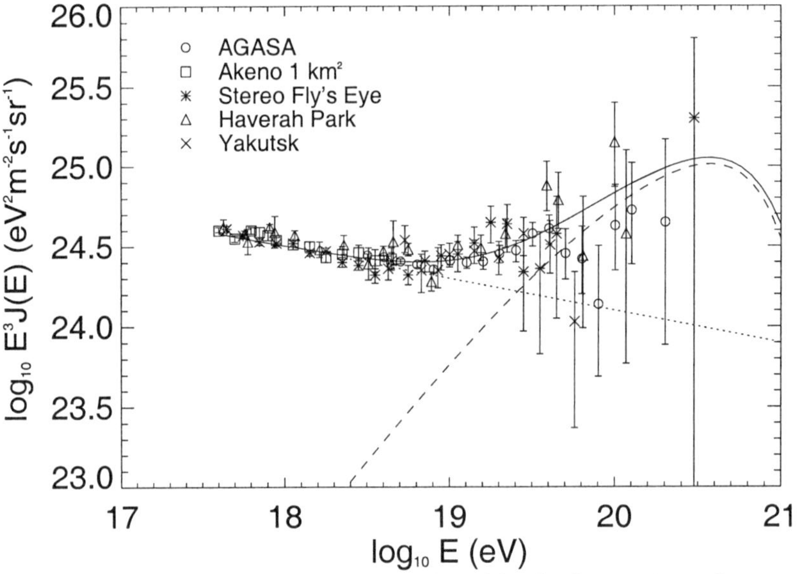

Figure 23. The spectrum of UHECRs can be explained by the decays of superheavy metastable particles such as cryptons [127].

density of superheavy particles that is large enough to be of interest for cosmology, without being excessive. As was discussed in Lecture 3, the plausible upper limit on the mass of a relic particle that was initially in thermal equilibrium is of the order of a TeV. However, equilibrium might have been violated in the early Universe, around the epoch of inflation, and various non-thermal production mechanisms have been proposed [128]. These include out-of-equilibrium processes at the end of the inflationary epoch, such as parametric resonance effects, and gravitational production as the scale factor of the Universe changes rapidly. It is certainly possible that superheavy relic particles might be produced with a significant fraction of the critical density.

We have seen that UHECRs could perhaps be due to the decays of metastable superheavy relic particles. They might have the appropriate abundance, their life-times might be long on a cosmological time-scale, and the decay spectrum might be compatible with the events seen. Pressure points on this interpretation of UHECRs include the composition of the UHECRs — there should be photons and possibly neutrinos, as well as protons, and no heavier nuclei; their isotropy — UHECRs from relic decays would exhibit a detectable galactic anisotropy; and clustering — this would certainly be expected in astrophysical source models, but is not excluded in the superheavy relic interpretation.

The Auger project currently under construction in Argentina should provide much greater statistics on UHECRs and be able to address many of these issues [129]. In the longer term, the EUSO project now being considered by ESA for installation on the International Space Station would provide even greater sensitivity to UHECRs [130]. Thus an experimental programme exists in outline that is capable of clarifying their nature and origin, telling us whether they are indeed due to new fundamental physics.

5.5. *Summary*

We have seen in these lectures that the Standard Model must underlie any description of the physics of the early Universe. Its extensions may provide the answers to many of the outstanding issues in cosmology, such as the nature of dark matter, the origin of the matter in the Universe, the size and age of the Universe, and the origins of the structures within it. Theories capable of resolving these issues abound, and include many new options not stressed in these lectures. Continued progress in understanding these issues will involve a complex interplay between particle physics and cosmology, involving experiments at new accelerators such as the LHC, as well as new observations.

References

1. C. Lineweaver, Lectures at this School.
2. For a recent review, see: K. A. Olive 2002, in *Astroparticle Physics. Proc. 1st NCTS Workshop*, eds H. Athar, G-L. Lin and K-W. Ng, (World Scientific).
3. E. W. Kolb and M. S. Turner, *The Early Universe* (Addison-Wesley, Redwood City, USA, 1990).
4. See the LHC home page: `http://lhc-new-homepage.web.cern.ch/lhc-new-homepage/`.
5. S.L. Glashow, *Nucl. Phys.* **22**, 579 (1961); S. Weinberg, *Phys. Rev. Lett.* **19**, 1264 (1967); A. Salam, Proc. 8th Nobel Symposium, Stockholm 1968, ed. N. Svartholm (Almqvist and Wiksells, Stockholm, 1968), p. 367.
6. LEP Electroweak Working Group, `http://lepewwg.web.cern.ch/LEPEWWG/Welcome.html`.
7. C. Bouchiat, J. Iliopoulos and Ph. Meyer, *Phys. Lett.* B **138**, 652 (1972) and references therein.
8. C. Quigg, *Gauge Theories of the Strong, Weak and Electromagnetic Interactions* (Benjamin-Cummings, Reading, 1983).
9. D. Brandt, H. Burkhardt, M. Lamont, S. Myers and J. Wenninger, *Rept. Prog. Phys.* **63**, 939 (2000).
10. M. Veltman, *Nucl. Phys.* B **123**, 89 (1977); M.S. Chanowitz, M. Furman and I. Hinchliffe, *Phys. Lett.* B **78**, 285 (1978).
11. M. Veltman, *Acta Phys.Pol.* **8**, 475 (1977).
12. J. R. Ellis, M. K. Gaillard and D. V. Nanopoulos, *Nucl. Phys.* B **106**, 292 (1976).
13. LEP Higgs Working Group for Higgs boson searches, OPAL Collaboration, ALEPH Collaboration, DELPHI Collaboration and L3 Collaboration, *Search for the Standard Model Higgs Boson at LEP*, CERN-EP/2003-011.
14. J. R. Ellis, Lectures at 1998 CERN Summer School, St. Andrews, *Beyond the Standard Model for Hillwalkers*, arXiv:hep-ph/9812235.
15. J. R. Ellis, Lectures at 2001 CERN Summer School, Beatenberg, *Supersymmetry for Alp hikers*, arXiv:hep-ph/0203114.

16. J. Scherk and J. H. Schwarz, *Nucl. Phys.* **B81**, 118 (1974); M. B. Green and J. H. Schwarz, *Phys. Lett.* **149B**, 117 (1984) and **151B**, 21 (1985); J. R. Ellis, *The Superstring: Theory Of Everything, Or Of Nothing?*, Nature **323**, 595 (1986).
17. K. Hagiwara *et al.* [Particle Data Group Collaboration], Phys. Rev. D **66**, 010001 (2002).
18. A. Osipowicz *et al.* [KATRIN Collaboration], arXiv:hep-ex/0109033.
19. O. Elgaroy *et al.*, *Phys. Rev. Lett.* **89**, 061301 (2002) [arXiv:astro-ph/0204152].
20. C. L. Bennett *et al.*, *ApJS* **148**,1 (2003); D. N. Spergel *et al.*, *ApJS* **148**,175 (2003); H. V. Peiris *et al.*, *ApJS* **148**, 213 (2003).
21. H. V. Klapdor-Kleingrothaus *et al.*, *Eur. Phys. J.* A **12**, 147 (2001) [arXiv:hep-ph/0103062]; see, however, H. V. Klapdor-Kleingrothaus *et al.*, *Mod. Phys. Lett.* A **16**, 2409 (2002) [arXiv:hep-ph/0201231].
22. Y. Fukuda *et al.* [Super-Kamiokande Collaboration], *Phys. Rev. Lett.* **81**, 1562 (1998) [arXiv:hep-ex/9807003].
23. Q. R. Ahmad *et al.* [SNO Collaboration], *Phys. Rev. Lett.* **89**, 011301 (2002) [arXiv:nucl-ex/0204008]; *Phys. Rev. Lett.* **89**, 011302 (2002) [arXiv:nucl-ex/0204009].
24. R. Barbieri, J. R. Ellis and M. K. Gaillard, *Phys. Lett.* B **90**, 249 (1980).
25. M. Gell-Mann, P. Ramond and R. Slansky, Proceedings of the Supergravity Stony Brook Workshop, New York, 1979, eds. P. Van Nieuwenhuizen and D. Freedman (North-Holland, Amsterdam); T. Yanagida, Proceedings of the Workshop on Unified Theories and Baryon Number in the Universe, Tsukuba, Japan 1979 (edited by A. Sawada and A. Sugamoto, KEK Report No. 79-18, Tsukuba); R. Mohapatra and G. Senjanovic, *Phys. Rev. Lett.* **44**, 912 (1980).
26. P. H. Frampton, S. L. Glashow and T. Yanagida, *Phys. Lett.* B **548**, 119 (2002).
27. T. Endoh, S. Kaneko, S. K. Kang, T. Morozumi and M. Tanimoto, *Phys. Rev. Lett.* **89**, 231601 (2002).
28. J. R. Ellis, J. S. Hagelin, S. Kelley and D. V. Nanopoulos, *Nucl. Phys.* B **311**, 1 (1988).
29. J. R. Ellis, M. E. Gómez, G. K. Leontaris, S. Lola and D. V. Nanopoulos, *Eur. Phys. J.* C **14**, 319 (2000).
30. J. R. Ellis, M. K. Gaillard and D. V. Nanopoulos, *Nucl. Phys.* B **109**, 213 (1976).
31. M. Kobayashi and T. Maskawa, *Prog. Theor. Phys.* **49**, 652 (1973).
32. Z. Maki, M. Nakagawa and S. Sakata, *Prog. Theor. Phys.* **28**, 870 (1962).
33. Y. Oyama, arXiv:hep-ex/0210030.
34. S. Fukuda *et al.* [Super-Kamiokande Collaboration], *Phys. Lett.* B **539**, 179 (2002) [arXiv:hep-ex/0205075].
35. S. A. Dazeley [KamLAND Collaboration], arXiv:hep-ex/0205041.
36. S. Pakvasa and J. W. Valle, *Proc. Indian Nat. Sci. Acad.* **70A**, 189 (2004).
37. H. Minakata and H. Sugiyama, *Phys. Lett.* B **567**, 305, (2003)
38. Chooz Collaboration, *Phys. Lett.* B **420**, 397 (1998).
39. A. De Rújula, M.B. Gavela and P. Hernández, *Nucl. Phys.* B **547**, 21 (1999) [arXive:hep-ph/9811390].
40. A. Cervera *et al.*, *Nucl. Phys.* B **579**, 17 (2000) [Erratum-ibid. B **593**, 731 (2001)].
41. B. Autin *et al.*, *Conceptual design of the SPL, a high-power superconducting H^- linac at CERN*, CERN-2000-012.
42. P. Zucchelli, *Phys. Lett.* B **532**, 166 (2002).
43. M. Apollonio *et al.*, *Oscillation physics with a neutrino factory*, arXiv:hep-ph/0210192; and references therein.
44. J. A. Casas and A. Ibarra, *Nucl. Phys.* B **618**, 171 (2001) [arXiv:hep-ph/0103065].
45. J. R. Ellis, J. Hisano, S. Lola and M. Raidal, *Nucl. Phys.* B **621**, 208 (2002) [arXiv:hep-

ph/0109125].
46. S. Davidson and A. Ibarra, JHEP **0109**, 013 (2001).
47. J. R. Ellis, J. Hisano, M. Raidal and Y. Shimizu, *Phys. Lett.* B **528**, 86 (2002) [arXiv:hep-ph/0111324].
48. J. R. Ellis, J. Hisano, M. Raidal and Y. Shimizu, *Phys. Rev.* D **66**, 115013 (2002) [arXiv:hep-ph/0206110].
49. M. Fukugita and T. Yanagida, *Phys. Lett.* B **174**, 45 (1986).
50. J. R. Ellis and M. Raidal, *Nucl. Phys.* B **643**, 229 (2002) [arXiv:hep-ph/0206174].
51. L. Maiani, *Proceedings of the 1979 Gif-sur-Yvette Summer School On Particle Physics*, 1; G. 't Hooft, in *Recent Developments in Gauge Theories, Proceedings of the Nato Advanced Study Institute, Cargese, 1979*, eds. G. 't Hooft *et al.*, (Plenum Press, NY, 1980); E. Witten, *Phys. Lett.* B **105**, 267 (1981).
52. S. Ferrara, J. Wess and B. Zumino, *Phys. Lett.* B **51**, 239 (1974); S. Ferrara, J. Iliopoulos and B. Zumino, *Nucl. Phys.* B **77**, 413 (1974).
53. P. Fayet, as reviewed in *Supersymmetry, Particle Physics And Gravitation*, CERN-TH-2864, published in *Proc. of Europhysics Study Conf. on Unification of Fundamental Interactions*, Erice, Italy, Mar 17-24, 1980, eds. S. Ferrara, J. Ellis, P. van Nieuwenhuizen (Plenum Press, 1980).
54. R. Haag, J. Lopuszánski and M. Sohnius, *Nucl. Phys.* B **88**, 257 (1975).
55. H. E. Haber and G. L. Kane, *Phys. Rep.* **117**, 75 (1985).
56. J. Ellis, S. Kelley and D. V. Nanopoulos, *Phys. Lett.* B **260**, 131 (1991); U. Amaldi, W. de Boer and H. Furstenau, *Phys. Lett.* B **260**, 447 (1991); P. Langacker and M. x. Luo, *Phys. Rev.* D **44**, 817 (1991); C. Giunti, C. W. Kim and U. W. Lee, *Mod. Phys. Lett.* A **6**, 1745 (1991).
57. H. Georgi, H. R. Quinn and S. Weinberg, *Phys. Rev. Lett.* **33**, 451 (1974).
58. Y. Okada, M. Yamaguchi and T. Yanagida, *Prog. Theor. Phys.* **85**, 1 (1991); J. R. Ellis, G. Ridolfi and F. Zwirner, *Phys. Lett.* B **257**, 83 (1991); H. E. Haber and R. Hempfling, *Phys. Rev. Lett.* **66**, 1815 (1991).
59. J. Ellis, J. S. Hagelin, D. V. Nanopoulos, K. A. Olive and M. Srednicki, *Nucl. Phys.* B **238**, 453 (1984).
60. H. Goldberg, *Phys. Rev. Lett.* **50**, 1419 (1983).
61. G. W. Bennett *et al.* [Muon g-2 Collaboration], *Phys. Rev. Lett.* **89**, 101804 (2002) [Erratum-ibid. **89**, 1219903 (2002)] [arXiv:hep-ex/0208001].
62. M. Davier, S. Eidelman, A. Hocker and Z. Zhang, *Eur. Phys. J.* **C27**, 497 (2003); see also K. Hagiwara, A. D. Martin, D. Nomura and T. Teubner, *Phys. Lett.* D **557**, 69 (2003); F. Jegerlehner, unpublished, as reported in M. Krawczyk, arXiv:hep-ph/0208076.
63. Joint LEP 2 Supersymmetry Working Group, *Combined LEP Chargino Results, up to 208 GeV*, `http://lepsusy.web.cern.ch/lepsusy/www/inos_moriond01/ charginos_pub.html`.
64. Joint LEP 2 Supersymmetry Working Group, *Combined LEP Selectron/Smuon/Stau Results, 183-208 GeV*, `http://lepsusy.web.cern.ch/lepsusy/www/sleptons_summer02/ slep_2002.html`.
65. J. Ellis, K. A. Olive, Y. Santoso and V. C. Spanos, *Phys. Lett.* B **565**, 176 (2003).
66. M. S. Alam *et al.*, [CLEO Collaboration], *Phys. Rev. Lett.* **74**, 2885 (1995), as updated in S. Ahmed *et al.*, CLEO CONF 99-10; BELLE Collaboration, BELLE-CONF-0003, contribution to the 30th International conference on High-Energy Physics, Osaka, 2000. See also K. Abe *et al.*, [Belle Collaboration], arXiv:hep-ex/0107065; L. Lista [BaBar Collaboration], arXiv:hep-ex/0110010; C. Degrassi, P. Gambino and G. F. Giudice, JHEP **0012**, 009 (2000) [arXiv:hep-ph/0009337]; M. Carena, D. Garcia,

U. Nierste and C. E. Wagner, *Phys. Lett.* B **499**, 141 (2001) [arXiv:hep-ph/0010003].
67. J. R. Ellis, G. Ganis, D. V. Nanopoulos and K. A. Olive, *Phys. Lett.* B **502**, 171 (2001) [arXiv:hep-ph/0009355].
68. H. N. Brown *et al.* [Muon g-2 Collaboration], *Phys. Rev. Lett.* **86**, 2227 (2001) [arXiv:hep-ex/0102017].
69. M. Knecht and A. Nyffeler, *Phys. Rev.* D **65**, 073034 (2002); M. Knecht, A. Nyffeler, M. Perrottet and E. De Rafael, *Phys. Rev. Lett.* **88**, 071802 (2002); M. Hayakawa and T. Kinoshita, arXiv:hep-ph/0112102; I. Blokland, A. Czarnecki and K. Melnikov, *Phys. Rev. Lett* **88**, 071803 (2002); J. Bijnens, E. Pallante and J. Prades, *Nucl. Phys.* B **626**, 410 (2002).
70. L. L. Everett, G. L. Kane, S. Rigolin and L. Wang, *Phys. Rev. Lett.* **86**, 3484 (2001) [arXiv:hep-ph/0102145]; J. L. Feng and K. T. Matchev, *Phys. Rev. Lett.* **86**, 3480 (2001) [arXiv:hep-ph/0102146]; E. A. Baltz and P. Gondolo, *Phys. Rev. Lett.* **86**, 5004 (2001) [arXiv:hep-ph/0102147]; U. Chattopadhyay and P. Nath, *Phys. Rev. Lett.* **86**, 5854 (2001) [arXiv:hep-ph/0102157]; S. Komine, T. Moroi and M. Yamaguchi, *Phys. Lett.* B **506**, 93 (2001) [arXiv:hep-ph/0102204]; J. Ellis, D. V. Nanopoulos and K. A. Olive, *Phys. Lett.* B **508**, 65 (2001) [arXiv:hep-ph/0102331]; R. Arnowitt, B. Dutta, B. Hu and Y. Santoso, *Phys. Lett.* B **505**, 177 (2001) [arXiv:hep-ph/0102344] S. P. Martin and J. D. Wells, *Phys. Rev.* D **64**, 035003 (2001) [arXiv:hep-ph/0103067]; H. Baer, C. Balazs, J. Ferrandis and X. Tata, *Phys. Rev.* D **64**, 035004 (2001) [arXiv:hep-ph/0103280].
71. S. Mizuta and M. Yamaguchi, *Phys. Lett.* B **298**, 120 (1993) [arXiv:hep-ph/9208251]; J. Edsjo and P. Gondolo, *Phys. Rev.* D **56**, 1879 (1997) [arXiv:hep-ph/9704361].
72. J. Ellis, T. Falk and K. A. Olive, *Phys. Lett.* B **444**, 367 (1998) [arXiv:hep-ph/9810360]; J. Ellis, T. Falk, K. A. Olive and M. Srednicki, *Astropart. Phys.* **13**, 181 (2000) [arXiv:hep-ph/9905481]; M. E. Gómez, G. Lazarides and C. Pallis, *Phys. Rev.* D **61**, 123512 (2000) [arXiv:hep-ph/9907261] and *Phys. Lett.* B **487**, 313 (2000) [arXiv:hep-ph/0004028]; R. Arnowitt, B. Dutta and Y. Santoso, *Nucl. Phys.* B **606**, 59 (2001) [arXiv:hep-ph/0102181].
73. M. Drees and M. M. Nojiri, *Phys. Rev.* D **47**, 376 (1993) [arXiv:hep-ph/9207234]; H. Baer and M. Brhlik, *Phys. Rev.* D **53**, 597 (1996) [arXiv:hep-ph/9508321] and *Phys. Rev.* D **57**, 567 (1998) [arXiv:hep-ph/9706509]; H. Baer, M. Brhlik, M. A. Diaz, J. Ferrandis, P. Mercadante, P. Quintana and X. Tata, *Phys. Rev.* D **63**, 015007 (2001) [arXiv:hep-ph/0005027]; A. B. Lahanas, D. V. Nanopoulos and V. C. Spanos, *Mod. Phys. Lett.* A **16**, 1229 (2001) [arXiv:hep-ph/0009065].
74. J. R. Ellis, T. Falk, G. Ganis, K. A. Olive and M. Srednicki, *Phys. Lett.* B **510**, 236 (2001) [arXiv:hep-ph/0102098].
75. J. L. Feng, K. T. Matchev and T. Moroi, *Phys. Rev. Lett.* **84**, 2322 (2000) [arXiv:hep-ph/9908309]; J. L. Feng, K. T. Matchev and T. Moroi, *Phys. Rev.* D **61**, 075005 (2000) [arXiv:hep-ph/9909334]; J. L. Feng, K. T. Matchev and F. Wilczek, *Phys. Lett.* B **482**, 388 (2000) [arXiv:hep-ph/0004043].
76. M. Battaglia *et al.*, *Eur. Phys. J.* C **22**, 535 (2001) [arXiv:hep-ph/0106204].
77. J. R. Ellis and K. A. Olive, *Phys. Lett.* B **514**, 114 (2001) [arXiv:hep-ph/0105004].
78. J. Ellis, K. Enqvist, D. V. Nanopoulos and F. Zwirner, *Mod. Phys. Lett.* A **1**, 57 (1986); R. Barbieri and G. F. Giudice, *Nucl. Phys.* B **306**, 63 (1988).
79. G. L. Kane, J. Lykken, S. Mrenna, B. D. Nelson, L. T. Wang and T. T. Wang, *Phys. Rev.* D **67**, 013001 (2003).
80. D. R. Tovey, *Phys. Lett.* B **498**, 1 (2001) [arXiv:hep-ph/0006276].
81. F. E. Paige, hep-ph/0211017.
82. ATLAS Collaboration, *ATLAS detector and physics performance Technical Design*

Report, CERN/LHCC 99-14/15 (1999); S. Abdullin *et al.* [CMS Collaboration], *J. Phys.* G **28**, 469 (2002); S. Abdullin and F. Charles, *Nucl. Phys.* B **547**, 60 (1999) [arXiv:hep-ph/9811402]; CMS Collaboration, Technical Proposal, CERN/LHCC 94-38 (1994).
83. I. Hinchliffe, F. E. Paige, M. D. Shapiro, J. Soderqvist and W. Yao, *Phys. Rev.* D **55**, 5520 (1997).
84. J. R. Ellis, G. Ganis and K. A. Olive, *Phys. Lett.* B **474**, 314 (2000) [arXiv:hep-ph/9912324].
85. J. Silk and M. Srednicki, *Phys. Rev. Lett.* **53**, 624 (1984).
86. J. Ellis, J. L. Feng, A. Ferstl, K. T. Matchev and K. A. Olive, *Eur. Phys. J.* C **24**, 311 (2002).
87. J. Silk, K. A. Olive and M. Srednicki, *Phys. Rev. Lett.* **55**, 257 (1985).
88. M. W. Goodman and E. Witten, *Phys. Rev.* D **31**, 3059 (1985).
89. R. Bernabei *et al.* [DAMA Collaboration], *Phys. Lett.* B **436**, 379 (1998).
90. D. Abrams *et al.* [CDMS Collaboration], *Phys. Rev.* D **66**, 12203 (2002); A. Benoit *et al.* [EDELWEISS Collaboration], *Phys. Lett.* B **513**, 15 (2001) [arXiv:astro-ph/0106094].
91. R. W. Schnee *et al.* [CDMS Collaboration], *Phys. Rept.* **307**, 283 (1998).
92. M. Bravin *et al.* [CRESST Collaboration], *Astropart. Phys.* **12**, 107 (1999) [arXiv:hep-ex/9904005].
93. H. V. Klapdor-Kleingrothaus, arXiv:hep-ph/0104028.
94. G. Jungman, M. Kamionkowski and K. Griest, *Phys. Rept.* **267**, 195 (1996) [arXiv:hep-ph/9506380]; `http://t8web.lanl.gov/people/jungman/neut-package.html`.
95. D. H. Lyth and A. Riotto, *Phys. Rept.* **314**, 1 (1999) [arXiv:hep-ph/9807278].
96. A. H. Guth, *Phys. Rev.* D **23**, 347 (1981).
97. A. G. Riess *et al.* [Supernova Search Team Collaboration], *Astron. J.* **116**, 1009 (1998) [arXiv:astro-ph/9805201]; S. Perlmutter *et al.* [Supernova Cosmology Project Collaboration], *Astrophys. J.* **517**, 565 (1999) [arXiv:astro-ph/9812133]; Perlmutter, S. & Schmidt, B. P. 2003 arXiv:astro-ph/0303428; J. L. Tonry *et al.*, *Astrophys. J.* **594**, 1 (2003).
98. N. A. Bahcall, J. P. Ostriker, S. Perlmutter and P. J. Steinhardt, *Science* **284**, 1481 (1999) [arXiv:astro-ph/9906463].
99. A. D. Linde, *Phys. Lett.* B **108**, 389 (1982).
100. A. D. Linde, *Phys. Lett.* B **129**, 177 (1983).
101. D. La and P. J. Steinhardt, *Phys. Rev. Lett.* **62**, 376 (1989) [Erratum-ibid. **62**, 1066 (1989)].
102. J. R. Ellis, D. V. Nanopoulos, K. A. Olive and K. Tamvakis, *Phys. Lett.* B **118**, 335 (1982) and *Nucl. Phys.* B **221**, 524 (1983).
103. J. M. Bardeen, P. J. Steinhardt and M. S. Turner, *Phys. Rev.* D **28**, 679 (1983).
104. W. H. Kinney, arXiv:astro-ph/0301448.
105. J. R. Ellis, M. Raidal and T. Yanagida, *Phys. Lett.* B **581**, 9 (2004).
106. V. Barger, H. S. Lee and D. Marfatia, *Phys. Lett.* B **565**, 33 (2003).
107. H. Murayama, H. Suzuki, T. Yanagida and J. Yokoyama, *Phys. Rev. Lett.* **70**, 1912 (1993); H. Murayama, H. Suzuki, T. Yanagida and J. Yokoyama, *Phys. Rev.* D **50**, 2356 (1994) [arXiv:hep-ph/9311326].
108. A. D. Sakharov, *Pisma Zh. Eksp. Teor. Fiz.* **5**, 32 (1967)
109. M. Takeda *et al.*, *Astropart. Phys.* **19**, 447; T. Abu-Zayyad *et al.* [High Resolution Fly's Eye Collaboration], *Phys. Rev. Lett.* **92**, 151101 (2004)and arXiv:astro-ph/0208301.
110. H. Georgi and S.L. Glashow, *Phys. Rev. Lett.* **32**, 438 (1974).

111. S. Dimopoulos and H. Georgi, *Nucl. Phys.* B **193**, 50 (1981); S. Dimopoulos, S. Raby and F. Wilczek, *Phys. Rev.* D **24**, 1681 (1981); L. Ibàñez and G. G. Ross, *Phys. Lett.* B **105**, 439 (1981).
112. J. Ellis, M. K. Gaillard and D. V. Nanopoulos, *Phys. Lett.* B **91**, 67 (1980).
113. A. J. Buras, J. Ellis, M. K. Gaillard and D. V. Nanopoulos, *Nucl. Phys.* B **135**, 66 (1978).
114. M. Shiozawa *et al.* [Super-Kamiokande collaboration], *Phys. Rev. Lett.* **81**, 3319 (1998).
115. J. Ellis, D. V. Nanopoulos and S. Rudaz, *Nucl. Phys.* B **202**, 43 (1982); S. Dimopoulos, S. Raby and F. Wilczek, *Phys. Lett.* B **112**, 133 (1982).
116. S. Weinberg, *Phys. Rev.* D **26**, 287 (1982); N. Sakai and T. Yanagida, *Nucl. Phys.* B **197**, 533 (1982).
117. M. Yoshimura, *Phys. Rev. Lett.* **41**, 281 (1978) [Erratum-ibid. **42**, 746 (1979)].
118. J. R. Ellis, M. K. Gaillard and D. V. Nanopoulos, *Phys. Lett. B* **80**, 360 (1979) [Erratum-ibid. B **82**, 464 (1979)].
119. K. Greisen, *Phys. Rev. Lett.* **16**, 748 (1966); G. T. Zatsepin and V. A. Kuzmin, *Pisma Zh. Eksp. Teor. Fiz.* **4**, 114 (1966).
120. P. G. Tinyakov and I. I. Tkachev, *Pisma Zh. Eksp. Teor. Fiz.* **74**, 3 (2001) [arXiv:astro-ph/0102101].
121. P. G. Tinyakov and I. I. Tkachev, *Pisma Zh. Eksp. Teor. Fiz.* **74**, 499 (2001) [arXiv:astro-ph/0102476] and arXiv:astro-ph/0301336. See, however, W. Evans, F. Ferrer and S. Sarkar, *Phys. Rev.* D **67**, 103005 (2003).
122. N. W. Evans, F. Ferrer and S. Sarkar, *Astropart. Phys.* **17**, 319 (2002) [arXiv:astro-ph/0103085].
123. M. G. Abadi, J. F. Navarro, M. Steinmetz and V. R. Eke, *Astrophys. J.* **591**, 499 (2003)and *Astrophys. J.* **597**, 21 (2003).
124. J. R. Ellis, J. L. Lopez and D. V. Nanopoulos, *Phys. Lett.* B **247**, 257 (1990); V. Berezinsky, M. Kachelriess and A. Vilenkin, *Phys. Rev. Lett.* **79**, 4302 (1997) [arXiv:astro-ph/9708217].
125. K. Benakli, J. R. Ellis and D. V. Nanopoulos, *Phys. Rev.* D **59**, 047301 (1999) [arXiv:hep-ph/9803333].
126. J. R. Ellis, G. B. Gelmini, J. L. Lopez, D. V. Nanopoulos and S. Sarkar, *Nucl. Phys.* B **373**, 399 (1992).
127. See, for example, M. Birkel and S. Sarkar, *Astropart. Phys.* **9**, 297 (1998) [arXiv:hep-ph/9804285].
128. See, for example, D. J. Chung, P. Crotty, E. W. Kolb and A. Riotto, *Phys. Rev.* D **64**, 043503 (2001) [arXiv:hep-ph/0104100].
129. A. Letessier-Selvon, arXiv:astro-ph/0208526; J. Cronin *et al.*, `http://www.auger.org/`.
130. L. Scarsi, *EUSO: Using high energy cosmic rays and neutrinos as messengers from the unknown universe*, in *Metepec 2000, Observing ultrahigh energy cosmic rays from space and earth*, p113.